Why We See What We Do Redux

T0337861

Why We See What We Do Redux

A WHOLLY EMPIRICAL THEORY OF VISION

Dale Purves
Department of Neurobiology
Duke University Medical Center

R. Beau Lotto
Institute of Ophthalmology
University College London

Sinauer Associates, Inc. • Publishers
Sunderland, Massachusetts, U.S.A.

About the cover
The Beacon is a 6 meter tower of colored Plexiglas, solar panels, and light that when seen at night creates color effects like some of those described in Chapter 3. The installation, created by Beau Lotto, produces its own electricity through an array of solar panels mounted at the top of the tower and through solar paving slabs made from photo voltaic cells embedded in recycled rubber. The Beacon invites passersby in London to explore how we come to understand the world around us, in this instance by showing that "illusions" are not parlor tricks, but ways of seeing the world that the brain finds useful. The brain has evolved perceptions to tell itself narratives that guide behavior, and the Beacon reminds people that our environment provides not only physical resources, but the basis of what we see and think.

Why We See What We Do Redux, A Wholly Empirical Theory of Vision
Copyright (©) 2011 by Sinauer Associates, Inc. All rights reserved.
This book may not be reproduced in whole or in part without permission from the publisher.

For information, address:
Sinauer Associates, Inc., 23 Plumtree Road, Sunderland, MA 01375 U.S.A.
Fax: 413-549-1118
Email: publish@sinauer.com
Internet: www.sinauer.com

Library of Congress Cataloging-in-Publication Data

Purves, Dale.
Why we see what we do redux : a wholly empirical theory of vision / Dale Purves and R. Beau Lotto.
p. cm.
ISBN 978-0-87893-596-3
1. Visual perception. 2. Vision. I. Lotto, R. Beau. II. Title.
BF241.P87 2010
152.14--dc22
2010044136

Printed in China
3 2

Contents

Preface

Although the ideas and evidence about the why we see what we do in the original version of this book were appreciated in many quarters, the reception in others was distinctly cool. The most strident critics concluded that a wholly empirical concept of vision was unbelievable, while, ironically, others felt that what we said was self-evident. Given this mixed reception, we felt duty-bound to try again. Our objective was, and remains, to present a framework that differs from the concepts that have driven this field for most of the last 50 years, has a strong biological rationale, and, most important, can explain the phenomenology of what we see.

The foundation of the argument is a different supposition about the information that is needed to generate useful behavior, and, as a result, what vision presents to percipients. In contrast, many modern investigators have taken the purpose of vision—usually implicitly—to be a representation of objects and conditions in the physical world. Indeed, it's hard to imagine how seeing the world other than "it really is" would be useful. In this conception, the relevant information is stimulus features that are detected and analyzed, either directly or statistically, by a series of postulated filters. The output of this processing is recombined in the higher stations of the visual system to create a mental model of external reality that represents the environment with reasonable accuracy, forming a basis for action and corresponding to what we see. This general understanding of vision, however, is undermined by the inverse optics problem (see Chapter 1), which precludes apprehending the world by any operation on images. How the visual system deals with this quandary is the central theme of the present account.

The argument we advance is that the visual system contends with the inevitably uncertain significance of sensory information for behavior by generating perceptions in a wholly empirical manner through trial-and-error interactions over phylogenetic and individual time with a world that is otherwise hidden. A corollary is that despite our overwhelming impression that we perceive the world as it is, what we in fact see on this basis is not a facsimile of the physical world that is occasionally misleading (causing visual "illusions"), but a subjective world fully determined by associations made between images and successful behavior over the course of species and individual history. As hard as it may be to understand and/or accept a framework based on historical data about the behavioral significance of sensory information rather than immediate "sense data" in images, evidence has continued to grow that what we see is generated in this way.

The wholly empirical theory we offer is in no way a restatement of the obvious fact that the organization of the visual system has been shaped by evolutionary and developmental demands, and is readily testable. The argument would be falsified if it were shown that (1) any other strategy of vision could contend with the direct unknowability of the physical world by means of retinal images; or (2) could predict the rich phenomenology of what we actually see.

We hope this revised version will encourage more readers to consider a wholly empirical concept of vision and its implications for interpreting, modeling, and ultimately understanding the structure and function of the human visual system.

Dale Purves and Beau Lotto
October, 2010

Acknowledgments

We reiterate our thanks to all those colleagues who gave us advice on the First Edition. Regarding this edition, further thanks are due Byron Boots, David Corney, Catherine Howe, Kyongje Sung, Bill Wojtach and Zhiyong Yang for new or ongoing collaborations that provided additional evidence that we discuss in this revision, as well as many helpful discussions. We are especially grateful to Catherine Howe, whose thesis forms the basis for much of Chapter 4, and to Bill Wojtach and Kyongje Sung whose work forms the basis for Chapter 6. Finally, we are indebted to Mike Hogan for reading the final manuscript and making lots of good suggestions, and to Andy Sinauer, Syd Carroll, Chelsea Holabird, Jean Zimmer, and the staff at Sinauer Associates for their fine job producing this new edition.

Preface to the First Edition

This appeal in *An Essay Towards a New Theory of Vision* betrays Berkeley's concern that readers might be diverted from the substance of his argument by inadequacies and/or infelicities in one or another of its component parts. We have a similar concern. There are certainly aspects of our thesis about vision that one can take issue with; indeed, if there were no strong reactions, we would deem our effort here a failure. Like Berkeley, however, we hope that whatever debate is generated will focus on the core of the argument and not on a "manner of expression" that might, for any of a variety of reasons, offend readers who have long entertained other ideas about vision and how it works.

There are several ways in which our intentions might be misread. The first is that we deem the physiological behavior of visual neurons described over the last four decades to be suspect, incorrect, or—worse still—irrelevant. This is certainly not so. The behavior of many types of visual neurons has been beautifully documented, and this work will inevitably form the foundation for whatever detailed understanding of the link between physiology and perception eventually emerges. The implication of our argument is not that this information is in any way unimportant, but that, on the basis of the phenomenology of what we actually see, there may be a better framework for understanding it.

A second possible misunderstanding that might arise concerns the evolution and development of the visual system and the role of the visual environment in these processes. Some readers might take our argument to imply that others have not sufficiently considered the degree to which visual circuitry must have been influenced by the structure of the world and that we are simply emphasizing this point. Any modern hypothesis about vision would presumably take for granted the importance of meeting the specific demands of the environment in both phylogeny and ontogeny. Our aim is not to underscore the fact that the visual system evolved (and develops) in accord with the demands of the natural world, but to indicate, again on the basis of perceptual phenomenology, the problem these processes are solving and what the manner of the solution implies about the organization of visual circuitry.

Finally, our intention is not to dismantle a current conception of vision and replace it with something entirely different. Since there is at present no consensus about the link between visual physiology and perception, there is nothing to dismantle, even if that were our aim. Indeed, having a well-defined "opponent" would have made prosecution of the argument a good deal easier. As it is, our intention is simply to build upon present knowledge a framework for understanding vision that is consistent with the perceptual evidence. Thus Berkeley's plea that judgment be based on the "truth" of one's "own experience" seems particularly apt. In each of the various aspects of vision that we consider—brightness, color, form, depth, and motion—the test of the strength of this framework is whether it explains and accurately predicts what people actually see.

Acknowledgments

We are grateful to Tim Andrews, David Coppola, Catherine Howe, Fuhui Long, Allison McCoy, Amita Shimpi, Surajit Nundy, Mark Williams, Len White and Zhiyong Yang whose collaboration, criticism and advice have been essential in generating both the ideas and the evidence discussed here. We also indebted to John Allman, Don Katz, Rolf Kuenhi, John Maunsell, Jim Schirillo, and Jim Voyvodic for extraordinarily helpful criticisms of earlier versions of the manuscript. Finally, we wish to thank to Robert Reynolds, who provided invaluable help over several years as a research assistant, tracking down numerous references, often obscure, with skill and insight, and to Tracey Madrid for her patient help formatting, proofreading and correcting what must have seemed far too many versions of the manuscript. Finally, we are deeply indebted to Mark Williams of Pyramis Studios (www.brainimages.com) for improving original illustrations and creating much additional art for the book. His thoughtful participation in both the artistic and intellectual aspects of the project has been enormously valuable.

D.P.
R.B.L.

1

Vision and Visual Perception

Introduction

Because images cannot specify the physical world that gives rise to them, the concept of vision we explore here is that what we see is an operational presentation of the world that links light stimuli with successful behavior, instantiated in neural circuitry determined by trial and error over evolutionary and individual time. In this counterintuitive framework, visual perceptions are not the result of retinal image analysis, do not correspond to physical reality, and cannot be understood in terms of logical computations. The merits of this way of understanding vision rest on its ability to contend with a directly unknowable world; the evidence that supports it is the ability to explain the peculiar details of what we actually see. The chapters that follow examine how this conception of vision operates for each of the basic qualities of human visual perception: lightness, brightness, color, form, distance and depth, and motion. The purpose of this first chapter, however, is to explain the rationale for this way of understanding vision in general terms, and to put it in historical context.

A Longstanding Puzzle

Seeing begins with images of the world projected onto the retina by the optics of eye. A strong intuition is thus that the visual brain generates perceptions by relying on features of the image to represent objects and conditions in the physical environment. That the visual system must be doing its job in this general way seems obvious.

Some observations, however, have always raised puzzling questions about the relationship between retinal images and perceptions of the visual world. In 1625, a Jesuit friar and astronomer named Christopher Scheiner looked through the back of a human eye stripped of its scleral coat and saw there an image of the world in miniature (Helmholtz, 1866/1924, vol. I, p. 91) (Figure 1.1A). René Descartes is said to have repeated this experiment for himself a few years later and made it a centerpiece of his theory of vision, concluding in his dualist scheme that the "seeing" of the retinal

(A) Eye

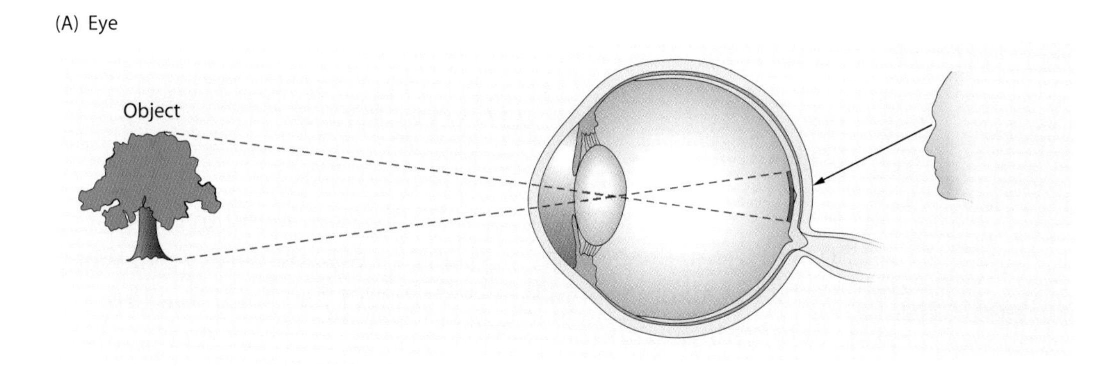

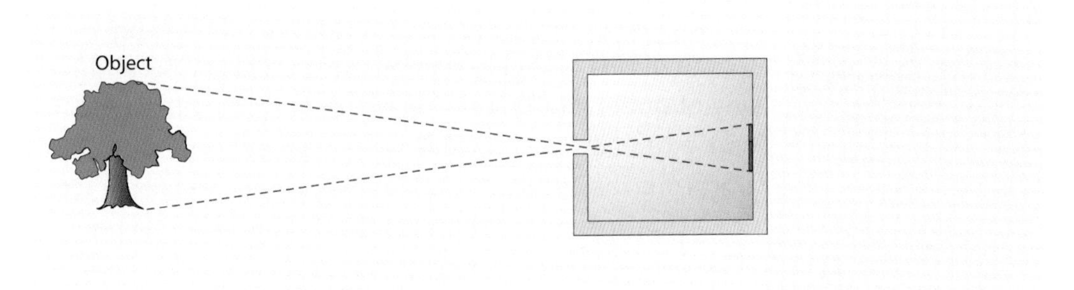

Figure 1.1 The formation of retinal images. (A) The observation that the eye generates an image on the retina that is inverted and shifted left for right was first made in the seventeenth century, although optics and optical projections were understood much earlier. (B) The process of image formation on a plane is the basis of photography, illustrated here by a pinhole camera. How retinal image features are analyzed has understandably dominated thinking about how the visual system works.

image must be done by an immaterial observer who stood above the biological fray (1637/1965; see also Polyak, 1957, pp. 100ff.). Whatever the interpretation, this observation made clear that we don't perceive the retinal image in a literal sense. After all, the projected world that Scheiner and others reported was, like the image formed by a pinhole camera (and for the same reasons), upside down, laterally reversed, and a great deal smaller compared with what the donor of the eye would have seen (Figure 1.1B).

In fact, there are many differences between retinal images and the perceptions the images ultimately give rise to. These differences include: (1) the perception of lightness, brightness, and color, even though these qualities don't exist as such in objects, their illumination, or the light reaching the eye; (2) the perception of distance and depth despite the absence of these features in two-dimensional projections; (3) the perception of objects as stationary despite continual movement of retinal images due to movements of the eyes, head, and body; and (4) the perception of motion when objects change position, even if a sequence of images is stationary (as in movies and video). Clearly, there is a great deal to explain.

A Conventional Framework

People interested in vision have naturally thought long and hard about these issues, what they imply about vision, and how the visual system must work. Distilling the upshot of these ideas is not easy since, in addition to neuroscientists and psychologists, philosophers, artists, and others have all weighed in on this debate over the last few hundred years or more. Despite this diversity of opinion, in recent decades the parties have generally agreed that the purpose of visual perception is to provide an accurate representation of the physical world (defined as what we are aware of seeing), or at least as accurate a representation as the limitations of biology allow. After all, to see the world other than the way it "really is" would presumably be maladaptive.

In neuroscientific terms, visual perception is taken to entail filtering and encoding retinal image features in signals carried centrally in the primary visual pathway, though exactly what low-level features are encoded is debated (if this pathway is unfamiliar, refer to the Appendix, which describes the organization of the visual system). Higher-order visual processing algorithms would then "decode" these signals, leading to a representation of the real-world sources of stimuli (see, for example, Barlow, 1961; Oja, 1982; Marr, 1982; Barlow, 1995; Edelman, 1999; Ullman, 1996; Ullman and Sali, 2000; Simoncelli and Olhausen, 2001; White and Wise, 1999; Wallis et al., 2001). In this way, the visual brain would construct a sufficiently accurate model of the external world to serve as a basis for motor actions and/or pertinent cognitive processes (e.g., attending to the visual scene, interpreting its meaning, responding to it emotionally, remembering it, and so on). Prior experience (often referred to as "top-down" information that can add to and/or modulate "bottom-up" information in retinal images) would then influence perception by, for example, bringing heuristics (i.e., simplifying rules) to bear.

A specific example of this sort of account is the conception of vision advocated by David Marr, one of the most influential vision theorists in the second half of the twentieth century. Marr was a brilliant synthesizer who had the persistence and courage to compile his ideas about vision in a book that has often served as a focal point for discussion since its debut more than 30 years ago (Marr, 1982). Although Marr died of cancer at age 35, just before his book was published, his ideas are regarded by many as having given rise to the field of computer vision. Marr, as most others, considered vision an information-processing task leading to a representation of the relevant information in the visual brain. For him, the goal of vision is "recovering from sensory information 'valid' properties of the external world," the idea being that "usually our perceptual processing...delivers a true description of what is there" (Marr, 1982, p. 29). Marr integrated into this framework much of what was then known about visual physiology, anatomy, and perception—the motivation for this ambitious task being his sense that "something was going wrong"—since the "discoveries of the 1950s and 1960s were not being followed by equally dramatic discoveries in the 1970s" (Marr, 1982, p. 14). The key problem, in his estimation, was that people were only describing vision and its neural underpinnings, but not explaining how visual stimuli, visual processing, and perception are related. It was into this breach that Marr threw himself.

Marr proposed that the information in the retinal image is subjected to algorithmic processing at three levels, which he referred to as the construction of the "primal sketch," the "2 1/2-D sketch," and the "3-D model representation" (the term "algorithmic" refers to following a set of logical instructions carried out either serially or in parallel). The purpose of the primal sketch was to make explicit information about the elemental geometries and intensities of the retinal image (including their statistical regularities), the latter being especially important for efficient coding (see below). The purpose of the 2 1/2-D sketch was to make explicit the orientation and approximate depth of surfaces in a viewer-centered frame of reference. Finally, the 3-D model represented shapes and their spatial organization in an object-centered frame of reference in more or less the way observers see them. For Marr, this overall process of internal representation was a "formal system for making explicit certain entities or types of information, together with a specification of how the system does this" (Marr, 1982, p. 20).

Marr's attempt to rationalize the corpus of vision science—which has grown enormously but has not changed in any fundamental way since Marr's death—in a comprehensive computational theory was remarkable. Although other syntheses in the same representational mode followed (e.g., Edelman, 1999; Ullman, 1996), Marr's effort stands as a pioneering attempt to explain vision in logical terms. Although not specifically due to Marr's influence, the general drift of thinking in the field has been focused on determining how the features of the retinal image are detected, processed, and represented according to a series of computational algorithms. As the psychologist Ken Nakayama put it, "Thanks to Marr's distinction of several levels, many researchers with very different interests and talents have been able to coexist almost peaceably, now recognizing their own efforts as part of a much larger endeavor" (Nakayama, 1998, p. 324). This conclusion notwithstanding, the theme of the present account is that if vision is ever to be explained, it must be on a basis fundamentally different from the detection, processing, and representation of images.

The Inverse Problem

What, then, is wrong with the seemingly sensible idea that the purpose of vision is to perceive the world more or less as it is; that this obviously beneficial goal is achieved by neuronal machinery that deals with image features; and that the visual system reconstructs from these a representation of the external world according to logical rules instantiated in visual processing circuitry? The fundamental problem with any scheme like this is that the real-world sources of a retinal stimulus—and thus their significance for subsequent action—cannot be determined by any operation on images as such. Any element of any visual stimulus could have arisen from many combinations of physical parameters in the world. The problem for visual animals is that knowing the specific parameters underlying a stimulus is obviously essential for successful behavior.

Since what we see in response to objects and conditions in the environment usually works quite well for us, the idea that visual stimuli cannot provide information about the physical world is extremely hard to appreciate. But despite our conviction that we see the properties of objects and their physical arrangement, that possibility is precluded by the fact that retinal images inevitably conflate the generative sources of light stimuli. Take, for example, the light reflected to the eye from the two target patches in Figure 1.2. When the identical patches on the page are under the same illuminant, measurements of the light coming from them and their projected forms are the same, and the two patches look the same. The retinal images of the physically identical patches in Figure 1.2 could, however, have been generated by many different sources and conditions in the world. The reason is that the illumination of the relevant objects, their surface reflectance properties, the transmittance of the intervening medium, and a host of other factors that determine the quantity and quality of light are intertwined in any visual stimulus. As a result, the light reaching the eye from the two patches in the figure could signify many different combinations of these variables, each requiring a different behavioral response. How the visual system recovers useful information in the face of this inherent uncertainty is called the *inverse optics problem.*

To make matters more confusing still, the appearance of any element in a scene can be changed, sometimes quite dramatically, by the context in which it is presented. For example, Figure 1.3 shows that when the two targets in Figure 1.2 are placed in a scene, their appearance is altered in lightness, color, shape, and spatial orientation. Moreover, the altered appearances are different for the left and right targets. These effects indicate that

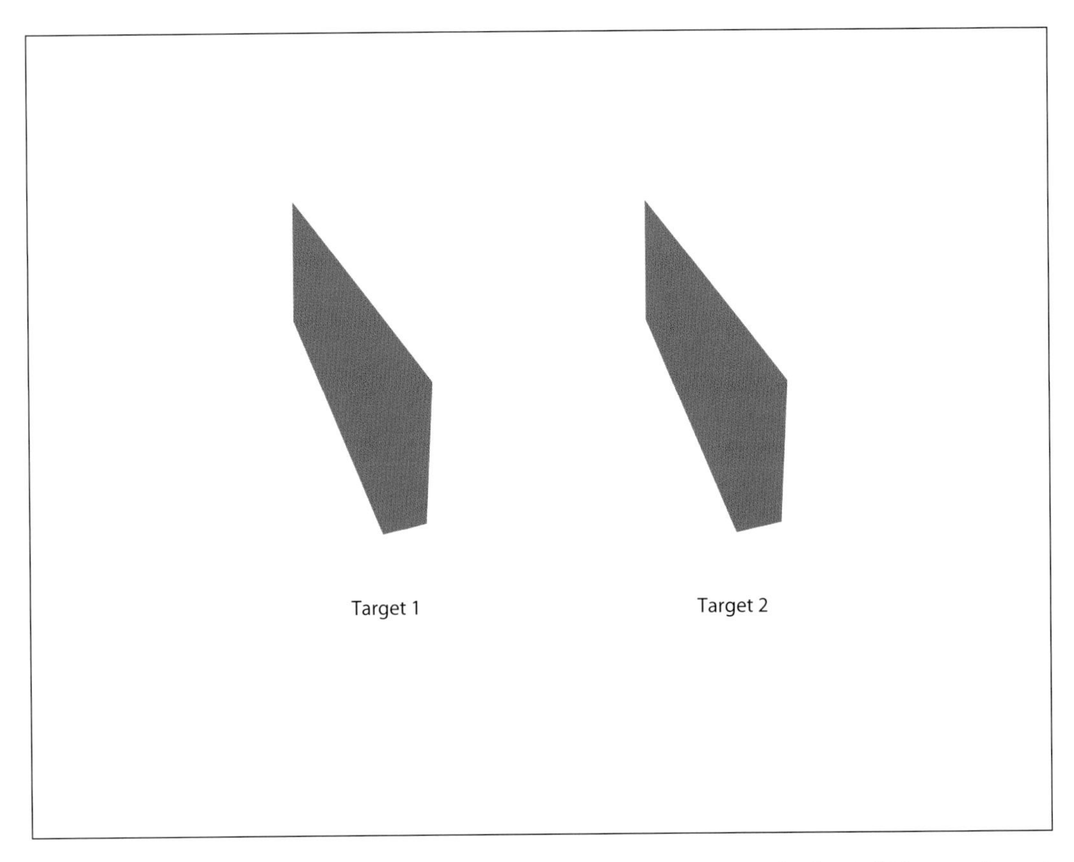

Figure 1.2 The two target patches shown here are physically identical: The quantity and quality of light they return to the eye, as well as the form of their projected contours, are the same. It is not surprising, therefore, that the two targets appear identical in all respects.

the way we see the world does not represent the physical features of the objects that are "out there" or the stimulus features they generate.

Scientists and others have long considered how these facts can be accommodated in theories of vision. One way is to call effects such as those in Figure 1.3 "illusions" and relegate them to the realm of curious mental phenomena best left to psychologists and philosophers. Another way of explaining away the peculiar way we see the world is to imagine that the inverse problem is only an issue when observers are presented with the relatively impoverished stimuli used in a vision science laboratory. This was in part the argument made by the American psychologist James Gibson in the 1970s (see below). Gibson imagined that when we view natural scenes, the rich information about the conditions and circumstances of any element in the scene is sufficient to generate perceptions of the underlying reality.

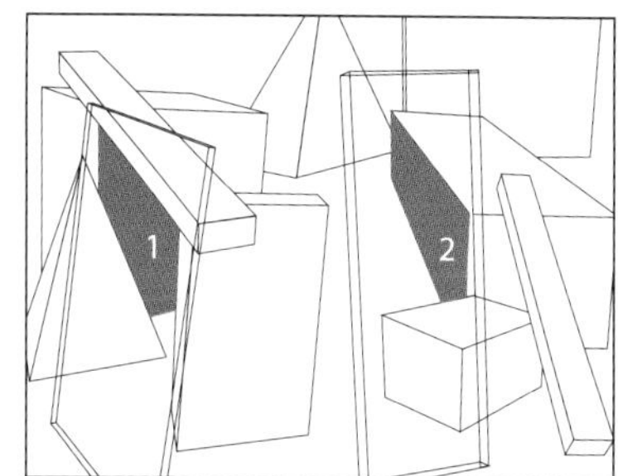

Figure 1.3 The same targets in Figure 1.2 (see inset) look quite different—and different from one another—when presented in a more natural scene. The two identical patches now appear to have different surface reflectance values, to be under different illuminants, and to be seen through differently transmitting media. Moreover, the two objects now seem to have different shapes, to be in different orientations, and to be at different distances.

Another way to deal with the inverse problem is to depend on statistical regularities in images, an idea that Marr took to be important in creating his primal sketch. The visual system would surely want to make use of this information, and much current thinking in vision science concerns image statistics. But however helpful image regularities may be in driving the evolution of visual processing, regularities cannot overcome the inability of retinal images to specify objects and conditions in the world. In short, the directly unknowable nature of the world by means of images as such remains the key obstacle in understanding vision.

An Empirical Strategy for Dealing with the Inverse Problem

Our argument is that vision and the visual system can only be understood in terms of an empirical process that links images with behavior by trial and error interactions with the environment. Given mechanisms of feedback and enough time, retinal light patterns can be associated with the success or failure of responses empirically. Such associations derive from history rather than logical analysis, and need not (indeed cannot) represent the actual properties of stimulus sources. Moreover, perceptual responses to stimuli are reflexive in that they arise from associative circuitry already in place rather than from on-line heuristic "computations." Whenever inherited visual circuitry with random variations made an empirical link a little more successfully than the circuitry of the competing members of a cohort, the improved reproductive fitness of the beneficiary would tend to increase the prevalence of that circuitry in the population as a whole. Over eons, the neural machinery underlying successful behavioral responses to visual stimuli would gradually wax and unsuccessful connectivity would wane, leading to the circuits and relative behavioral success that humans and other visual species enjoy today. Experience-dependent neural plasticity during an animal's lifetime would further refine the visual system according to the mechanisms of activity-dependent plasticity.

Testing the merits of the idea that the way we see the world has been fully determined by species and individual history is not as difficult as it might seem. If conditions are kept simple so that vision is the main contributor to perception (thus getting around the multiplicity of other inputs relevant to any perception and ensuing behavior), it is possible to ask how well the psychophysical characteristics of the basic visual qualities are predicted by human experience with the sorts of images generated by the world in which we live.

A Simple Analogy

To make this alternative strategy of vision clearer, consider how a player could behave successfully in a game of dice when direct analysis (e.g., weighing the dice, inspecting their structure by cutting them up, etc.) is precluded as a way to determine whether the dice are fair or "loaded." This scenario presents a problem similar to the one confronting vision, where direct information about the physical world is likewise unavailable.

Despite the exclusion of direct analysis, an operational evaluation of the dice and the implications for behavior (betting advantageously) could nonetheless be made by tallying the frequency of occurrence of the numbers that come up over the course of many throws. Just as the properties of the physical world determine the frequency of occurrence of different retinal stimuli, so the physical properties of the dice determine the frequency of occurrence of the different numbers that are rolled. If the numbers come up about equally, then a player should behave as if the dice are fair and bet accordingly; conversely, if some numbers appear more frequently than others, the dice are likely to be "loaded" and, to succeed, the player's behavior should change.

This information about overall biases, however, provides a limited guide to behavior. More specific information is needed for successful action and, again, empirical evaluation presents a way forward. Since there are two dice, the number on one die is relevant to the behavioral significance of the number on the other. In the game of "craps," for instance, the significance of a 1 is contingent on the number on the other die (if it is a 6, you win; if a 1, you lose). One die can therefore be thought of as a "target" in a stimulus pattern and the other the "context." If any particular number on one of the die is taken as a "target," then tallying how often that number comes up together with each of the possible numbers on the other die provides a more refined guide to behavior. For example, if a 5 on the "target" die is often associated with a 2 on the "context" die, then betting on 7 would be a good strategy. The more frequently a target number on one die is associated with a contextual number on the other, the greater the chance that a given behavior in response to the combination will succeed.

The human environment is like loaded dice in that the complex physical properties of the world ensure that stimuli, visual or otherwise, do not come up equally over time. Despite the inverse optics problem, appropriate behavior can be generated by progressive modification of the connectivity of the visual brain according to the frequency of occurrence of stimulus targets in the myriad contexts that nature provides. In vision or any other sensory modality, useful behaviors can therefore be generated without representing the generative physical sources of stimuli.

However, relying on the behaviorally relevant associations in large numbers of stimuli rather than on the (unavailable) properties of objects themselves means that visual perceptions will never correspond to the properties of the physical world, just as the associations of the numbers on the dice do not correspond to the physical properties of the dice. In this framework, then, what we see is not objects or conditions as such, but historically determined perceptions that lead to behaviors that worked in the past and are thus likely to work in the future. Given the inverse problem, this mode of operation seems inevitable. Subsequent chapters describe how this idea has been evaluated by asking how well the frequency of occurrence of visual stimuli predicts what we see.

Presentation Versus Representation

This concept of vision raises another issue: If the actual properties of objects in the physical world are not represented, what then is the proper way to describe what we see? Although more a philosophical question than a scientific one, the answer is critical to the argument we are making here.

The concept of representation is inappropriate when discussing vision or, for that matter, any perception. Take the perception of pain, for example. It seems obvious that pain does not exist in the physical world, and that speaking of pain as if it were a physical aspect of the world that could be "represented" would be nonsensical. It is a subjective quality that we evolved to perceive because of its usefulness in behavior and, ultimately, survival. Other perceptual qualities—those pertinent to gustation, olfaction, and audition, for example—are similar to pain: Although they arise from obvious antecedent causes in the physical world, tastes, odors, speech sounds, and so on do not "represent" physical things in any conventional sense of the word. Appearances to the contrary, visual qualities are just the same. Although we routinely attribute these qualities to objects and conditions in the world, our experiences of lightness, brightness, color, form, and motion are likewise subjective constructs that have arisen for reasons of behavioral advantage. In consequence, these qualities are better thought of as "presented" to the perceiver, rather than represented, the latter term implying that we have access to the physical properties of objects through the senses, when in fact we do not.

If vision is not representational, then neither are its underlying physiological and anatomical mechanisms, which must be thought of, examined, and tested in different terms. Analysis (e.g., direct analysis of the physical properties of the dice in the analogy above) is very different in principle from an empirical evaluation of what works based on accumulated history. The fundamental difference between logical analysis and history as a means of generating perceptions and behaviors is why we refer to the argument here as "wholly empirical."

Notice that in this account of vision experience with a series of images alone is insufficient. Since the mechanisms for instantiating the needed visual circuitry are natural selection and activity-dependent neural plasticity, feedback from the success or failure of behavior in natural environments is essential to complete the biological loop underlying a wholly empirical visual strategy.

Precedents for Understanding Vision in Empirical Terms

One of the first thinkers to explicitly consider the inverse optics problem and the possibility of an empirical solution was the Irish philosopher and Anglican cleric George Berkeley. In a remarkable treatise, *An Essay Towards a New Theory of Vision* published in 1709, Berkeley pointed out "the estimate we make of the distance of objects… [is] an act of judgment grounded on experience [rather] than sense…. It is evident that when the mind perceives any idea… it must be by means of some other idea" (pp. 7–8). His concern was based chiefly on the difficulty of imagining how the physical structure of the world can be conveyed by light, given the sort of uncertainty illustrated in Figure 1.4. Berkeley recognized that solving this problem could not be done on the basis of the information present in a retinal image as such, and his insights established him as a leading vision theorist.

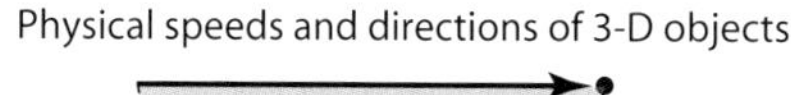

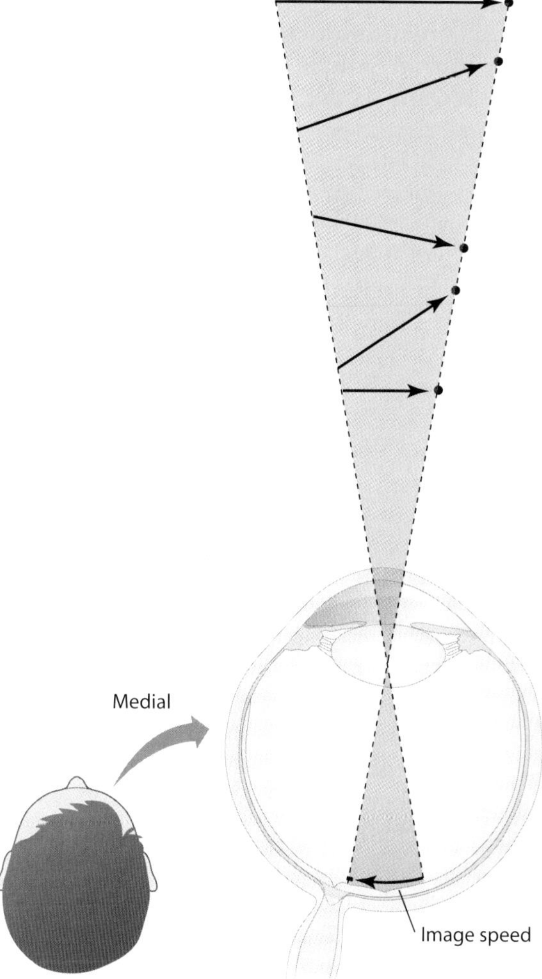

Figure 1.4 The inherent uncertainty of the real-world significance of visual stimuli. Berkeley pointed out that the nature of real-world sources of visual stimuli cannot be specified by retinal images. The reason is that objects with very different properties—physical speeds and directions, in this example—can project the same image or image sequence on the retina. (After Wojtach et al., 2008.)

Reckoning with the influence of experience on visual perception has played some part in theories of vision ever since. The preeminent nineteenth century advocate of experience as a factor influencing how we see the world was Hermann von Helmholtz, a dominant figure in vision and general German science of that era (he also made fundamental contributions to studies of audition, the physics of sound, and thermodynamics). Helmholtz based his thinking on a distinction he drew between sensation and perception, a dichotomy that persists in vision science today. (At least three current undergraduate textbooks on the subject are titled *Sensation and Perception*, and a recent review of "illusions" states that such "misperceptions"are useful "because they allow us to distinguish neural responses that correlate with basic sensory features from those that correlate with perception" [Eagleman, 2002, p. 922].) At the outset of the third volume of his remarkable exposition of nearly every aspect of vision science that could be pursued in the nineteenth century, Helmholtz remarked that it is "hard to say how much of [perception] derived by the sense of light is due directly to sensation, and how much...is due to experience and training" (Helmholtz, 1866/1924–25, vol. III, p. 10). He went on to define the "sensory" aspect of vision as perceptions that experience cannot change or "overcome" by "unconscious inference." Helmholtz's contemporary and rival Ewald Hering (1868/1964) had concluded that perception derives primarily from inherited physiology, setting in motion an ongoing controversy about the relative contributions of nature and nurture to visual perception (see Turner, 1994).

The role of experience in perception was also a focus in the long-running dispute between the behaviorist and gestalt schools of psychology. The "behaviorist school," founded by John Watson in the early twentieth century, argued that studies of subjective phenomena ("mental processes") should be avoided in science because they cannot be observed and measured. "Psychology as the behaviorist views it is a purely objective branch of natural science," claimed Watson (1913, p. 158). "Introspection forms no essential part of its methods." Frustrated by the contentious and relatively unscientific nature of perceptual psychology in his time, Watson and his successors—preeminently, B. Frederick Skinner at Harvard—made great strides, leading to a variety of conceptual and technical advances (e.g., the concept of operant conditioning and, in Skinner's skilled hands, the means to study it). The price paid, however, was ignoring what people perceived, felt, or thought. In consequence, the behaviorist school eventually gave way to cognitive psychology and, more recently, to the burgeoning field of cognitive neuroscience. Despite the unwarranted proscription of perceptual phenomenology, the heart of the behaviorist argument accords with the idea that the operation of the brain (or at least the visual part of it) is best thought of in terms of reflexes rather than ill-defined "top-down" cognitive processes.

The gestalt school took the opposing position that perception should and could be understood, but only in its own terms. The school, founded shortly after the turn of the nineteenth century by Max Wertheimer, took root with the work of his disciples, preeminently Kurt Koffka and Wolfgang Kohler. The word *gestalt* means "form" or "shape" in German and was used to suggest that the overall organization of a stimulus possessed a significance that transcended the parts. At the core of gestalt theory is the idea that the "units of experience go with the functional units in the underlying physiological processes" (Kohler, 1947, p. 63). The "law of *präganz*" (meaning, literally, the law of conciseness) expressed the idea that any stimulus would be seen by observers as the simplest possible real-world source.

One of the inheritors of the gestalt mantle in the second half of the twentieth century was James Gibson, a psychologist at Smith College and, later, Cornell University. Much like the gestalt psychologists, Gibson eschewed attempts to understand perception in terms of physiology, focusing on the objects and circumstances that observers encounter in natural scenes (see above). By his own admission, he had moved on from a concern with the retinal image in his first exposition in 1950 to an interest in what he called "the ambient optical array" in his last book on the subject 30 years later. Gibson's unorthodox proposal was that the brain directly perceives what he called "invariances" in the visual environment (Gibson, 1979). By this he meant that the consistent behavior of objects and conditions (e.g., commonly encountered ratios, textures, patterns, and the like) provided the fundamental categories that dictate visual stimuli. In this conception, vision doesn't generate a perception of the retinal stimulus as such but rather "the invariant structure of ambient light" (op. cit., p. 58). The visual system, he argued, had evolved to interpret these "invariances" without explicitly reconstructing the relevant qualities of the retinal image, a process that in Gibson's view entailed "perceptual systems" rather than "sensing systems." In the case of size and distance, for example, he argued that the ratio of the retinal extent of the projection of objects with respect to background textures provided the kind of invariant information that would allow an observer to make sense of otherwise ambiguous size–distance relationships (see Figure 1.4). The mechanism of this apprehension was taken to be "resonance," a concept borrowed from the sympathetic physical resonances of vibrating bodies.

Although Gibson's ideas today seem somewhat mystical and out of step with the detailed information about the organization of the visual system that he could have used, his insistence that an analysis of retinal images as such is not the way to understand vision and that the "invariant" relationships between elements of images that underlie perception has a good deal of merit. The present argument that visual perceptions depend on trial-and-error interactions with the world over evolutionary and individual time that reflexively link images and sources bears some similarity to Gibson's idea of "resonances."

The most notable empirical approach in the last decade or two is Bayesian decision theory. Bayes's theorem (Thomas Bayes was an eighteenth century minister and amateur mathematician) indicates the probability of one of a set of hypotheses being true (the posterior distribution), given another set of incomplete empirical evidence (the prior distribution). Examples of its use in medicine are deciding on a diagnosis given a battery of test results that are not definitive, or determining the best therapeutic strategy to follow based on a set of clinical trials. In the case of vision, the theorem has been used to determine the most likely cause of an ambiguous or noisy stimulus (Knill et al., 1996; Kersten, 2000; Rao et al., 2002; Weiss et al., 2002; Kersten and Yuille, 2003; Knill and Pouget, 2004). Thus, the posterior distribution represents the relative probability that a given retinal image has been generated by one or another set of possible physical sources. Perception would then be predicted by the highest value in the posterior distribution.

The application of Bayesian theory, which usefully formalizes Helmholtz's qualitative proposal about "unconscious inferences" as a means of contending with ambiguous or noisy visual stimuli, would be fine if disambiguating images were the problem confronting visual perception. But because the characteristics of the physical world can't be conveyed by images, the problem for vision is a very different one. To the extent that a Bayesian perspective remains wedded to the idea that the goal of vision is to determine the real-world properties underlying a retinal stimulus, it is untenable as a basis for understanding perception. Since there is, of course, nothing wrong with Bayes's equation, a wholly empirical theory could presumably be put in Bayesian terms, but this exercise would be superfluous.

Conclusion

Despite extraordinary knowledge about the anatomy and physiology of the visual system, the way in which we generate perceptions remains deeply uncertain. The central problem is how observers can respond so effectively to objects and conditions in the physical world when the sources of visual stimuli cannot be determined by means of any operation on retinal images as such. As a result, why we see what we do remains in many ways as puzzling today as it was for the natural philosophers who thought about these issues centuries ago.

Perception has been given relatively short shrift in vision research. A case in point is the otherwise marvelous book *Brain and Visual Perception* by David Hubel and Torsten Wiesel (2005), the two most influential vision scientists in the second half of the twentieth century. The title notwithstanding, the content includes almost nothing about perception, for which there is no entry in either the glossary or index. It is simply assumed that understanding what we see will eventually emerge from sufficiently detailed studies of the properties of visual neurons and their connectivity. In fact, as ever more refined anatomical and physiological studies of the visual system have emerged, the initial promise of this approach has been replaced by a sense that proceeding along this path may not be as enlightening as was first imagined.

The alternative idea—that what we see is a presentation based on history rather than a representation of the world derived from an analysis of images—is hard to accept. This concept of vision is made even less palatable by the rudimentary sense at present of how an empirical strategy of vision could be understood in terms of the physiological and anatomical "nuts and bolts" of the human visual system. Nonetheless, what follows is an effort to make the case for this wholly empirical framework, focusing on visual perception with all its quirky details. Our argument (or any other) stands or falls on how well what we actually see is explained.

2

Seeing Lightness and Brightness

Introduction

The activity of neurons that respond to visual stimuli is of course the basis for what we see, and describing their properties and interactions is essential. But this information has not explained how the visual system produces perceptions or successful visually guided behavior. If the strategy the visual system uses is wholly empirical, then as outlined in Chapter 1, the visual brain does not represent the properties of stimuli or the physical world discovered inferentially. Rather, what we see and do as a result is determined by the links forged between particular patterns of light on the retina and behaviors that have been empirically successful in responding to those stimuli in the past. What we see thus needs to be understood on this "non-representational" basis, which has implications for rationalizing the underlying physiology of perception. A good place to begin exploring this counterintuitive conception of vision is with the qualities of lightness and brightness generated by the intensity of light reaching the eyes.

Light, Luminance, and the Perception of Light Intensity

The visual brain processes neural activity generated by a small portion of electromagnetic radiation in the terrestrial environment that humans inhabit. This bandwidth, called "light," comprises photons vibrating at energies that interact with human photoreceptors, initiating the process that leads to what we see. Although it is more appropriate in physics to consider light in terms of the vibrational frequencies of photons (expressed in terahertz), discussing light in terms of wavelengths is conventional in biology (see Rodieck, 1998, pp. 513ff. for a discussion of these issues). The wavelengths of light ranges from about 400 to 700 nanometers (Figure 2.1); there is some leeway in these values, since wavelengths a little shorter than 400 nm and little longer than 700 nm can be perceived by most people in special circumstances. Light is therefore defined by the human visual system, which evolved to process radiation over this particular spectral range. Strictly speaking, then, the biblical injunction should have been "Let there be electromagnetic radiation" rather than the anthropocentric admonition "Let there be light." In any event, the physical input to the visual system is light patterns, which are generally referred to as "retinal images."

Luminance is a measure of the intensity of a light stimulus determined with a photometer and expressed in units such as candelas per square

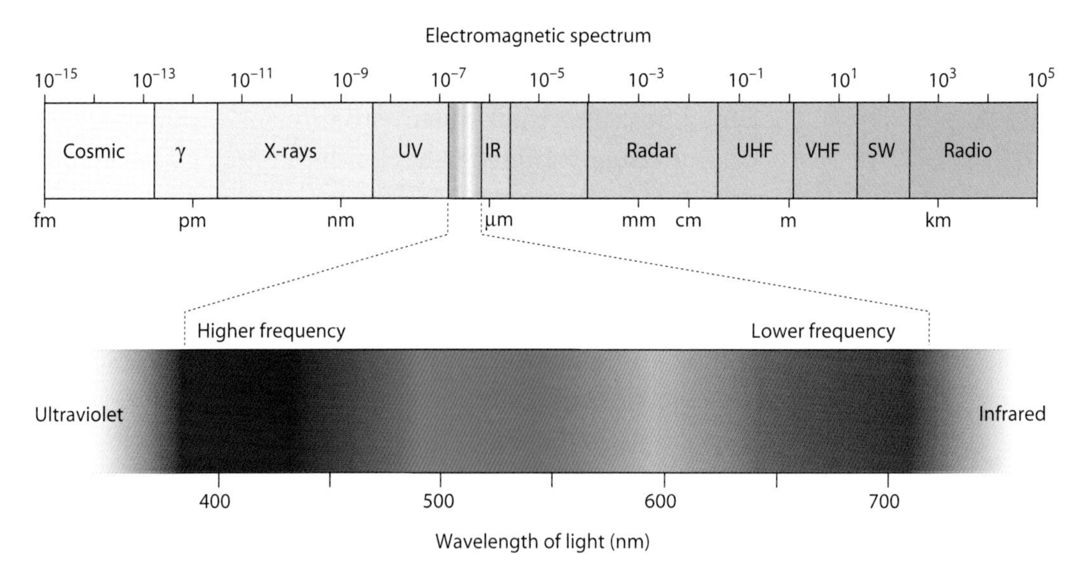

Figure 2.1 The electromagnetic spectrum, showing the relation of light to other forms of radiation. Colors represent the approximate sensations of hue elicited by monochromatic light at the indicated wavelengths, although, as emphasized in the text, light is in no sense "colored." The wavelengths of light (i.e., the distance from one peak to the next, in nanometers) are inversely related to the frequency of the waves. Thus, the smaller the wavelength of light, the more energetic it is.

meter. In fact, measuring the intensity of light and the units that describe it is a complicated business that depends on the practical goal (see Wandell, 1995 for a discussion of these issues). The photometer used in the context of human vision measures radiant energy with a filter that mimics the sensitivity of the average human observer, thus specifically measuring the narrow band of electromagnetic radiation that human photoreceptors have evolved to respond to. In dealing with visual perception it would make little sense to neglect the fact that receptors have particular sensitivities that together determine responses to light intensity.

The perceptual qualities elicited by luminance are called lightness and brightness. Lightness refers to the lighter or darker appearance of object surfaces, whereas brightness refers to the brighter or dimmer appearance of objects that are sources of light (e.g., the sun, fire, lightbulbs, etc.). Like all other percepts, lightness and brightness are not subject to direct measurement; they can only be evaluated by asking observers to report their threshold for light detection, the just noticeable difference between two stimuli, or, more commonly, the appearance of one surface or light source relative to the appearance of another.

The Peculiar Relationship between Lightness/Brightness and Luminance

In thinking about the generation of lightness and brightness, a simple expectation is that measurements of light intensity and perception would be proportional. After all, increasing the luminance of a stimulus increases the number of photons captured by photoreceptors and thus the output (action potentials per unit time) of the retina at any given level of background luminance. A corollary is that two objects in a scene that return the same amount of light to the eye should appear equally light or bright. It has long been known, however, that perceptions of lightness and brightness fail to meet these expectations. For example, two patches of equal photometric intensity are perceived as being differently light or bright when placed on backgrounds that have different luminance values (Figure 2.2A): A patch on a background of relatively low luminance looks lighter/brighter than the same patch on a background of higher luminance, a phenomenon called "simultaneous brightness contrast."

Until recently, the textbook explanation of this effect was predicated on the properties of neurons at the input level of the visual system, in particular the local interactions among retinal neurons that are important determinants of the information ganglion cells convey centrally (e.g., Coren et al., 1999, pp. 109ff.; Levine, 2000, pp. 77ff.; Goldstein, 2000, pp. 58ff.). The foundation of such accounts is evidence that, presumably as a means of enhancing the detection of contrasting luminance boundaries (the edges of things and related stimulus features such as spatial frequency and motion), the receptive fields of lower order visual neurons are organized in a center-surround fashion with the two components having opposite functional

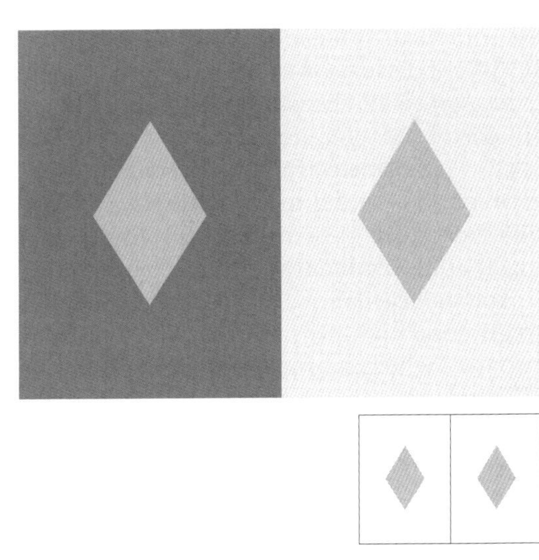

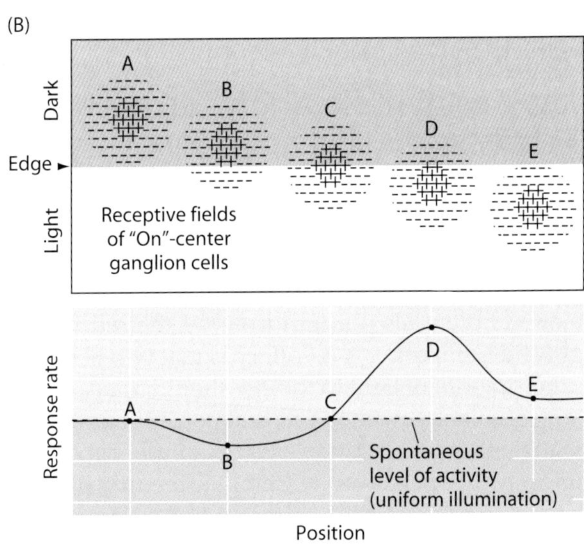

Figure 2.2 Standard demonstration of simultaneous brightness contrast. (A) A central target (the diamond) on a less luminant background (left) is perceived as being lighter/brighter than the same target on a more luminant background (right), even though the two targets are physically identical; if both are presented on the same background, they appear to have the same lightness/brightness (inset). (B) A possible explanation of this phenomenon, based on the center-surround receptive field properties of retinal ganglion cells (see text for explanation).

polarities (Hartline and Graham, 1932; Hartline, 1940; Kuffler, 1953; Barlow, 1953; Wiesel and Hubel, 1966; Kuffler, 1973; see the Appendix if the concept of receptive fields is unfamiliar). The firing rate of neurons whose receptive fields intersect a contrast boundary will therefore differ from the activity of neurons whose receptive fields fall entirely on one side of the boundary or the other (Figure 2.2B). For example, neurons whose receptive field centers lie just within the diamond on the dark background in Figure 2.2A will fire at a higher rate than the neurons whose receptive field centers lie just within the diamond on the light background (because the former are less inhibited by their oppositely disposed receptive field surrounds than the latter; see the lower part of Figure 2.2B). As a result, so the argument goes, the patch on the dark background should look lighter/brighter than the patch on the lighter background because of the difference in the retinal output elicited by the two identical targets in different surrounds.

To the extent that this argument emphasizes the significance of edges and the heightened responsiveness of the retina to contrast boundaries, it seems appropriate. Both electrical recording from neurons (e.g., Shapley, 2005) and psychophysical (eye-tracking) experiments (e.g., Yarbus, 1959) indicate that the most important information in any visual scene derives from regions where there are contrast boundaries. It makes sense, then, that the visual system would want to emphasize the information derived from boundaries and that the properties of the relevant neurons would reflect this goal.

Why This Explanation Is Wrong

It doesn't follow, however, that simultaneous brightness contrast is a price paid for achieving another goal, such as better edge detection. One reason to doubt that the different appearance of the equiluminant patches in Figure 2.2A is an incidental consequence of the organization of retinal receptive fields comes from the perceptual consequences of other patterns of contrasting luminance. For instance, in the stimulus depicted in Figure 2.3A, two equally luminant gray targets associated with a black cross differ in lightness/brightness despite the absence of any differences in local luminance contrast (both triangles are flanked by black on two sides and white on the other) (Wertheimer, 1912; Benary, 1924). Even more remarkable is White's illusion, in which the target patches on the left in Figure 2.3B are surrounded on average by a greater area of higher luminance territory and yet appear brighter/lighter than the targets on the right, which are surrounded on average by lower luminance territory (the converse argument applies to the patches on the right) (White, 1979). Although the disposition of the surrounds in White's stimulus is effectively the opposite of that in the standard presentation in Figure 2.2A, the brightness difference elicited is about the same in direction and magnitude (this sort of anomaly was first noted by Wilhelm von Bezold in the arabesques he studied in the nineteenth century to illustrate his so called "spreading effect," in which a black surround likewise makes a target look darker when compared with the effect of a white surround; see Evans, 1948, p. 181 and Plate XI).

In general, imagining that what we see is predicated on physiological constraints is not a good approach to understanding perception. Of course, such constraints provide limits; for example, sensations of lightness or brightness are generated only by photons whose wavelengths lie within the narrow range of electromagnetic radiation that humans have evolved to see. Although some phenomena arise from unusual stimuli that exceed physiological limits or uses ("after images" experienced following a photo flash are an example), many other stimuli have been devised over the years that are difficult to explain in such terms (e.g., O'Brien, 1958; Cornsweet, 1970; Gilchrist, 1977; Land, 1986; Knill and Kersten, 1991; Adelson, 1993, 2000; Gilchrist, 1994; Todorovic, 1997; Williams et al., 1998a,b; Lotto and Purves,

(A)

(B)

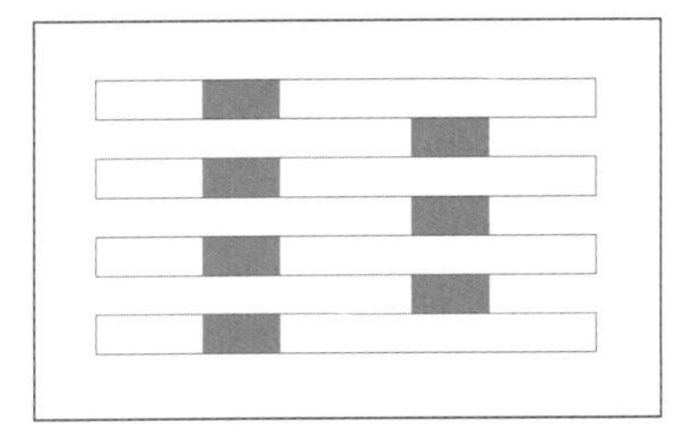

◀ **Figure 2.3** Some reasons why the model in Figure 2.2B fails to explain simultaneous brightness contrast. (A) In the Wertheimer-Benary stimulus, two equiluminant targets (the gray triangles) elicit different sensations of lightness/brightness despite having the same local contrast relationships (the upper triangle looks slightly lighter/brighter than the lower one; see inset for comparison). (B) White's stimulus generates a perception of relative lightness/brightness similar to the sensations elicited in Figure 2.2A despite the fact that the local contrast of the gray patches is now more or less opposite the contrasts in a standard brightness contrast stimulus. In this instance, the targets that appear lighter/brighter (the four patches in the left column; see inset) are mainly surrounded by areas of higher luminance, whereas the targets that appear darker (the three patches in the right column) are surrounded mainly by areas of lower luminance.

1999; Purves et al., 1999; Lotto and Purves, 2001). Although various mathematical models can account for some of these effects, these theories fail to explain the full range of complex effects that have been observed; they also depend on adjustable weighting factors and lack a biological rationale.

Other approaches have been based on loosely defined rules (heuristics). An example is the model proposed by Alan Gilchrist that divides visual stimuli into multiple "local frameworks" based on junction analysis, coplanarity, and other factors (Gilchrist et al., 1999; Gilchrist, 2006). The ratio of each patch's intensity and the maximum intensity of the local surround is used to predict surface reflectance by combining a "bright is white" and a "large is white" rule. Although these rules are effective in predicting simple stimuli, they do not explain the effects of the stimuli in Figure 2.3, much less natural scenes under spatially heterogeneous illumination (see below).

If the visual circuitry underlying perception is indeed determined historically to contend with the inverse problem (as well as the need to incorporate into perception many non-visual factors relevant to successful behavior), it is not surprising that models based on logical analysis fail to explain the range of these perceptual phenomena.

An Empirical Explanation of Lightness/Brightness

If explanations of lightness and brightness based on distortions arising from retinal processing, as well as models with adjustable parameters or gestalt-like rules, fail to account for the effects of stimuli such as those in Figures 2.2.A and 2.3, what does?

The answer is an empirical strategy, since this is the only way the visual system can contend with the uncertain provenance of light intensities falling on the retina. In most natural circumstances, three key aspects of the physical world determine the luminance at each retinal locus: the illuminant, the reflectance of object surfaces, and the transmittance of the space between the objects in a scene and the observer. As indicated in Figure 2.4, these factors are inevitably entangled in the retinal image. As a result of this conflation, many different combinations of illumination, reflectance, and transmittance could have given rise to any given retinal luminance value. There is no way

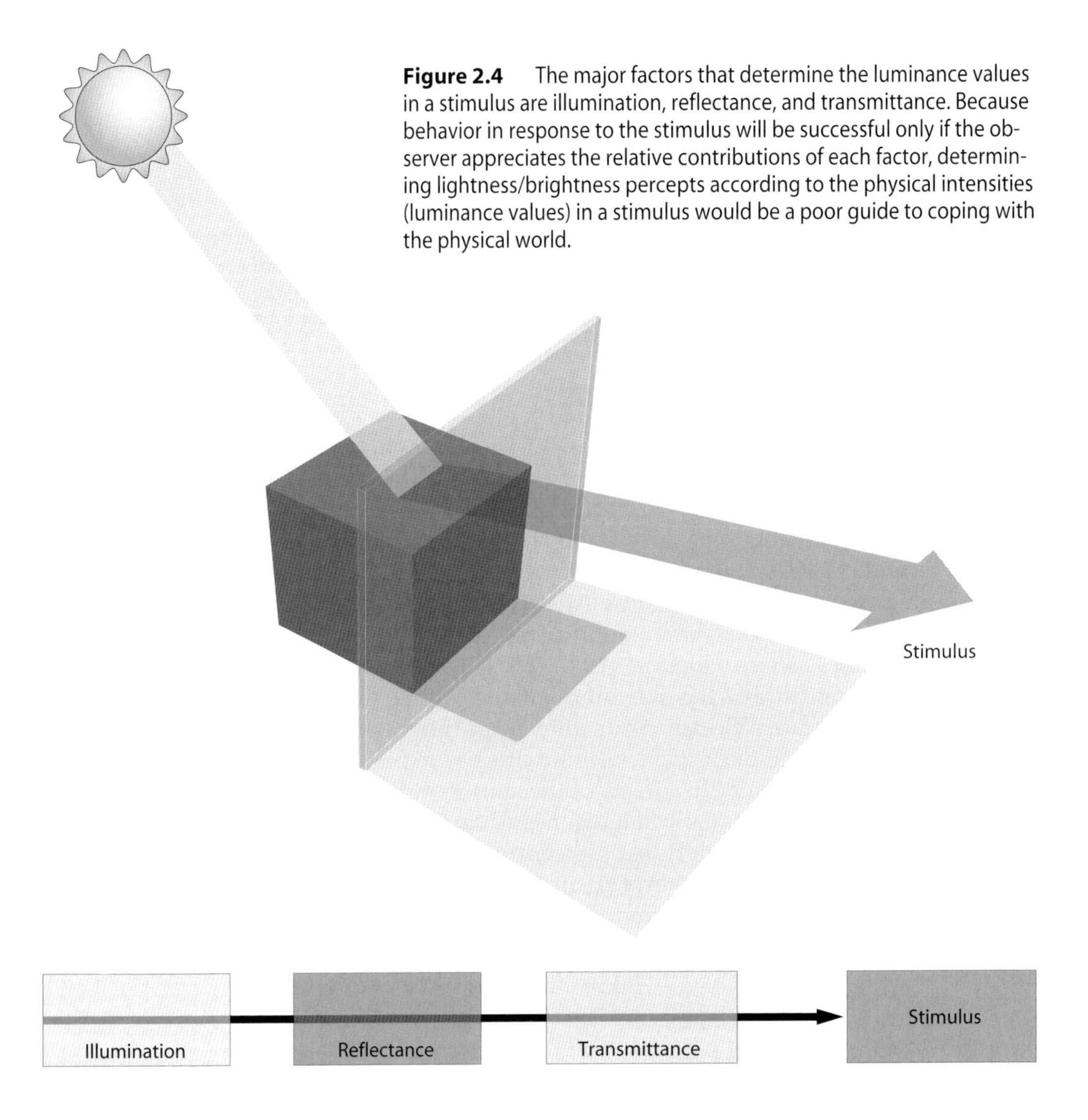

Figure 2.4 The major factors that determine the luminance values in a stimulus are illumination, reflectance, and transmittance. Because behavior in response to the stimulus will be successful only if the observer appreciates the relative contributions of each factor, determining lightness/brightness percepts according to the physical intensities (luminance values) in a stimulus would be a poor guide to coping with the physical world.

the visual system can determine how these factors have been combined to generate the luminance values in an image as such, and thus no way to inform a visual animal about what it should see or how it should react. Since survival requires successful behavior in the face of this uncertainty, perceptions based on retinal luminance values would be a useless strategy. Vision must therefore rely on some other way to glean information from images about how to behave in the world. As in the dice analogy in Chapter 1, the only plausible contender seems to be accumulated experience.

As discussed in Chapter 1, lots of earlier work recognized the importance of experience in perception and attempted to explain some aspects of what we see in empirical terms. For the most part, these explanations—beginning with Helmholtz—have allowed that the experience of individuals helps in "disambiguating" retinal images. While this perspective rightly assigns a role for experience in determining what we see, it wrongly attributes the routine success of vision to a neural representation of retinal images and/or the underlying features of objects and conditions in the world. Given the uncertain relationship of images to the physical world (see Figure 2.4), this possibility is ruled out. Experience isn't just a contributing factor to what we see; it is the only factor.

Although images as such are insufficient, visual animals could gain the information needed to survive by tallying up the success or failure of past responses to the different the combinations of illumination, reflectance, and transmittance generated by the world. In this scenario, however, the sense of lightness or brightness elicited by a stimulus would not correspond to underlying physical features or conditions in the world. In principle, visual perceptions would correspond to the operational link between the stimulus and a successful behavioral response, discovered over the course of evolutionary and individual time. If this idea about the genesis of the lightness or brightness is right, then these and other aspects of visual perceptions should vary predictably according to accumulated past experience.

Understanding Lightness/Brightness in Empirical Terms

To get a feel for how this framework can explain lightness and brightness, consider the conventional simultaneous brightness contrast stimulus shown in Figure 2.5A. The stimulus is consistent with the two equiluminant targets being similarly reflective surfaces under similar illuminants (Figure 2.5B) or differently reflective surfaces under different illuminants (Figure 2.5C). In an empirical framework, these and all other possibilities would have determined the lightness or brightness seen in proportion to the respective frequencies of occurrence of the relevant images in past human experience. As a result of the instantiation of this information in the visual system by the gradual modification of neural connectivity over evolutionary and individual time, the lightness/brightness of the two diamonds in the stimulus would arise reflexively.

The idea that the effects perceived in response to the standard stimulus in Figures 2.2A and 2.5A are generated in this way is very different from the idea that what we see is a representation of the physical intensities of the pattern of light falling on the retina at any moment. Nevertheless, by using the accumulated feedback from trial-and-error behavior over phylogenetic and ontogenetic time, observers can contend with the inverse problem as it pertains to lightness and brightness (see Figure 2.4). At the same time, this historical strategy provides a way of integrating the importance of all non-visual information relevant to successful behavior into the perceptual outcome. As a result, measured luminance values in a stimulus will always differ from the perceptual qualities accorded them—sometimes greatly (Figure 2.6).

Figure 2.5 Basis for simultaneous brightness contrast effects in empirical terms. (A) A standard simultaneous brightness contrast stimulus, much like the stimulus in Figure 2.2A. (B, C) Cartoons illustrating two commonly experienced but different sources of the stimulus in (A). (See text for explanation.)

(A) Standard simultaneous brightness contrast stimulus

(B) Similar surfaces under similar illuminants

(C) Different surfaces under different illuminants

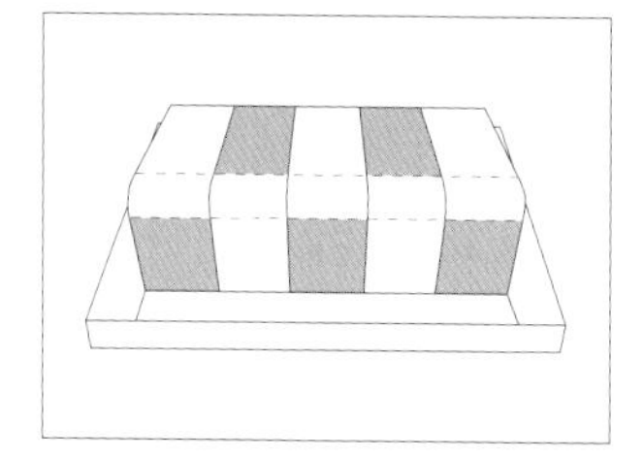

Figure 2.6 Although, as shown in the inset, the gray patches all have the same luminance, a striking perceptual difference among them is generated by the wealth of mutually consistent information indicating that one set of equiluminant patches is in deep shadow (the patches on the riser of the step) and the other set is in intense light (the patches on the surface of the step).

These discrepancies have given rise to the ill-conceived idea of "illusions" that differ from perceptual representations of the world that are routinely accurate. If the argument here is correct, *all* perceptions of lightness and brightness are empirical constructs generated on the same physiological basis for the same biological purpose of behavioral success. By the same token, the lightness and brightness values we see never correspond to physical reality (or do so only coincidentally).

Qualitative Evidence that Lightness and Brightness Are Empirically Determined

If the lightness/brightness values seen are generated empirically, they should increase or decrease as the information in a scene is varied. It is easy enough to test this prediction, even when the luminance of targets and their immediate surrounds remains unchanged. For instance, when the dark and light surrounds in the standard brightness contrast stimulus in Figure 2.2A are presented with the target diamonds being under different illuminants, as in Figure 2.7A, the apparent difference in the reported lightness/brightness of the targets increases (Williams et al., 1998a,b). Conversely, when viewing a depiction of the targets and their surrounds under the same

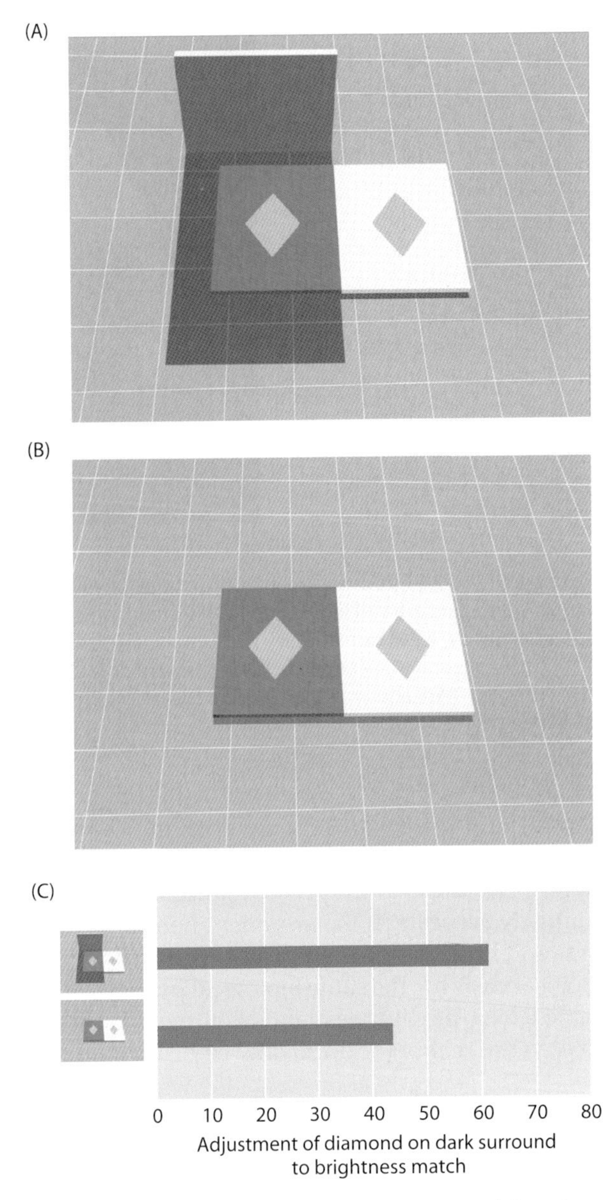

Figure 2.7 The effect of changing the depicted illumination of test targets while maintaining the luminance values of both the targets and their local surrounds. (A) Test targets and surrounds from Figure 2.2A presented such that the left diamond lies in an apparent shadow cast by an object between the light source and the surface. (B) The test targets and surrounds presented as intrinsic components of a uniformly illuminated surface. (C) Difference in the perceived lightness/brightness of the two identical test targets reported by observers. The upper bar shows the average adjustments made to equate the lightness/brightness of the diamond on the dark surround with that of the diamond on the light surround in the scene shown in (A) (as indicated on the left). The lower bar shows the adjustment made to equalize the appearance of the two diamonds on the unshadowed card in (B). (After Williams et al., 1998b.)

illuminant, as in Figure 2.7B, the perceived difference is diminished (the average responses of observers are shown in Figure 2.7C).

In empirical terms the explanation is as follows. Whenever two surfaces returning the same amount of light to the eye are under different intensities of illumination (as in Figure 2.7A), the surface in less illumination (shadow) would, in past interactions with the world, have required a perceptual and behavioral response different from that required by the surface under more intense illumination. Based on the influence of this accumulated experience on visual system connectivity, the target in implied shadow is seen as being lighter than the one in the implied illumination. This perceptual difference occurs not because the visual system is making inferences or in some way "decoding" these different luminance patterns, but simply because responding differently to the targets in this way facilitated successful behavior with physical sources in the past. This same reasoning explains why the perceived lightness difference is diminished in Figure 2.7B.

A more subtle aspect of illumination that can be manipulated to demonstrate the influence of experience on lightness/brightness is penumbras, the hazy borders that occur at the edges of shadows cast by the obstruction of light from the sun and most other sources. Since light sources are generally extended (the sun subtends about a degree of visual angle), the light rays reaching the shadow-casting object do not arise from a single point in space. As a result, the edges of natural shadows are blurred to an extent that depends on the diameter of the source, the distance of the shadow-casting object from the shadowed surface, the clarity of the atmosphere, and a variety of other factors (Minnaert, 1937; Lynch and Livingston, 1995). Thus, penumbras typically signify that the luminance of the target in question entails shadowing rather surface reflectance. Since shadows cast by occluding objects will always have been adorned by penumbras (only point or collimated sources of light fail to generate them), a luminance gradient abutting a contrast boundary should elicit a different perception of lightness/brightness and thus a different behavioral response.

In Figure 2.8A, for instance, there is little or no information to indicate that the dark surround of the left diamond is a shadow. In Figure 2.8B, however, the probability of a shadow lying across the left side of a billboard has been increased by a luminance pattern that is more consistent with this possibility. Nonetheless, because the ends of the dark region are coextensive with the edges of the billboard, the dark bar could also be a surface feature of the sign. In Figure 2.8C, a penumbra has been added to the lateral borders of the bar. By enhancing the likelihood of a shadow, the penumbra increases the probability that the diamond on the left lies in a region with diminished illumination. All these changes elicit the empirically expected differences in the lightness/brightness of the test diamonds (Figure 2.8D).

Although the descriptions of these phenomena might be taken to imply cognitive or unconscious inferential processes are being carried out by the visual brain, in an empirical framework the effects seen are simply reflex responses generated by circuitry that, over time, has linked retinal light patterns to behaviors that have worked in the past.

(A)

Not since Paradise
have apples tasted so good ...
BLUE
DEVIL
CIDER
A DEVIL of a drink.

(B)

Not since Paradise
have apples tasted so good ...
BLUE
DEVIL
CIDER
A DEVIL of a drink.

(C)

Not since Paradise
have apples tasted so good ...
BLUE
DEVIL
CIDER
A DEVIL of a drink.

(D)

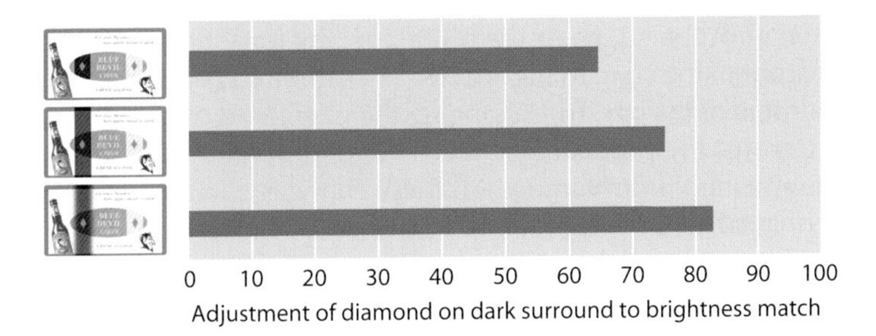
0
10
20
30
40
50
60
70
80
90
100
Adjustment of diamond on dark surround to brightness match

◀ **Figure 2.8** The effect of a penumbra on lightness/brightness perception. (A) Scene in which the dark and light surrounds of the left and right diamonds are presented as surface features of a billboard. (B) Scene in which a bar extending across a portion of the sign is presented in a manner that is to some degree consistent with a shadow. (C) The addition of a penumbra to the possible shadow in (B), thus enhancing the likelihood that the bar is indeed a shadow. (D) Graph showing the average adjustment made by observers to equalize the appearance of the two test diamonds in the three scenes. (After Williams et al., 1998a.)

Empirical Explanations of Other Lightness/ Brightness Phenomena

Qualitative explanations in empirical terms can also be given for other well-known lightness/brightness phenomena. An example is the Cornsweet edge effect, named after the American psychologist Thomas Cornsweet. In Figures 2.9A and B, the territory that adjoins the lighter gradient appears lighter/brighter than the territory adjoining the darker gradient, as shown in Figure 2.9C. This perception is a specific instance of a broader class of edge effects first described by Kenneth Craik in the 1940s (Craik, 1948/1966; see also O'Brien, 1959). The basis of the effect resides in the nature of the edge that separates the two territories that look differently light or bright, since blocking out this portion of the stimulus (a finger works well) immediately abolishes the perceptual difference between the flanking regions (Figure 2.9D). Much like White's effect in Figure 2.3B, the Cornsweet effect belies explanations of lightness/brightness based on distorted retinal output arising from local contrast, since the territory that looks darker is next to the less luminant gradient, and vice versa. Although Cornsweet and others invoked lateral interactions between retinal neurons to account for the effect (Cornsweet, 1970, 1985), there is no generally accepted explanation. The empirical alternative is that the Cornsweet effect is the consequence of experience with what such edges (gradients) in visual stimuli would have required for successful behavior in past experience (Purves et al., 1999).

Luminance gradients are typically produced in one of two ways: (1) by systematic variation in the reflectance properties of surfaces or (2) by systematic variation in the illumination of surfaces (Figure 2.10; see also the previous section on shadows). A luminance gradient arising from the reflectance properties of an object will usually have been associated with adjacent surfaces that are under the same illumination. A luminance gradient that arises from illumination, on the other hand, usually signifies different amounts of light falling on adjacent surfaces that have similar reflectance properties. Thus equiluminant territories adjoining gradients produced by reflectance will typically have been similarly reflective surfaces under the same illuminant, whereas territories adjoining gradients of illumination will typically have been differently reflective surfaces under different amounts of illumination. Since successful responses to these

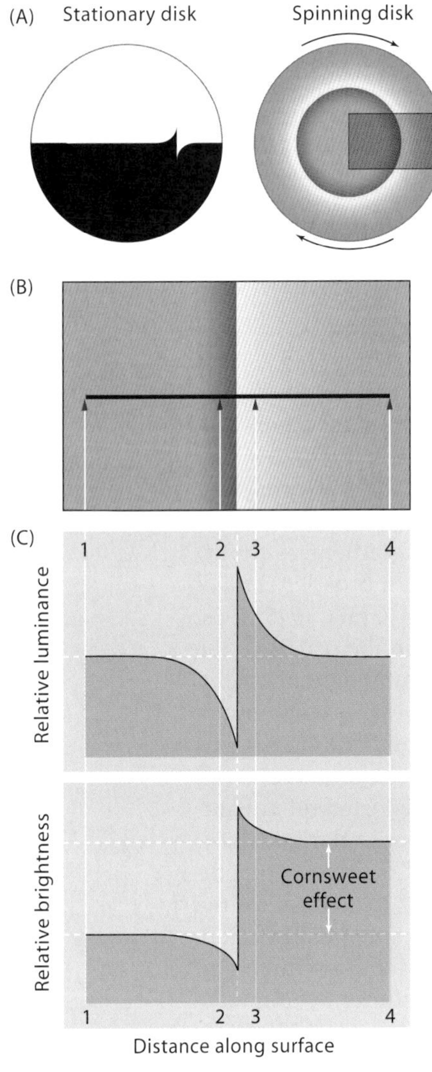

Figure 2.9 The Cornsweet edge effect. (A) Diagram of the spinning disk used by Cornsweet to demonstrate that when two equiluminant regions are separated by an edge comprising a pair of oppositely disposed luminance gradients, the adjoining territories appear differently light/bright. (B) Standard presentation of the Cornsweet stimulus, shown here as a blowup of a portion of the rotating disk (for the sake of simplicity, the edge has been straightened). (C) Comparison of the photometric and perceptual profiles of the stimulus in (B). Despite the equal luminance values of the territories adjoining the two gradients, the territory (1) to the left of the dark gradient (2) looks darker than the territory (4) to the right of the light gradient (3). (D) This effect is abolished by covering up the opposing luminance gradients, as indicated by the inset. (After Purves et al., 1999.)

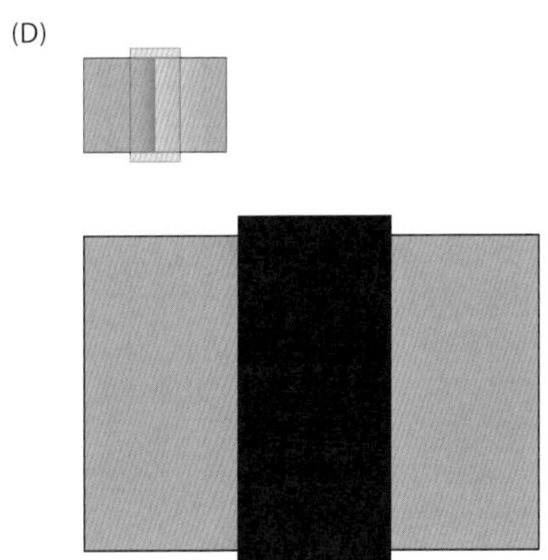

two circumstances will have entailed different behaviors, the associated perceptions of lightness/brightness also differ. In the Cornsweet edge effect, the surfaces abutting the gradients look differently light because of this experience, particularly with differently reflective surfaces under different amounts of light (see the bottom panel in Figure 2.10).

As in lightness/brightness perceptions elicited by simpler stimuli, an empirical framework predicts how changes in the presentation of the Cornsweet stimulus should affect what is seen. For instance, if the information in the stimulus is made more consistent with gradients arising from variations in surface properties (the top

Figure 2.10 Physical sources of Cornsweet edge stimuli. The luminance gradients in the standard Cornsweet stimulus could arise from gradual changes in surface reflectance adjacent to territories having the same reflectance properties and observed under the same illuminant (top panel) or from gradual changes in illumination of two surfaces that have different reflectance properties and are under different illuminants (bottom panel). (Although the illuminated side of the darker cube and the shadowed side of the lighter one in the bottom panel look differently bright, they are actually equiluminant; the perceptual effects under discussion here can't be avoided, even in a didactic illustration.) (After Purves et al., 1999.) ▶

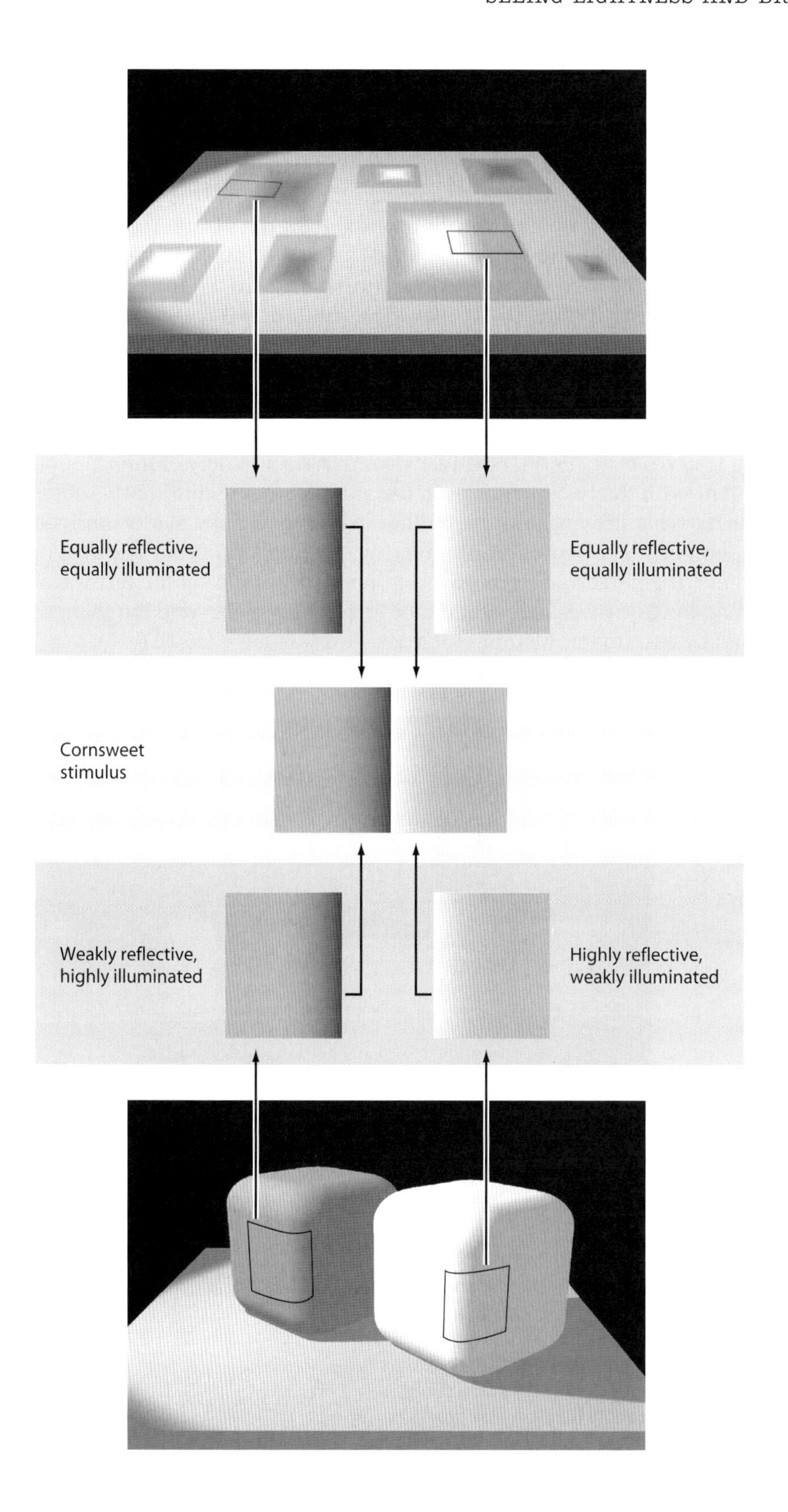
Equally reflective,
equally illuminated
Equally reflective,
equally illuminated
Cornsweet
stimulus
Weakly reflective,
highly illuminated
Highly reflective,
weakly illuminated

panel in Figure 2.10), the perceived difference in the lightness/brightness of the equiluminant regions should decrease. The reason is that the equiluminance of the adjoining territories will, in this case, have signified two surfaces with more or less the same material properties under more or less the same illumination. Sources that are the same require the same behavior and thus should, on empirical grounds, look the same. Accordingly, the Cornsweet effect should no longer be seen. On the other hand, if the information in the scene is made more consistent with a transition in illumination, making it more likely that the equiluminant regions adjoining the gradients are differently reflective surfaces under different amounts of illumination (see the bottom panel in Figure 2.10), then the perceived lightness/brightness difference of the adjacent equiluminant territories should be enhanced. Both of these predictions are borne out in psychophysical testing (Purves et al., 1999). Finally, by combining a variety of information consistent with the two territories in the stimulus being differently reflective surfaces in different amounts of illumination (e.g., information derived from perspective, orientation, shadowing, texture, and other features of the scene), the perceived lightness/brightness difference of the territories adjoining the Cornsweet edge should be increased well beyond the changes induced by each manipulation separately—as it is (Figure 2.11).

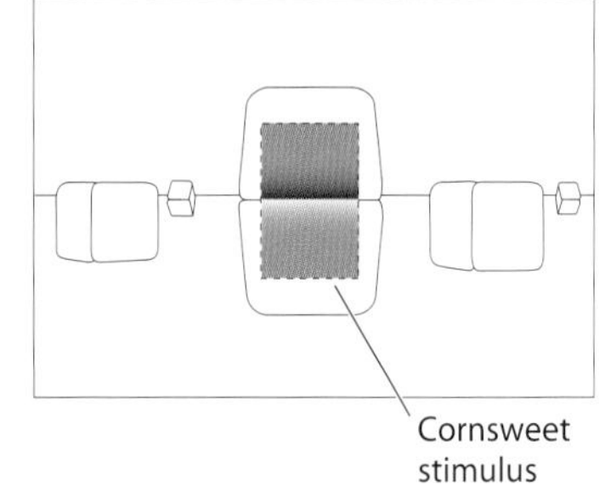

Figure 2.11 Enhancement of the Cornsweet effect by concordant information that increases the likelihood of the territories adjoining the opposing luminance gradients arising from two differently reflective surfaces under different levels of illumination. The inset shows the location of standard Cornsweet stimulus in the scene. By combining a variety of mutually reinforcing information in a scene, the Cornsweet effect can be enhanced more than ten-fold over the standard presentation in Figure 2.9B. (After Purves et al., 1999.)

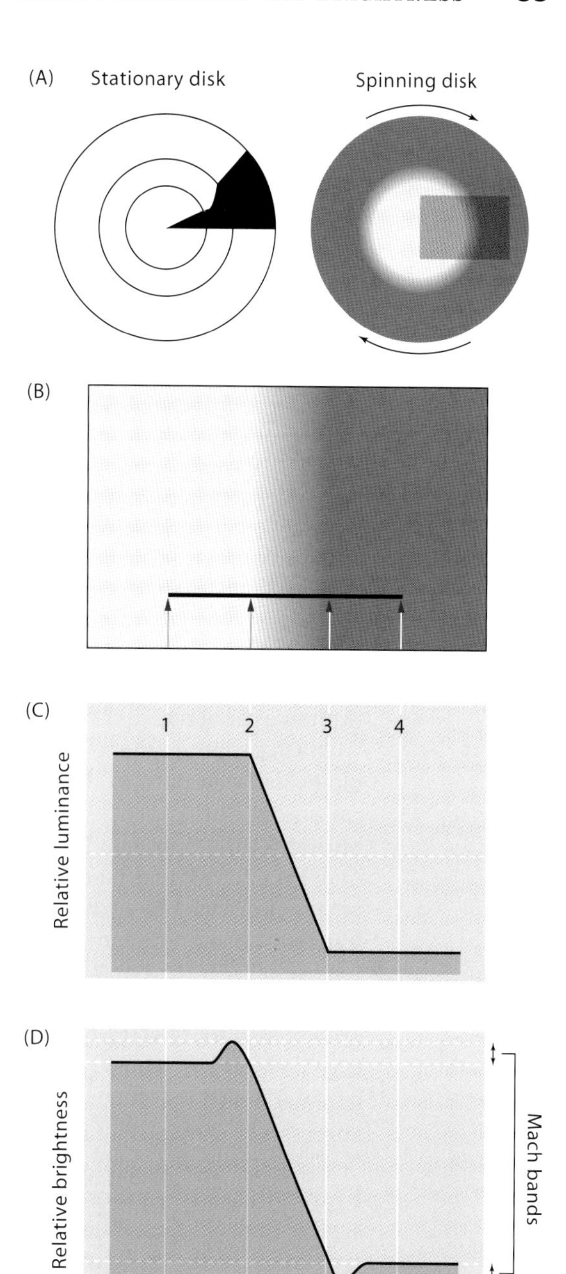

Figure 2.12 Mach bands. (A) The stimulus used by Mach. When this disk is spun, a luminance gradient is established between the uniformly lighter center of the disk and the uniformly darker region at its periphery. (B) Blowup of a portion of the spinning stimulus in (A), indicating the nature and position of Mach bands (the curvature of the disk has been removed for simplicity). The stimulus elicits a band of perceived maximum lightness at position (2) and a band of maximum darkness at position (3), neither of which is present in photometric measurements. (C) Because the portion of the black sector between points (2) and (3) in (A) is a segment of an Archimedean spiral, the luminance gradient generated between the corresponding points on the spinning disk is linear, as indicated by photometric measurement along the line in (B). (D) Graph of the relative lightness/brightness seen by observers, indicating the illusory lightness maximum just before the initiation of the linear gradient (2) and the illusory minimum just after its termination (3). These graphs represent the actual luminance values in (B). (After Lotto et al., 1999a.)

Another challenge for any explanation of lightness/brightness effects is Mach bands. These bands, first reported by the physicist Ernst Mach in 1865, refer to regions of spuriously increased darkness and lightness seen at the onset and offset, respectively, of luminance gradients (Figure 2.12) (Mach, 1866/1959; Ratliff, 1965). The stimulus used by Mach was a disk with black and white sectors that, when spun, generated a linear gradient that linked a uniformly lighter region occupying the center of the disk to a uniformly darker region occupying the periphery (Figure 2.12A–C). In response, observers perceive a band of maximum lightness at the initiation of the gradient and a band of maximum darkness at its termination (Figure 2.12D). The peculiar character of Mach bands and the wealth of literature seeking to explain these phenomena (summarized in Floyd Ratliff's centennial volume on the subject in 1965) presents a daunting test of the idea that

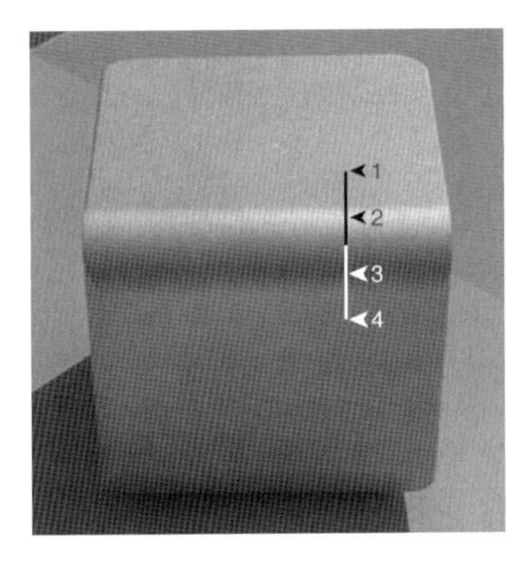

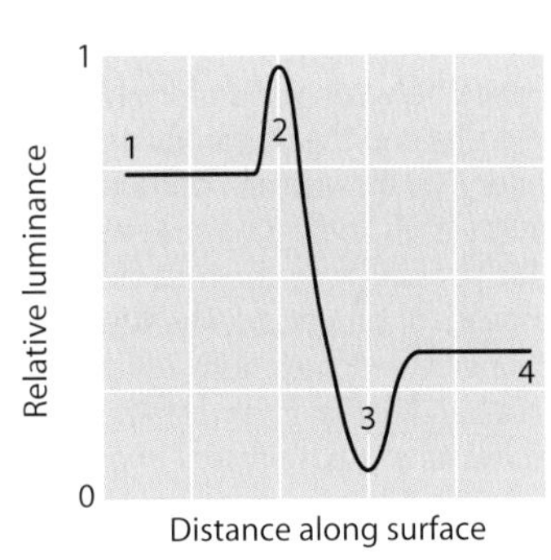

Figure 2.13 Photograph of an aluminum cube in sunlight. The highlights and lowlights that adorn the surfaces of many real-world objects under typical conditions of reflectance and illumination are apparent in both the picture and the accompanying photometric measurement. Numbers indicate corresponding points on the surface of the cube and the photometric profile. (After Lotto et al., 1999a.)

all lightness and brightness perceptions can be understood in terms of the human experience that has linked visual stimuli to successful behavior.

Mach proposed that the illusory bands are a direct consequence of physiological interactions in the retina, and he elaborated a detailed mathematical model of this process based on reciprocal inhibition between neighboring retinal points, a theory that he continued to modify and improve over several decades (see Ratliff, 1965 for English translations of this series of papers). Despite a great deal of additional work on Mach bands during the subsequent century, Mach's reasoning, at least in general terms, has remained the conventional explanation of this striking effect (e.g., Ratliff, 1965; Cornsweet, 1970, 1985; Coren et al., 1999, pp.110ff.; Arend and Goldstein, 1987; Grossberg, 1987; Goldstein, 2000, pp. 66ff.; Eagleman, 2001). However, Mach bands, like the Cornsweet edge effect, can also be explained as a result of human experience with the responses that lead to successful behavior in the face of uncertain sources of luminance gradients.

The relative frequency of occurrence of luminance patterns determined by the physics of light coming to the eye from natural scenes works equally well in rationalizing Mach bands. As for other edge effects, human perception and behavior will have been molded by how often luminance gradients have arisen from the graded illumination of curved surfaces, the changing reflectance values of surfaces, and the penumbral gradients generated by cast shadows (see Figures 2.7, 2.8, and 2.10). If the visual system evolved and developed to respond to these gradients on a wholly empirical basis, then the perceptual response to a Mach band stimulus should accord with this experience and not the characteristics of the stimulus as such. Among the sources of luminance gradients are those produced by curved surfaces, which have physical maxima and minima in the exact location of Mach bands (Figure 2.13). It follows that any perceptual response to a luminance gradient will be influenced by this fact in proportion to the frequency of occurrence of images in which this association was helpful in generating a successful behavioral response. Thus in empirical terms, Mach bands are simply another manifestation of lightness and brightness perceptions having been shaped according to the relative success of behavioral responses to patterns of retinal luminance.

The Empirical Effects of Experience with Transparency

Understanding edge effects and Mach bands concerns experience with the way illumination gradients and surface properties have informed successful behavior. Another empirical influence on lightness/brightness perceptions is transmittance, which changes the amount and quality of light reaching the eye (see Figure 2.4). One would thus predict other empirically explainable effects on perception arising from this aspect of experience.

A phenomenon that confirms this prediction is the Chubb effect (Chubb et al., 1989). The relevant stimulus entails the differential lightness/brightness of randomly arranged elements in a patterned target. The perceptual effect is that the apparent contrast between the target elements is less when they are embedded in a high-contrast background that has the same spatial frequency as the target (Figure 2.14A) compared to the same target pattern presented in a uniform background (Figure 2.14B). This perception occurs despite the fact that the target patterns are identical and that the average luminance of the surrounding territory in the two presentations is the same.

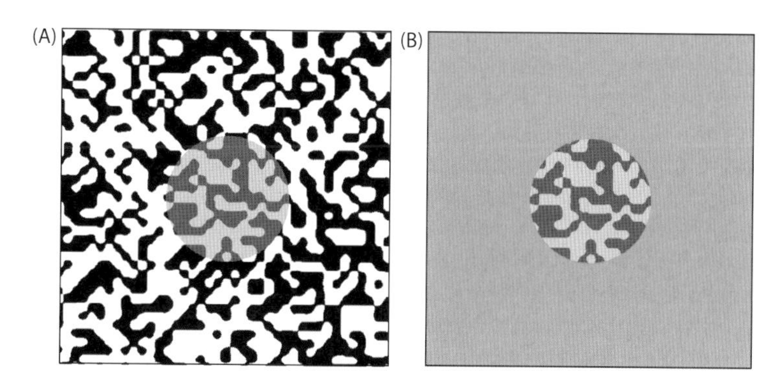

Figure 2.14 The Chubb effect. A patterned target (the circular patch) embedded in a surround in which the same pattern has a higher luminance contrast (A) appears to have less contrast than an identical target embedded in a uniform surround of the same average luminance (B). The inset below shows the patterned elements of the two identical targets when presented without their respective surrounds. (After Lotto and Purves, 2001.)

The interpretation of this effect by investigators who studied it was based on inhibitory interactions among cortical neurons "tuned" to similar spatial frequencies (Chubb et al., 1989; see also Olzak and Laurinen, 1999), and thus an anomalous by-product of what the authors call "contrast gain control" (reviewed by D'Zmura, 1998). The presumed response of these neurons to the contrasting target elements in Figure 2.14A was taken to be diminished in comparison with the response to Figure 2.14B because the higher contrast of the elements in the surround in Figure 2.14A more vigorously activated inhibitory connections between the relevant cells than did the stimulus pattern in Figure 2.14B (in which the surround has the same average luminance but no spatial frequency component). If perceptions of contrast were a more or less direct manifestation of the relative activity of such spatial contrast frequency "detectors," then the lightness/brightness values of the target elements would be expected to appear more similar in Figure 2.14A than in Figure 2.14B, as they do.

There is, however, a simpler empirical explanation of the Chubb effect. All natural scenes are seen through semitransparent media (e.g., the atmosphere, water vapor, water, glass, etc.). And, to a greater or lesser degree, the medium affects the amount of light that reaches the eyes from objects in the world. This influence depends on transmittance, defined as the amount of light reaching a detector compared to the amount of light reflected from object surfaces or generated by objects, expressed as a percentage. The relative clarity of the atmosphere means that transmittance in most circumstances is slight. However, viewing objects at a distance, nearby objects in fog or smog, or objects through semitransparent liquids (water) or solids (glass) are all common—and consequential—determinants of the amount and quality of light that falls on the retina and initiates visual perception.

These consequences of imperfect transmittance are illustrated in Figure 2.15A. If, for example, two target surfaces reflect, respectively, 80 percent and 30 percent of the incident light, the light returned from the more reflective surface in perfectly transmitting conditions will be greater than that returned from the less reflective surface by a ratio of 8:3. If, however, the light returned from same surfaces passes through an imperfectly transmitting medium, this ratio is decreased. Although the interposition of such a filter reduces the amount of light coming from the two surfaces in question, some light is also added to the luminance values attributable to the surfaces because the material of the medium reflects light. Since this reflected light is added equally to any return from a surface viewed through the medium, the result is that the luminance attributable to the less reflective target surface is increased to a greater degree than the luminance associated with the more reflective surface. Consequently, the difference in the luminance of the two target surfaces is reduced, in this example from a ratio of 8:3 to about 7:5 (see also Metelli, 1970, 1974; Metelli et al., 1985). In short, an imperfectly transmitting medium, irrespective of its particular properties, typically reduces the luminance differences between differently reflective surfaces seen through the medium (Figure 2.15B).

(A)

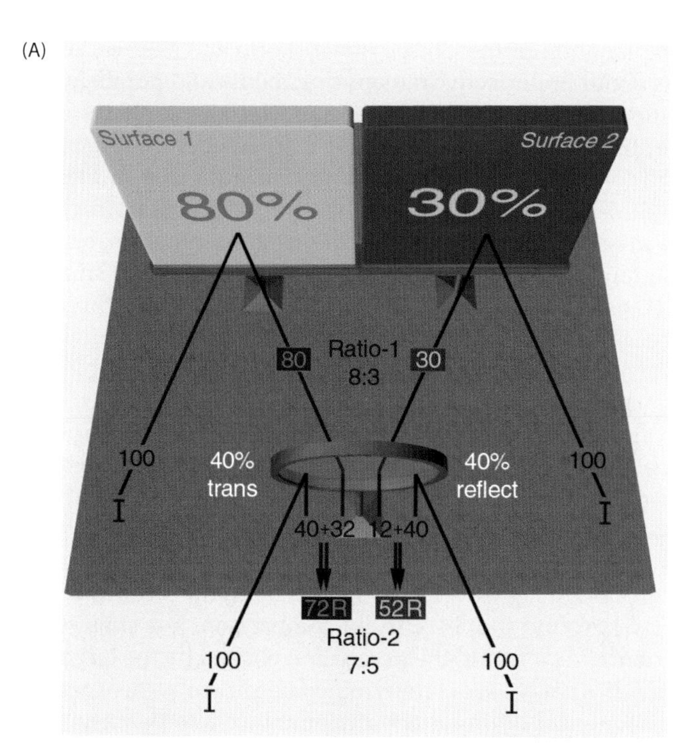

(B)

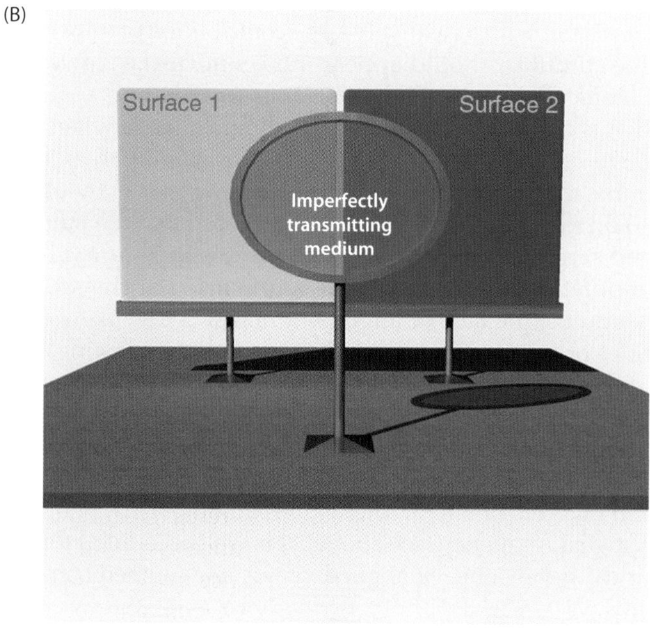

Figure 2.15 The physical effects of transmittance and their perceptual consequences. (A) An imperfectly transmitting medium interposed between object and observer reduces the luminance differences between the light reaching the eye from surfaces seen through that medium (see text for further explanation). (B) Illustration showing how the conditions diagrammed in (A) would typically change the appearance of the relevant surfaces. (After Lotto and Purves, 2001.)

An empirical explanation of the Chubb effect (Lotto and Purves, 2001) rests on these facts about imperfectly transmitting media and parallels the empirical explanations of the Cornsweet edge and Mach band effects. The information in the stimuli in Figures 2.14A and B are inherently uncertain with respect to the contribution of transmittance: as indicated in Figure 2.4, the identical luminance patterns from the central targets in the two panels could have arisen from many combinations of illumination, reflectance, and transmittance. However, past experience with the two stimuli makes imperfect transmittance a more likely contributor to the stimulus depicted in Figure 2.14A than to the stimulus depicted in Figure 2.14B, and thus more likely to have been the determinant of successful behavior in response. This increased likelihood arises because: (1) the patterned elements of the surround in Figure 2.14A are continuous across an implicit border with the patterned elements of the target and (2) the diminished luminance values of the target elements compared with the elements in the surround accord with values that would have occurred if the central target were being viewed through an imperfectly transmitting medium. The uniform background in Figure 2.14B, although it has the same average luminance as the background in 2.14A, is, by comparison, less consistent with human experience. As a result, the perception elicited by the target in Figure 2.14A should, if what we see is determined empirically, incorporate the behavioral consequences of imperfect transmittance into the response elicited by the stimulus in Figure 2.14B. Thus, the elements of the target in 2.14A, which are consistent with a relatively large contribution of imperfect transmittance to the stimulus, should appear more similar (i.e., lower in contrast) than the target elements in Figure 2.14B, as they do.

If this explanation is correct, the Chubb effect should change when the stimuli like the one in Figure 2.14A are made less consistent with experience viewing textures through imperfectly transmitting media by other means. One way to do this it to rotate the central target so that the pattern is no longer aligned across the target boundary, as it is in Figure 2.14A. Whereas this manipulation alters neither the luminance nor the spatial frequency of the stimulus, the lack of alignment makes it less likely that the luminance values of the target pattern are generated by viewing this part of the scene through a filter, and more likely that they arise directly from differences in surface reflectance properties. The apparent contrast of the target elements is indeed increased when the alignment is abolished (see www.purveslab.net for a demonstration). Although this result has no obvious explanation in terms of lateral interactions among visual neurons similarly tuned to spatial frequency (see above), it is well accounted for by the different empirical significance of aligned versus non-aligned components in the stimulus.

Another way of making the Chubb stimulus less consistent with experience viewing textures through imperfectly transmitting media is changing the luminance relationships between the targets and surrounds. The luminance difference between the elements of the surround in Figure 2.16A, for example, is 3 times greater than the luminance difference between

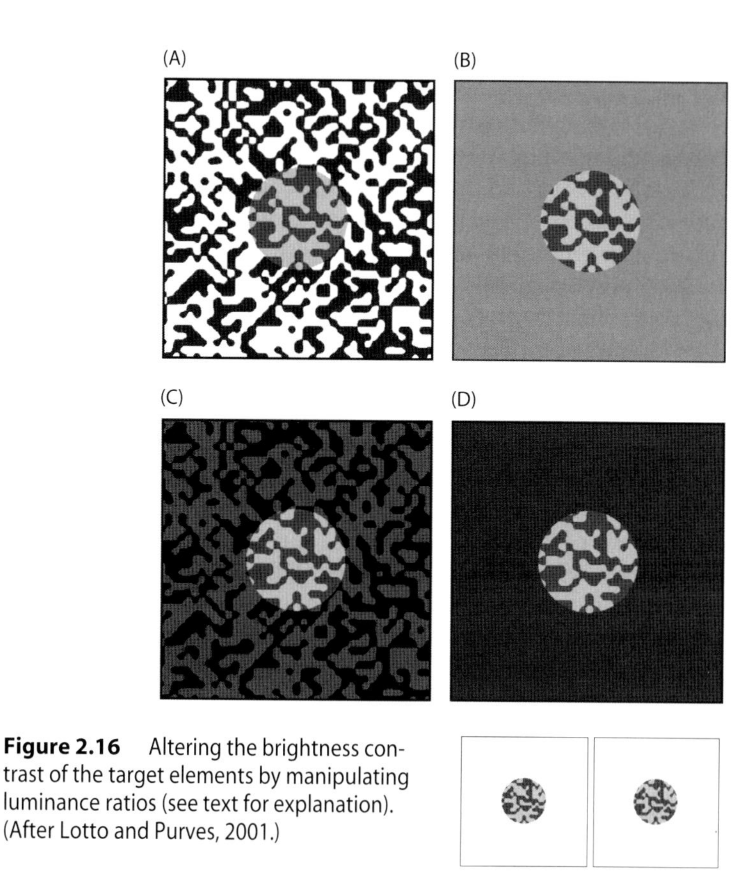

Figure 2.16 Altering the brightness contrast of the target elements by manipulating luminance ratios (see text for explanation). (After Lotto and Purves, 2001.)

the elements of the target. As in Figure 2.14A, this ratio of luminance values is consistent with the stimulus arising from an imperfectly transmitting medium interposed between the observer and the central target and gives rise to the same effect (the stimuli in Figure 2.16A and B differ from those Figure 2.14A and B only in that all the luminance values have been adjusted for the purposes of this demonstration). A quite different result is obtained, however, when the average luminance of the two surrounds are both decreased, as in Figures 2.16C and D (the target luminance values remaining the same). Although the luminance contrast of the elements in the surround remains 3 times greater than that of the elements in the target, this manipulation makes the stimulus in Figure 2.16C less consistent (indeed, incompatible) with a contribution of transmittance to the light returned from the target (the condition represented is the empirical equivalent of shining a spotlight on the target). This inconsistency arises because an imperfectly transmitting medium overlying the central region of the surrounding pattern could not increase the luminance values of both the light and dark elements of the pattern in the surround to provide the luminance values that are now being returned to the eye from the target. As a result, the contrast of the target elements in Figure 2.16C actually appears somewhat greater than that of the same target on a uniform surround in Figure 2.16D. Although this result is expected on empirical grounds, it is again difficult to rationalize in other terms.

These several observations contradict explanations of the Chubb effect based on the anomalous activation of inhibitory connections between cortical or other neurons similarly "tuned" to detect spatial contrast frequencies in the stimulus. The phenomena are also inconsistent with interpretations based on "contour junctions" (see Adelson, 2000), since such junctions are not explicit in the Chubb stimulus (as pointed out by Anderson, 1998). The common denominator of all these effects is simply the empirical significance of the stimulus for successful behavior.

The Effects of Luminance Scaling in Empirical Terms

In addition to rationalizing these specific phenomena, an empirical framework can also explain the general relationship between luminance and lightness/brightness observed in experimental paradigms referred to as "scaling." A longstanding objective in neuroscience has been to understand the relationship between the physical characteristics of any class of stimuli and the overall "shape" of the relevant perceptual space (i.e., the subjective organization of perceptual qualities such as lightness and brightness). This issue was first taken up in the nineteenth century by Ernst Weber and Gustav Fechner (Weber, 1834/1996; Fechner, 1860/1966; see also Laming, 1997), and carried forward in the twentieth century by investigators such as Harvard University psychologist Stanley Stevens (1966, 1975). The approach Stevens took was to present a full range of stimuli in the same simple context (e.g., a light source whose intensity could be varied, or pieces of paper ranging in reflectance from dark to light) and ask subjects to rate their perceptions on a subjective scale.

As shown in Figure 2.17, Stevens and others found as many peculiarities in scaling tests as in responses to complex stimuli such as the Cornsweet edge, the Mach band stimulus, or the Chubb stimulus. In particular, doubling the intensity of light coming from a source or reflected from a surface does not simply double perceived brightness. For increments of the test light above the background luminance (the portions of the functions to the right of each of the dashed lines in the figure), the lightness/brightness seen is greater than expected for relatively small increases in the target luminance, but less than expected as target luminance values increase (Stevens, 1966; Stevens, 1975; Laming, 1997). This relationship between luminance and lightness/brightness measured under standard conditions in the laboratory is often called "Stevens' Law." Another peculiarity is that this relationship varies according to the circumstances of testing. When, for instance, the light sources of different intensities are presented in a dark background, the exponent of the Stevens Law function is approximately 0.5; if, however, the target intensities arise from the reflectance values of

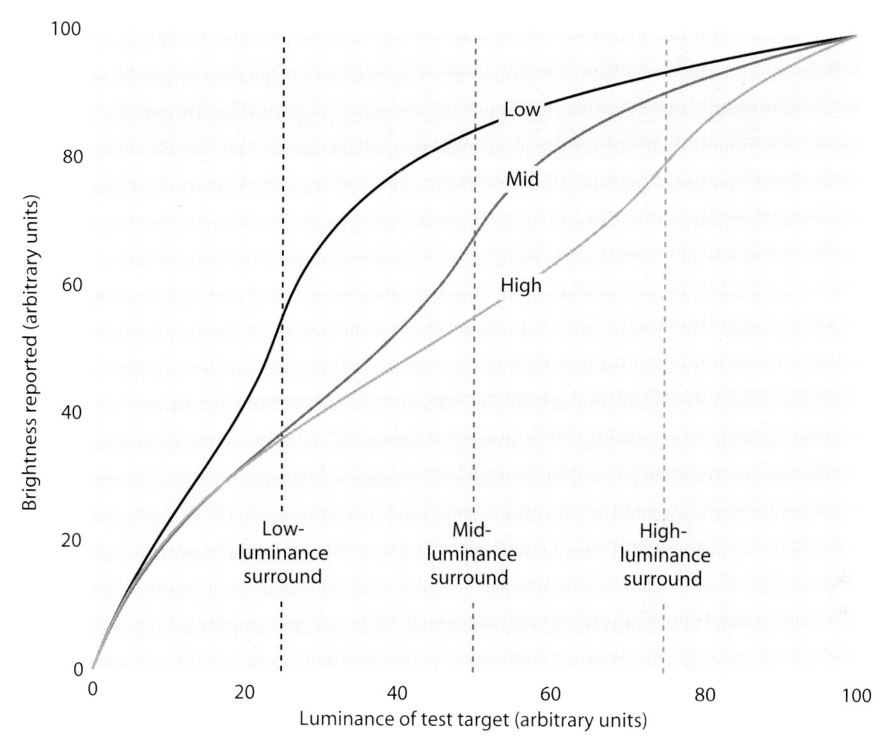

Figure 2.17 The major features of lightness/brightness scaling. The graph shows scaling tests at three levels of background luminance; the three vertical dashed lines indicate the point at which the luminance of the target is the same as the background. The features include: (1) the non-linearity of the relationship, (2) the different form of the function for increments and decrements, (3) the different shape of the relationship as a function of the background luminance, and (4) the steeper slope of the relationship when the luminance of the target is near that of the background. The units are arbitrary. (After Nundy and Purves, 2002.)

a range of gray papers, then the exponent of the psychophysical function approaches 1. Other anomalies are that the slope of the relationship is more or less opposite for increments and decrements, that the relationship varies as a function of the background luminance, and that the slope is greatest when the luminance of the test target is similar to the luminance of the background. There is no generally accepted explanation of this phenomenology, although various models have been proposed (Whittle, 1992; Laming, 1997; Gilchrist, 2006).

The empirical explanation of these observations is shown in Figure 2.18. Illumination (*I*) and reflectance (*R*) are plotted as the orthogonal dimensions of a graph that (very crudely) represents human experience with the relative contributions of these two major factors that commonly determine the luminance of a target (the less typical circumstance of luminance generated by a light source is taken up below). Each yellow line in the figure

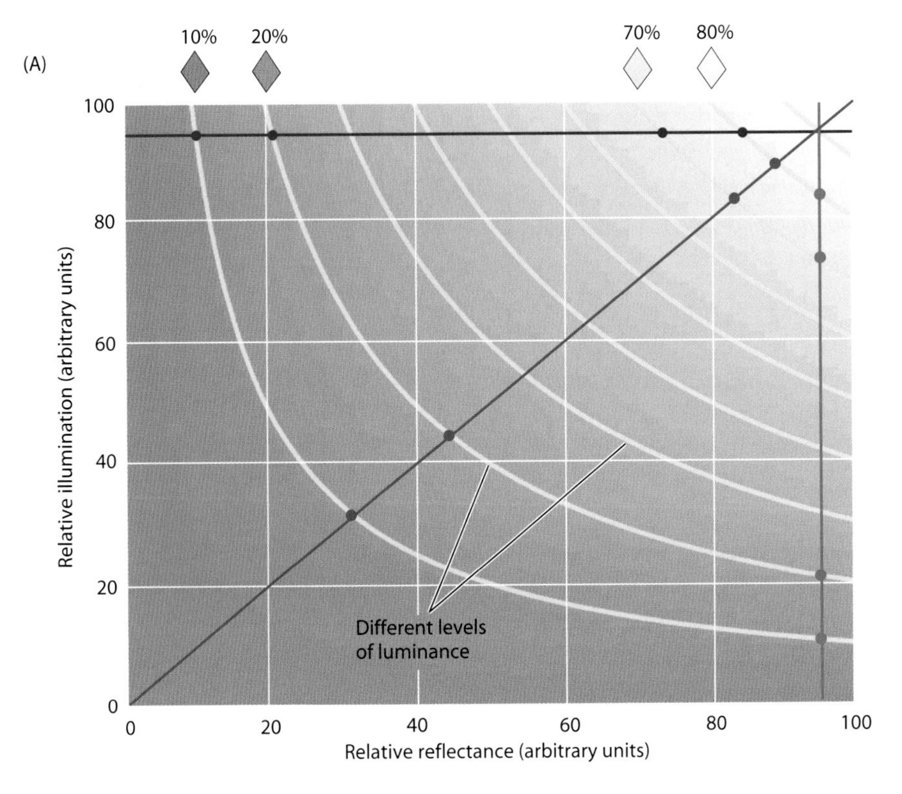
(A)
10%
20%
70%
80%
Relative illumination (arbitrary units)
Relative reflectance (arbitrary units)
Different levels of luminance
0
20
40
60
80
100

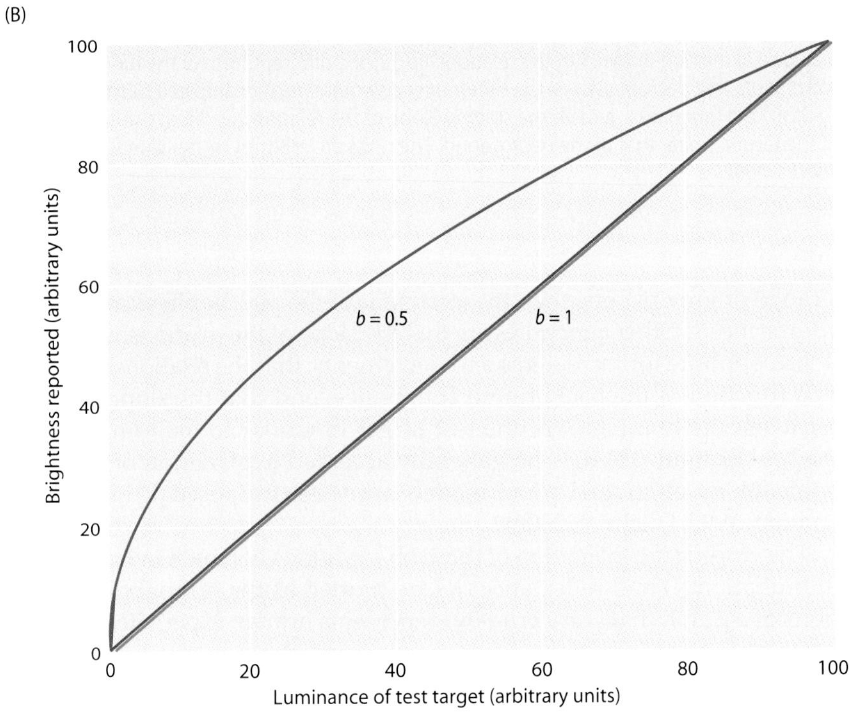
(B)
Brightness reported (arbitrary units)
Luminance of test target (arbitrary units)
b = 0.5
b = 1
0
20
40
60
80
100

◀ **Figure 2.18** Understanding luminance scaling in empirical terms. (A) Graph showing the possible combinations of illumination and reflectance that could have generated particular values of luminance. The relative reflectance (*x* axis) and illumination (*y* axis) values are plotted in arbitrary units relative to the average values of these parameters in a hypothetical scene. The three differently colored lines show the effects on lightness/brightness scaling expected as the empirical significance of the stimulus varies (see text for explanation). (B) The scaling relationships between luminance and lightness/brightness predicted by the empirical framework in (A). If the information in a stimulus is consistent with the target luminance arising predominantly from illumination *or* reflectance, the exponent of the function should approach 1 (i.e., the psychophysical function should track the evenly spaced intervals along the green or the blue line in [A] and [B]); if, on the other hand, the information in the stimulus is consistent with a more equal contribution of illumination and reflectance, then the exponent should approach 0.5 (i.e., the function should track along the red lines in [A] and [B]). (After Nundy and Purves, 2002.)

represents all the possible (I, R) combinations that could have given rise to that particular luminance value. The likelihood that any particular (I, R) combination along one of these isoluminance lines is the generative source of the luminance an observer must respond to is a function of the probability distribution of the (I, R) values over the yellow lines and can be thought of as a third dimension of the graph (i.e., coming up out of the page). The diagram thus provides a rough guide to how the luminance–lightness/brightness relationships would, on empirical grounds, be expected to vary. If, for example, the conditions in a given brightness scaling test changed the probable contribution of either illumination or reflectance to the luminance of the target, then the probability distribution of possible (I, R) values along any isoluminance line in Figure 2.18A would change accordingly. If the testing conditions increased the probability that the range of target luminance values derives from changes in reflectance, then the central tendency of the probability distribution of possible (I, R) combinations that could have given rise to a particular target luminance would shift toward the blue line. If, on the other hand, the conditions increased the probability that the luminance of the targets derived predominantly from changes in illumination, then the central tendency of probability distributions of the (I, R) values would shift toward the green line. Finally, if the experimental conditions were consistent with past experience that provided little or no bias informing behavior directed about the sources of the luminance values, or if experience indicated that the likely contributions of illumination and reflectance were about the same, then the distribution of the possible (I, R) combinations underlying any luminance value would track more centrally with respect to the isoluminance lines, as indicated by the red line in Figure 2.18A.

The scaling functions predicted by this empirical argument are shown in Figure 2.18B, again as rough approximations. If perceptions of luminance values are determined by behaviors pertinent to dealing successfully with generative sources that cannot be known from luminance as such, they should vary as a function of the distances between the isoluminance lines in Figure 2.18A. Thus, if a testing paradigm biases the probable sources of luminance values toward a predominant contribution from either illumination or reflectance, the differences in target brightness reported by observers should vary according to the smaller and more equal distances between isoluminant values along the blue or green lines shown in Figure 2.18A. The resulting scaling function would thus tend to follow the correspondingly colored lines in Figure 2.18B. Conversely, if the test conditions are consistent with experience due to a more equal contribution of illumination and reflection to luminance, then the brightness difference elicited by different luminance values would track along the red line in the Figure 2.18A, and the scaling function should tend to correspond to the red curve in Figure 2.18B.

This framework explains why Stevens and others observed different results when papers were used as test objects instead of a variable source of light. Recall that when Stevens tested light sources in his scaling paradigm, the psychophysical function of brightness that subjects reported tended to have an exponent of approximately 0.5 (see the red line in Figure 2.18B). The empirical explanation is that experience with light sources would have shown that their intensity is not much affected by illumination or reflectance. As a result, the perception of luminance values arising from light sources should also track along the red line and curve, respectively, in Figure 2.18A and B.

Rudimentary as it is, an empirical framework indicates how the accumulation of experience instantiated in visual circuitry according to feedback from the relative success of behavior in the world can explain what people report seeing in scaling tests of lightness and brightness.

Evidence Derived from Natural Scene Data

Ultimately, the way to validate a wholly empirical framework of perception is to predict what people see based on the retinal light patterns humans have always experienced in natural settings and used to link visual stimuli with useful perceptions and behavior. Although this sort of analysis remains difficult, it can be carried out in limited ways for a basic category of visual perception such as lightness/brightness using a database of natural scenes—black and white scenes, in this instance (Figure 2.19). A large collection of such scenes gathered today can be reasonably taken to represent luminance patterns much the same as those our ancestors would have experienced. Since luminance is a physical measure, the frequency of occurrence of different luminance patterns indicates the relative value of the trial-and-error responses that would have linked the patterns in images to apposite behaviors. It may be helpful to refer back to the dice analogy in Chapter 1 to be reminded of how this empirical linkage works based on tallying up the frequency of occurrence of different patterns (or targets in different contexts).

Figure 2.19 Examples of natural scenes that, in large enough numbers, can serve as a proxy for human experience with luminance patterns. (From Van Hateren and Van der Schaaf, 1998.)

The frequency of occurrence of different target luminance values in dark and light surrounds, for instance, can be determined by applying an appropriate template millions of times to thousands of natural scenes such as those depicted in Figure 2.19 (Figure 2.20A) (Yang and Purves, 2004). For perceptual qualities such as lightness and brightness, the percept seen should accord with the percentile rank of the target luminance among all possible luminance values that co-occur naturally in that pattern. A percentile target rank of 0 corresponds to the perception of maximum darkness, a rank of 100 to maximum lightness or brightness.

To understand how this approach can be used to predict what people should see on empirical grounds, consider a standard simultaneous lightness/brightness stimulus in which equiluminant patches are presented in the dark and light surrounds (Figure 2.20A). As shown in Figure 2.20B, in nature less luminant targets occur far more often in dark surrounds than in light surrounds and vice versa, simply because surfaces made of the same "stuff" in the same illumination are more common. Figure 2.20C indicates

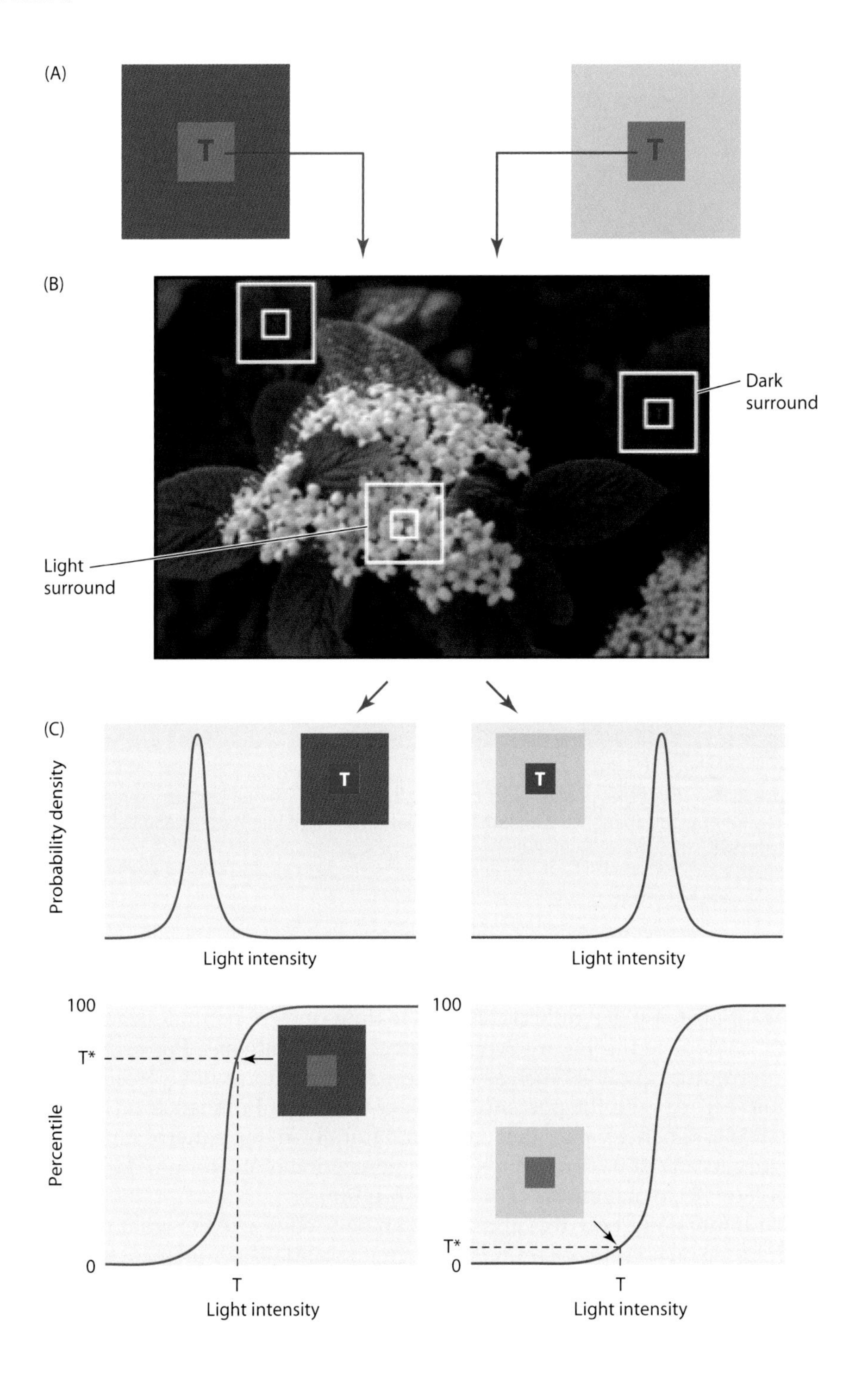
(A)
T
T
(B)
Dark
surround
Light
surround
(C)
Probability density
T
T
Light intensity
Light intensity
100
T*
Percentile
0
T
Light intensity
100
T*
0
T
Light intensity

◀ **Figure 2.20** Using natural scenes to predict the lightness/brightness of the same target in different contexts. The stimulus configurations used as templates (A) to repeatedly sample natural scenes (B). In this way, the probability of occurrences of the luminance of the target (T) in the different contexts—a light surround versus a dark surround, in this example—can be determined. The graphs in (C) show that the lightness/brightness percepts elicited by the luminance of the targets in (A) are predicted by the percentile rank of that luminance in the probability distribution functions (i.e., the integrals of the probability densities). (After Yang and Purves, 2004.)

that for any target luminance value within the limits set by the values of the two surrounds, many more images will have been experienced in which the luminance of the central target in the dark surround is relatively low; conversely, target luminance values in the lighter surround will more often have been relatively high. Thus, in terms of frequency of occurrence, the percentile rank of the target in the dark surround will be much higher than the rank of the same target in lighter surround. Since, in empirical terms, behavioral success will have depended on these ranks, the target in the dark surround should be seen as lighter/brighter than the target in the light surround. This perceptual effect occurs not because seeing the targets in this way better represents reality (it obviously does not), but as a signature of the empirical strategy that vision is using to resolve the inverse problem.

The empirical information gleaned in this way using other template patterns predicts many other lightness/brightness effects, including the responses elicited by the brightness contrast stimuli in Figure 2.3 (see Yang and Purves, 2004).

Evidence from Autonomous Agents Behaving in Virtual Environments

Another way of validating an empirical explanation of lightness/brightness is to ask whether artificial neural networks with visual input that are trained (or evolved) in virtual environments that simulate the relevant aspects of the real-world respond to luminance values in the same general way we do (Corney and Lotto, 2007; Boots, et al., 2007). An advantage of this approach is the possibility, in principle, of exploring how the organization of networks in trained or evolved agents might be relevant to understanding circuits in the visual brain.

Figure 2.21 shows an example of luminance patterns that mimic a variety of natural surfaces under heterogeneous illumination, which artificial neural networks can be trained to distinguish (Corney and Lotto, 2007). When presented with two identical stimuli in light and dark surrounds, the networks respond to the targets in the darker surrounds as if they were more reflective surfaces than the targets in the lighter surrounds.

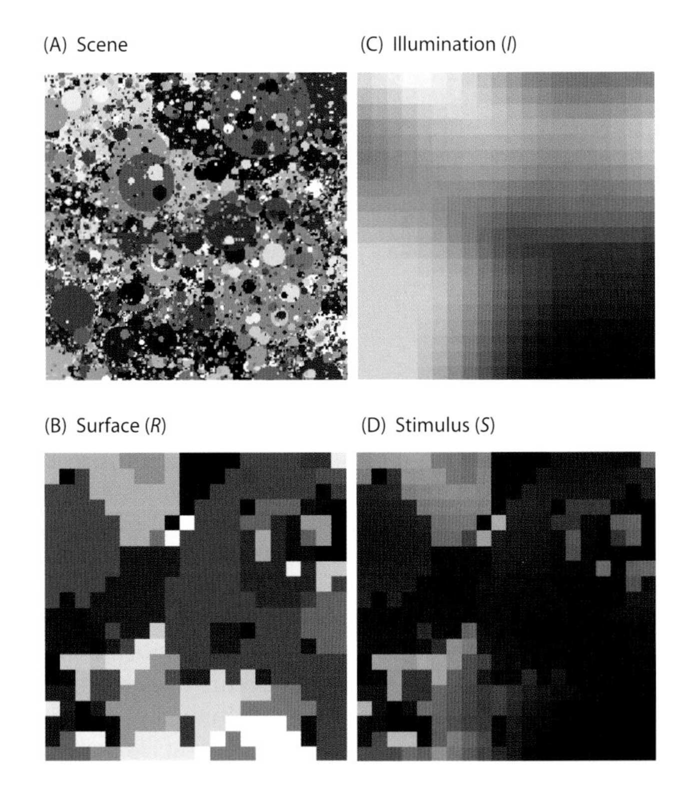

Figure 2.21 Training artificial neural networks on luminance patterns. (A) Example of a large area of surfaces ("Scene") used to train agents. The scene comprises overlapping circular disks with intensities drawn independently from a uniform distribution, which results in images that have much the same statistics as natural images. (B) A 20 × 20 pixel matrix with reflectance values (*R*) chosen randomly from the scene in (A); 40–60 disks are present in such samples. (C) The illumination (*I*) falling on a typical surface modeled here to represent gradual changes across the stimulus. (D) The stimulus (*S*) based on the product of the (*R*) and (*I*) values in (B) and (C). See text for explanation. (From Corney and Lotto, 2007.)

Moreover, the networks show asymmetrical responses to darker versus lighter backgrounds, which is again similar to what humans perceive (see Figure 2.17). The presumptive reason is that a lower luminance surround changes the behavioral significance of the target relative to the lighter surround, which is likely to be a surface that reflects more light. As a result, the dark surround in a brightness contrast stimulus has a stronger effect on the perception of its central target than does the light surround (see Blakeslee and McCourt, 1999; Gilchrist et al., 1999; Nundy and Purves, 2002). Furthermore, when multiple elements in a scene are made mutually consistent with a given empirical significance, the responses of the artificial networks indicate they further "overestimate" the lightness of a target in a darker surround compared with the way they respond to the same target in a lighter surround, much as we do.

Further evidence comes from the responses of trained networks to variations in the number of surfaces in a stimulus. When identical targets are presented on light and dark surrounds that comprise a patchwork of elements, human observers report an increase in the contrast between the targets (Anderson and Rosenfeld, 2000; Maloney and Schirillo, 2002; Gilchrist, 2006). Again, the behavioral responses made by trained networks are much the same as those made by human observers (Corney and Lotto, 2007). Such networks also show variations in behavior as a function of stimulus structure, including contrast and luminance gradients (Figure 2.22). Thus,

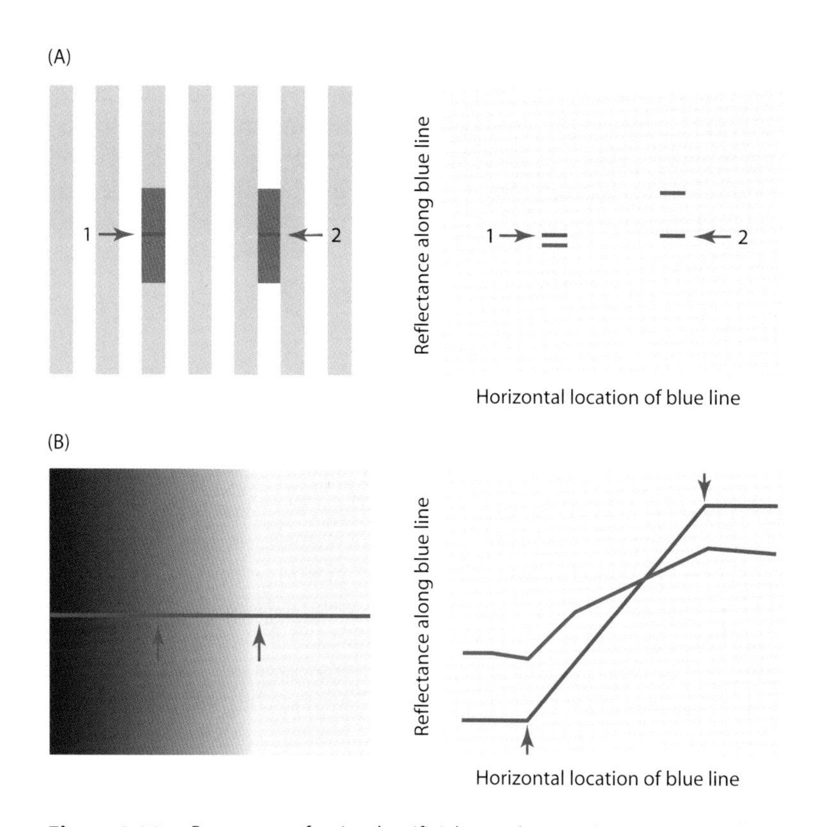

Figure 2.22 Responses of trained artificial neural network to more complex stimuli. The left panels are the test stimuli and the right panels the judgments made by the trained networks. Blue lines in the panels indicate luminance values; the red lines show the reflectance values predicted by the behavioral responses of the networks. The blue arrows indicate the components of stimulus that generate the relevant perceptions in human subjects. (A) Responses to White's stimulus. Although the intensity of the stimulus at the locations of the two blue arrows is the same, the networks predict these regions to have different reflectances, specifically that the area indicated by arrow 2 is more reflective than the area indicated by arrow 1. This is much the same way we see such stimuli. (B) Responses to a Mach band stimulus. The predicted responses of the networks to the stimulus at the locations of the arrows is not flat, as expected from the stimulus intensity at that location. Instead there is a dip in the predicted reflectance "seen" by the networks at the site of the left arrow, and a slight rise in the predicted reflectance at the site of the right arrow. These responses again correspond to the perceptions of human subjects. (From Corney and Lotto, 2007.)

their responses track human perceptions produced by White's stimulus (Figure 2.22A; see also Figure 2.3B) and Mach band stimuli (Figure 2.22B; see also Figure 2.12).

Finally, artificial neural networks can be evolved in virtual environments in which new generations are produced by parent networks according to fitness determined by the relative success of, for example, their ability to avoid collisions with the walls of an arena (Schlessinger et al., 2005; Boots and Purves, 2007). In this paradigm a genetic algorithm preserves the architecture of the most fit networks in each generation. The result is that the artificial networks explore the environment more and more efficiently as evolution progresses. This improvement presumably occurs because the selected networks increasingly associate the stimulus patterns projected onto their visual sensory arrays with source-appropriate behavior based on trial and error.

Conclusion

Because luminance conflates the factors that determine the physical intensity of the light that reaches the eye, the visual system must link light patterns to behaviors that have worked in the past rather than encoding and analyzing this stimulus feature. As a result, lightness and brightness perceptions can't be understood in terms of the physical characteristics of the world, logical algorithms, heuristics, or cognitive inferences about reality. The evidence for this different concept of vision is that many of the peculiar relationships between luminance and lightness/brightness can be explained in empirical terms. In this framework, the seemingly odd effects elicited by luminance patterns are not aberrations or "illusions," but signatures of the fundamental strategy of vision. The chapters that follow provide evidence that the other basic qualities used to describe what we see are determined in this same way.

3

Seeing Color

Introduction

Evidence that perceptions of lightness and brightness are signatures of the way successful behavior is generated to contend with the inverse problem implies that color perceptions are also determined according to accumulated experience with the spectral information humans have always witnessed. After all, the same challenge confronts the perception of luminance and perceptions elicited by the distribution of spectral intensities in light, which is the basis of color vision. In both cases, the light that reaches the eye is a conflation of illumination, reflectance, transmittance and other factors; accordingly, there is no analytical way of unraveling the relative contributions of these and other factors to the generation of appropriate visually guided responses to light spectra. This chapter examines whether the strategy of vision that underlies lightness and brightness can also rationalize the phenomenology of color perception. The evidence described indicates that it can.

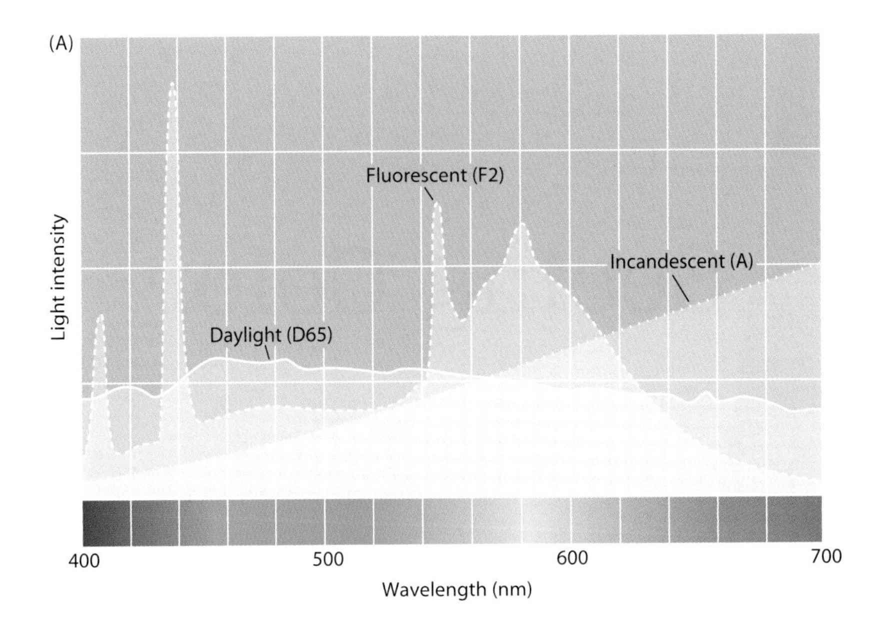
(A)
Light intensity
Fluorescent (F2)
Incandescent (A)
Daylight (D65)
400
500
600
700
Wavelength (nm)

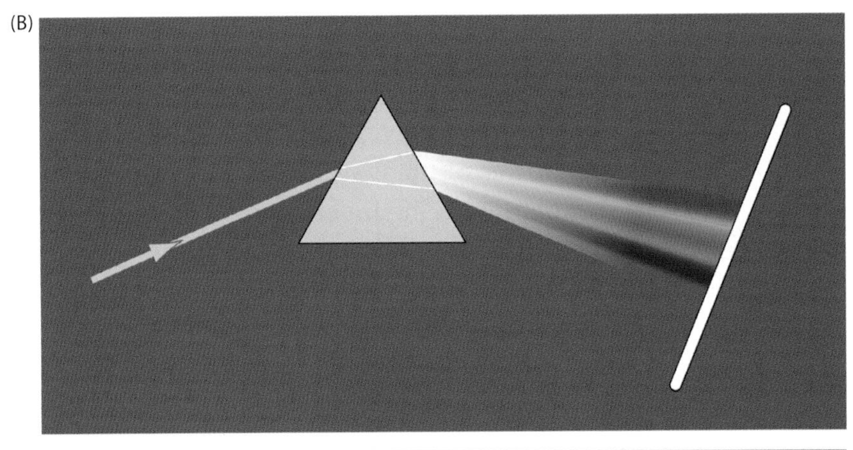
(B)

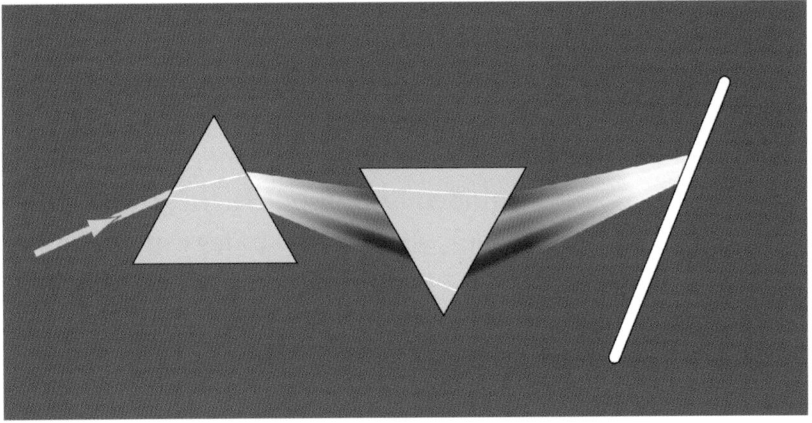

What Is Color?

Although color perceptions seem to indicate the properties of objects, in the absence of an observer there is no such thing as color (or any other sensory quality, for that matter). Although a few "color realists" can still be found, most people who think about vision recognize that colors are not properties of objects or conditions in the world, but subjective qualities generated by the brain for the behavioral advantages they confer. The perception of colors is obviously initiated by the radiant energy reaching the eye from a scene, but it is not determined by light spectra as such. Explaining this puzzle is the central problem in color vision.

As mentioned in Chapter 2, the human visual system responds to radiation with wavelengths of about 400 to 700 nanometers, which is a very small fraction of the electromagnetic spectrum. "White light," which loosely refers to spectral characteristics of light from the sun, incandescent filaments, fluorescent lights, and similar sources, contains a broad but by no means uniform distribution of these wavelengths (Figure 3.1A; in principle, white light, like white noise, should contain all frequencies in equal measure). The common denominator of white light stimuli is that the collection of wavelengths (or frequencies of photon vibrations) activates the three cone types in the human retina to about the same degree. As a result, these spectra are seen as some shade of gray on a perceptual scale that ranges from white to black.

In the late seventeenth century, Isaac Newton famously showed that white light can be decomposed into subsidiary spectra that evoke sensations of color and explained why this is so (Figure 3.1B) (Newton, 1704/1952). Prior to Newton, the prevailing idea was that white light represented "unmixed" or "pure" light, whereas colors represented "mixed" or "impure" light (see, for instance, Hooke, 1665). Given the religious significance then attached to this interpretation (i.e., that white represents purity), a theory suggesting otherwise was a bolder departure than it seems today. In any event, this was the intellectual environment in the late 1660s and early 1670s when Newton, then a relatively unknown professor at Cambridge, conducted a series of experiments that marked a new beginning in the study of light and color. Newton's principal experiment on color was to project a narrow beam of white light onto the wall of his study through a prism. René Descartes, Francesco Grimaldi, Robert Hooke, and

◀ **Figure 3.1** "White light." (A) The spectra of sunlight, fluorescent light, and light from a tungsten filament are all considered "white light," although the wavelength composition from these several sources is neither evenly distributed nor even very similar. This fact is well known to photographers, who must use different films and/or filters to compensate for the distortions that would otherwise occur when scenes are photographed in different "white light" illuminants (e.g., the "bluish" light from the sky in typical outdoor conditions, or the "yellowish" light from the sun or tungsten filaments). (B) As Newton showed, white light comprises spectral components that elicit perceptions of color when they interact with the three human cone types (see text for explanation).

Robert Boyle had all done similar experiments. In earlier work, however, the investigators had always placed a prism quite close to the projection surface or used a spherical prism that was less effective (Westfall, 1976). What Newton did differently was to use a triangular prism and increase the distance to the projection surface. Thus, whereas others had noted only a region of white flanked by a narrow strip of red on one side and blue on the other, Newton saw the full color spectrum. This observation led him to suggest correctly that spectra eliciting sensations of color are actually components of "white light," the perception of white arising from their admixture. Newton did several further experiments to confirm his hypothesis, the most important of which was to place a second, oppositely oriented prism in the path of the light emerging from the first prism (see the lower panel in Figure 3.1B). If "white light" is a mixture of more elemental "colored lights," then recombining these elements should again produce white light, which of course it did.

Newton's observation that white light can be decomposed into subsidiary wavelength bands that look colored has been a source of confusion as well as inspiration. To many, both then and even now, his findings suggested that the perception of color has primarily to do with the wavelengths of the light reflected (or emitted) by objects or that objects have some property appropriately referred to as "their color." Whereas Newton did not believe that either light or objects were "colored," he nonetheless thought that "Colours in the Object are ... a disposition to reflect this or that sort of Rays more copiously than the rest" (Newton, 1704, p. 125). This assertion, however, is not quite right. Although the perception of color is initiated by different mixtures of wavelengths, the colors seen are not a direct consequence of any specific spectral composition of light stimuli. As in the relationship of lightness/brightness and luminance, there is no simple connection between light spectra and the colors seen. Thus color perceptions cannot be explained by the reflectance properties of object surfaces or by the light spectra that reach the eye.

The Generation of Color Percepts

To understand the controversies that have arisen in attempts to explain color vision, some basic facts about the generation of color percepts are needed. As described in more detail in the Appendix, humans and most other mammals have two basic types of photoreceptors: rods and cones. The rods, about 125 million in each human retina, play relatively little part in color vision, being the initiating elements for the achromatic visual sensations at very low levels of light intensity (the so-called scotopic levels characteristic of starlight or moonlight; see Appendix Figure A.4). In light of sufficient intensity to evoke the full range of color sensations (i.e., at photopic levels), the rods are for the most part irrelevant because the large number of photons impinging on the retina saturates the ability of rods to respond. In contrast, the 6 million or so cones in each human retina generate signals even in very intense light and are inactivated only briefly.

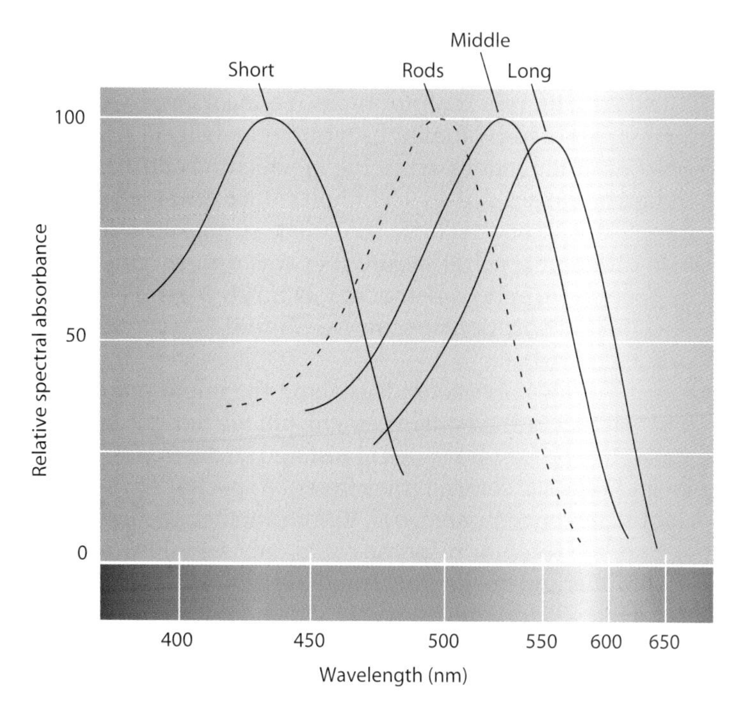

Figure 3.2 The spectral sensitivity of photopigments in the human retina. The solid curves represent the absorption spectra of the three different cone types (short- middle-, and long-wavelength cones), each distinguished by a different opsin pigment. The dashed curve is the absorption spectrum of rhodopsin, the rod pigment. Absorbance is defined as the difference between the amount of light falling on a surface (or sample of a substance) and the light reflected (or transmitted by the sample). (After Schnapf et al., 1987.)

Human color perception depends on three distinct types of cones characterized by photopigments called cone opsins. Each cone type has a different absorption spectrum and therefore responds best to a different portion of the visible spectrum (roughly speaking, to long, middle, and short wavelengths, respectively) (Figure 3.2). That three different types of receptors underlie human color vision was first suggested by one "George Palmer" in 1777 (see Walls, 1956 for an account of this pseudonymous genius) and, much more famously, a few decades later by Thomas Young. Young, of course, knew nothing about cone pigments, but he nevertheless contended in two lectures given to the Royal Society that there must be three different classes of receptive "particles" (Young, 1801/1923; 1802). Young's (and Palmer's) argument was based on the phenomenology of what humans perceive when lights having different spectra are mixed (a technique made possible by Newton's discoveries a century earlier). His key observation was that most color sensations could be produced by mixing appropriate amounts of long-, middle-, and short-wavelength light (mixing spectral lights is called "color addition" and is different from, and not to be confused with, mixing pigments, which effectively subtracts light of different wavelengths by absorption).

Young explained the ability of a mixture of three such lights to elicit most color sensations by postulating that three receptive elements—one for "red light," one for "green light," and one for "blue light"—are the basis of color vision (a modern version of his statement would refer to long-, middle-, and short-wavelength light, since the stimuli in question are not themselves colored, nor do they necessarily evoke red, green, and blue sensations). The perception of different colors, Young argued, arises as a result of the relative activities of the three different receptor types. Young's theory—which was a relatively minor part of his voluminous writings on natural science—was largely ignored until the latter part of the nineteenth century, when it was revived and greatly extended by Helmholtz (1866/1924–25, vol. II, pp. 61ff.) and James Clerk Maxwell (1855, 1861). The supposition of three receptor types with different spectral sensitivities was directly confirmed only in the 1960s, when spectrophotometry demonstrated the three human cone pigments and their absorption spectra (see Figure 3.2) (Brown and Wald, 1964; Marks et al., 1964).

The importance of Young's prescient ideas about the initiation of color vision is not simply that they turned out to be right, but the recognition that, in principle, a comparative process is needed to distinguish differences across the light spectrum and thus see colors. If the effect of a spectral stimulus on one receptor type could not be compared with that on another, then, as in the case of the rod system, the receptor response could only signal the relative intensity of a light stimulus and not its wavelength distribution. Achromatic experience—seeing a black-and-white movie or video, for instance—arises from the more or less equal stimulation of the three cone types.

The fact that humans have three types of cones with absorption spectra that peak in long-, middle-, and short-wavelength ranges respectively is referred to as trichromacy. The further hypothesis that the relative activation of the three cone types by light explains the colors actually seen is called the trichromacy theory (or the Young-Helmholtz theory). In fact, trichromacy theory as an explanation of color perception works quite well in accounting for the sensations that arise from mixing lights under restricted conditions (Judd, 1933; Evans, 1948; Hurvich, 1981). Studies carried out in this way, referred to as colorimetry, entail presenting monochromatic light within a surround that neither emits nor reflects much light, thus minimizing the contextual richness of natural stimuli (Figure 3.3). In such experiments, a "matching" stimulus is typically produced by three independent sources

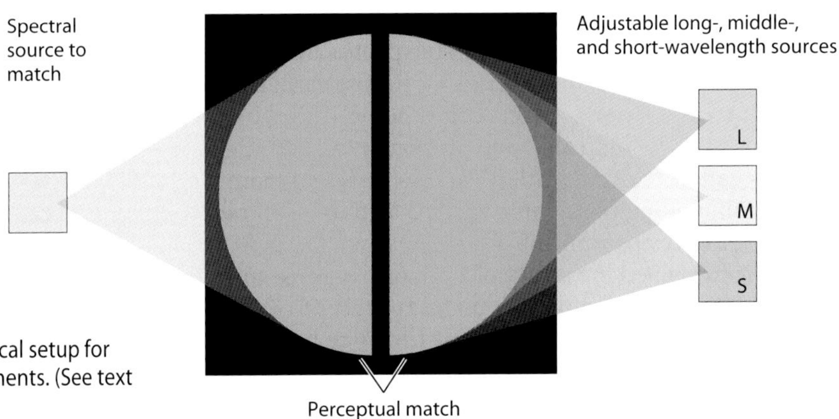

Figure 3.3 A typical setup for colorimetry experiments. (See text for explanation.)

producing long-, middle- and short-wavelength light superimposed onto half of a bipartite frosted glass diffuser. The "test" light is projected onto the other half of the diffuser, and the subject is asked to adjust the intensity of the three light sources until the color of two sides appears the same.

The responses of subjects in such matching tests are well predicted by the functional characteristics of the three cone receptor types (e.g., Wyzecki and Stiles, 1982). Thus, the results of colorimetry suggest that, to a first approximation, color sensations are indeed determined by the absorption spectra of the receptors, as "Palmer," Young, Helmholtz, and Maxwell had surmised.

Problems with the Trichromatic Theory of Color Vision

Although successful in accounting for many aspects of color perception, trichromacy theory as an explanation of color vision ran into trouble almost immediately. In addition to the fact that some color sensations cannot be generated by light mixing (see below), Helmholtz's contemporary Ewald Hering pointed out that humans perceive red to be an opponent color to green, and blue to be an opponent color to yellow (see Figure 3.5) (Hering, 1868/1964). By this he meant that while observers can see and/or imagine a transition from red to yellow through an incremental series of intermediates without entertaining any other primary color sensation, there is no parallel perception or conception of how to get from red to green, or from blue to yellow except through gray. Moreover, humans perceive a certain hue of red, green, blue, and yellow to be unique; that is, we see one color value in each of these categories as not being a mixture of any other colors (in contrast to the way we see orange as a mixture of red and yellow, or purple as a mixture of blue and red). As a consequence, there are no readily perceived (or conceived) "reddish-green" or "bluish-yellow" intermediates (although see Crane and Piantanida, 1983; Billock et al., 2001; Billock and Tsou, 2010). Since trichromacy theory offers no explanation of these perceptual phenomena, Hering concluded that trichromacy is at best an incomplete account of how color sensations are generated.

The importance of color opponency has been confirmed by electrophysiological studies of wavelength-sensitive neurons at different stations of the visual system of nonhuman primates, as well as in other species with color vision. The majority of spectrally sensitive neurons in the retina and lateral geniculate nucleus of the thalamus have receptive fields that are organized in a color opponent fashion (Svaetichin and MacNichol, 1958; Gouras, 1968; reviewed in Hurvich, 1981; Daw, 1984; Lennie and Fairchild, 1994). Such cells are excited by light of one wavelength (e.g., long or "red") illuminating the center of their receptive field and inhibited (or "opposed") by another wavelength (e.g., middle or "green") falling in the region surrounding the center of the receptive field (Figure 3.4). In macaque monkeys, most (but not all) color opponent cells are antagonistic with respect to wavelengths

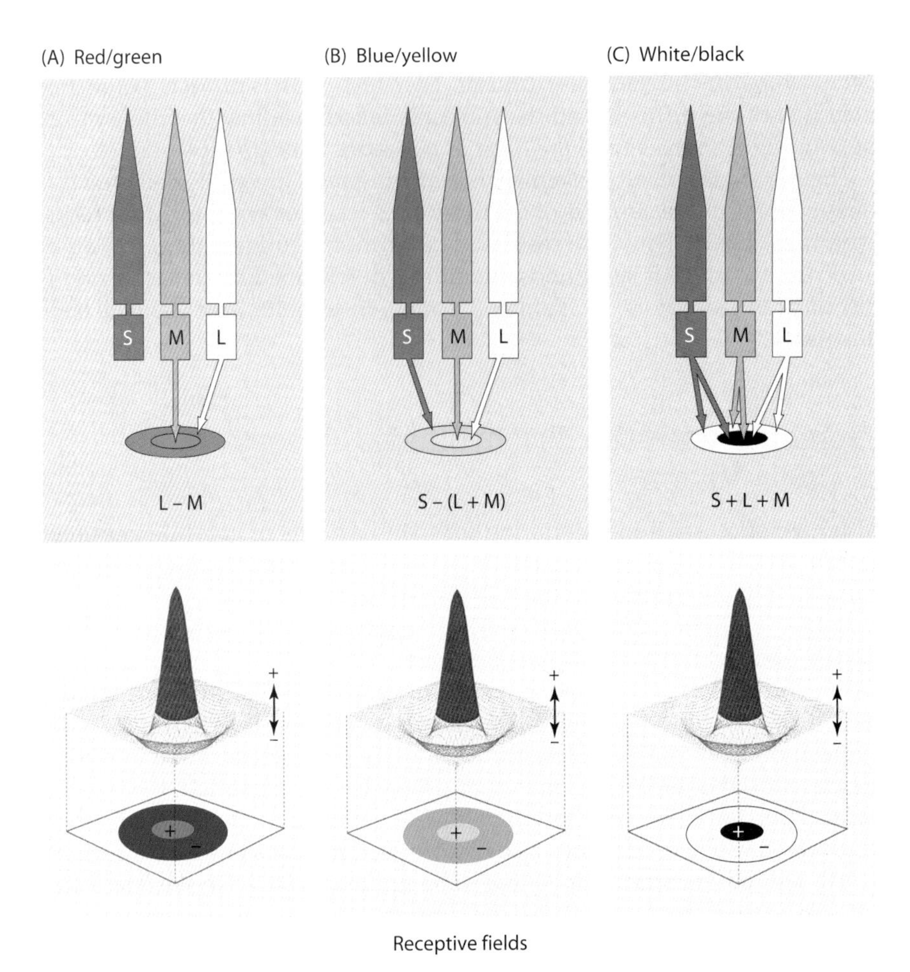

Figure 3.4 Diagram of the retinal connectivity that gives rise to postulated color opponent "channels" in the human visual system. (A) The "red/green" channel generated by two of the three human cone types. (B) The "blue/yellow" channel. (C) The "white/black" channel (the quotation marks indicate that the opponency is between long and middle wavelength spectral responses, not between colors). The lower panels show the activity patterns resulting from color opponent receptive fields.

that appear red and green, or blue and yellow. In addition to red/green and blue/yellow classes of opponent cells, other neurons are insensitive to differences in wavelength, being activated only by differences in the overall intensity of the stimulus. These cells are considered "white/black" opponent neurons. In light of this evidence, the explanation usually given for the perceptual phenomena Hering noted is that perceptions of color are elicited by neurons comprising three "channels" that operate in a push/pull fashion: For example, when the neurons responsible for seeing red are excited, those responsible for green are inhibited, and vice versa.

Beyond the level of the thalamus, however, simple "red/green" and "blue/yellow" opponent cell types have not been found. The neuronal receptive fields of spectrally sensitive cells at higher stations in the visual pathway are more complex, including "double opponent" neurons (i.e., neurons in which the surround is also inhibited by the spectrum that activates the center), neurons that cover a much larger range of opponent combinations than those evident in the properties of retinal or thalamic neurons, and spectrally sensitive neurons that show no obvious opponency (Schein and Desimone, 1990; Webster and Mollon, 1991; Komatsu, 1998; Conway, 2001). Thus, the idea of color "channels" should not be taken too literally.

The function of color opponency in perception also remains unclear, though some possible reasons have been proposed: to maximize the efficiency of color vision according to the demands of information theory (Buchsbaum and Gottschalk, 1983; Gottschalk and Buchsbaum, 1983; Ruderman et al., 1998; Hoyer and Hyvärinen, 2000) and/or to promote the perception of "color constancy" (see below) (D'Zmura and Lennie, 1986; Maloney, 1986; Shepard, 1992).

The Biological Advantages of Color Vision

Seeing in color requires populating the retina with receptors having different spectral sensitivities, which necessarily reduces visual acuity. Moreover, the circuitry at every station in the visual system must be modified to incorporate this perceptual quality. Given these biological costs, a reasonable question is: What benefit does color add to seeing scenes in shades of gray?

All sighted mammals (mole-rats and some species of moles and bats are effectively blind) can readily distinguish boundaries demarcated by luminance differences (Evans, 1948; Mollon, 1991; Jacobs, 1993; Allman, 1999; Mollon and Regan, 1999). Animals with little or no color vision (the majority), however, are unable to distinguish a further set of boundaries that arise from differences in the spectral composition of light reaching the eye. Thus, a visual system that uses both these qualities of light will generate more effective behavior than a system operating on the basis of light intensity alone.

Another advantage concerns the fact that the intensity of natural light varies by a factor of a billion or more; thus, the less-variable spectral qualities of light may provide a better guide to behavior than luminance, further encouraging the evolution of color vision. In accord with this idea, artificial neural networks, operating in virtual environments in which they must discriminate different surfaces to survive, evolve the equivalent of a single receptor type if the surfaces are under the same steady illuminant. However, if the illumination of the surfaces varies, as would be the case in nature, the majority of networks evolve two or more receptor types (i.e., the sort of computational infrastructure needed for "color vision") (Schlessinger et al., 2005).

Despite these advantages, the cost of imperfect or absent color vision for humans is relatively modest. The most common cause of color deficiency is a defect in one of the three cone photopigments. Human dichromats (so-called because they need to adjust only two instead of three variable sources of long-, middle-, and short-wavelength light to match spectral stimuli; see Figure 3.3) are unable to distinguish reds and greens or, less commonly, blues and yellows, depending on which of the three X-linked photopigment genes is defective. Such individuals (about 5 percent of the male population in the United States) are incapable of making some of the cone activity comparisons needed to generate the full range of color sensations. As a result, they are at a disadvantage in discriminating objects on the basis of spectral qualities (Shepard and Cooper, 1992; Cole, 1993). Although the actuarial consequences for color deficient humans are not substantial today, this lack presumably would have been more significant in earlier times; without color vision, edibles such as fruits; prey; predators; and other critical aspects of the environment would have been more difficult to respond to, since such objects are often camouflaged by variegated or dappled backgrounds (see Morgan et al., 1992; Mollon, 1995; Mollon and Regan, 1999; Sumner and Mollon, 2000). In short, animals whose visual systems categorize according to the distribution of power in light spectra as well as overall intensity should be more successful in behavior.

Perceptual "Color Space"

Attempts to represent human subjective color experience have led to numerous descriptions over the centuries (Kuehni and Schwarz, 2008). Many modern representations entail a series of planes, each corresponding to a different level of perceived intensity (Figure 3.5). At any particular intensity level—which corresponds to changes of lightness or brightness—movements around the perimeter of the plane correspond to changes in hue (i.e., changes in the apparent contribution of red, green, blue, or yellow to the percept), whereas movements along the radial axis correspond to changes in saturation (i.e., changes in the approximation of the color appearance to a neutral gray).

A presumptive requirement of color space (or any perceptual space) is that perceptions relate in some way to physical similarities and differences in the world pertinent to behavior, which is not the same as saying that perceptions actually represent those physical attributes. As in rationalizing perceptual responses to luminance in Chapter 2, stimulus targets that require the same behavioral response should look the same, whereas targets that have a different significance for behavior should look different, even if the targets are physically identical (Shepard and Carroll, 1966; Lotto and Purves, 2002).

The degree to which spectra are similar or different can be determined by multidimensional scaling, a technique that orders items such that their relative positions on a map represent the relative similarities and differences among them (Torgerson, 1952; Shepard, 1962; Kruskal, 1964; Kruskal and Wish, 1977). Figure 3.6 shows that progressively changing the shape of a spectrum without changing its variance entails moving around the perimeter of the map, which corresponds, in human color space, to changes in hue (see Figure 3.5). Conversely, moving between spectra that differ progressively in variance entails migration toward the center of the map, which corresponds to changes in saturation. Finally, the center of the map represents a spectrum of approximately equal intensity at all wavelengths, and thus a uniform relationship to all the other spectra in the plane (like that of gray to other color sensations at the same level of intensity). The overall similarity of the multidimensional scaling map in Figure 3.6 and the perceptual color space planes in Figure 3.5 supports the idea that color perceptions are based on spectral relationships rather than absolute values. This conclusion makes sense in empirical terms: Given the inverse problem, percepts based on absolute values would not abet successful behavior, whereas percepts based on spectral relationships could.

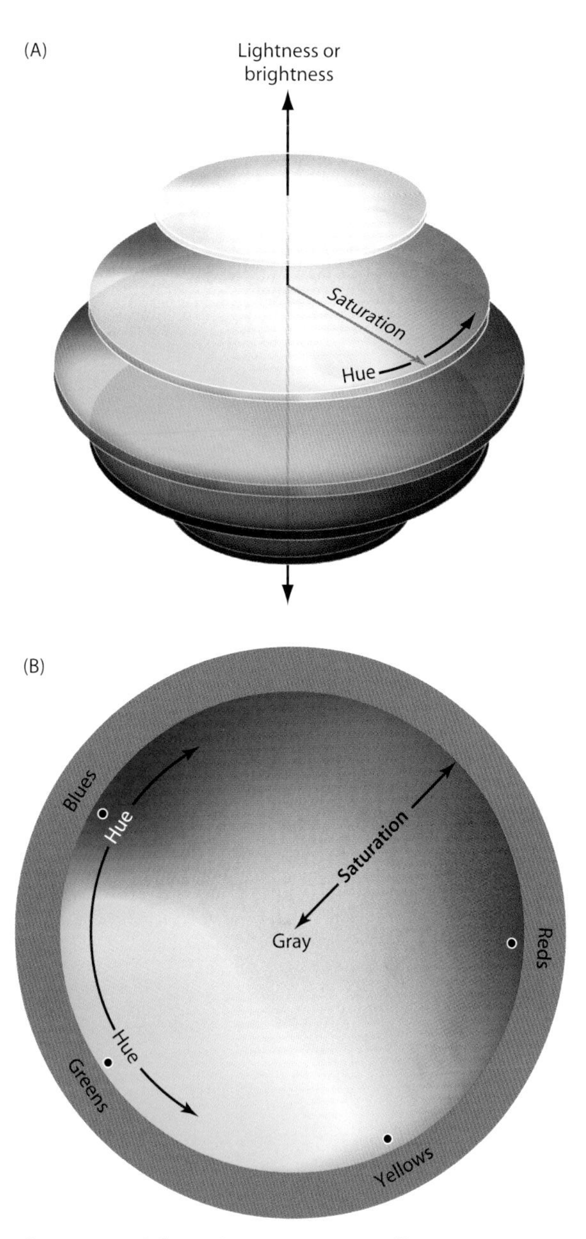

Figure 3.5 Schematic representation of human color space. (A) and (B) are different views of the overall organization of the perceptual qualities elicited by the spectral distribution of light. The black dots indicate the approximate locations of the unique hues that define each of the four primary color categories. (See text for further explanation.)

Another feature of color space is the organization hue according to four primary categories—red, green, blue, and yellow—each characterized by a unique hue that has no apparent admixture of the other three (see

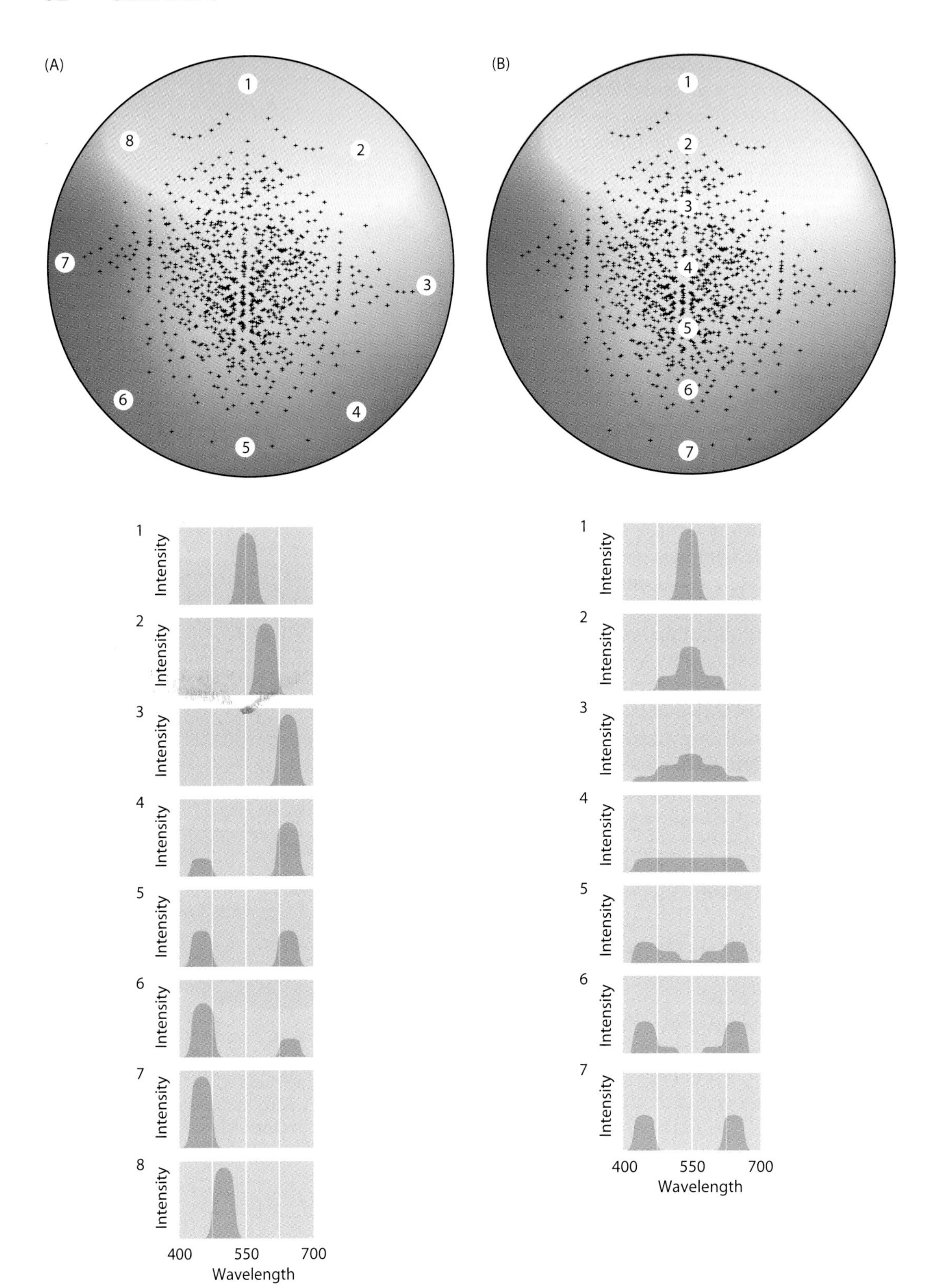
(A)
1
2
3
4
5
6
7
8
Intensity
400
550
700
Wavelength
(B)
1
2
3
4
5
6
7
Intensity
400
550
700
Wavelength

◀ **Figure 3.6** Multidimensional scaling map of representative spectra. (A) Crosses indicate the relative location of 1000 spectra; the spectra that occupy the numbered points around the perimeter of the map are illustrated below. Circular movements around the perimeter of the map identify spectra that would elicit progressive differences in perceived hue (see Figure 3.5). (B) The same map, but now indicating a series of spectra that define movement across the space from a spectrum to its physical opposite. Movements in this dimension correspond to spectral differences that would elicit different perceptions of saturation, the central axis being the location of the most nearly uniform spectrum. (After Lotto and Purves, 2002.)

Figure 3.5). Secondary color groupings such as purples, oranges, cyans, and yellow-greens are seen as mixtures of these four primaries. These categories have no apparent basis in the spectra themselves and are not predicted by the absorption characteristics of the three cone types or from what is known about opponent processing. As the color scientist John Mollon put it, the "special phenomenal status of the four pure hues is perhaps the chief unsolved mystery of colour science" (Mollon, 1995, p. 146).

One possible solution is that four categories of color sensation are, for visually guided behavior, a solution to the challenge of the so-called "four-color map problem" (Figure 3.7). Cartographers had known for centuries that four colors are needed to unambiguously distinguish territories on a geographical map. Based on this observation, the four-color map problem—the conjecture that "four colors are sufficient to color any map drawn in a plane or on a sphere so that no two regions with a common boundary [other than a point] are colored with the same color"—was formally posed in the latter part of the nineteenth century (Cayley, 1878) and proven formally about a hundred years later (Appel and Haken, 1976).

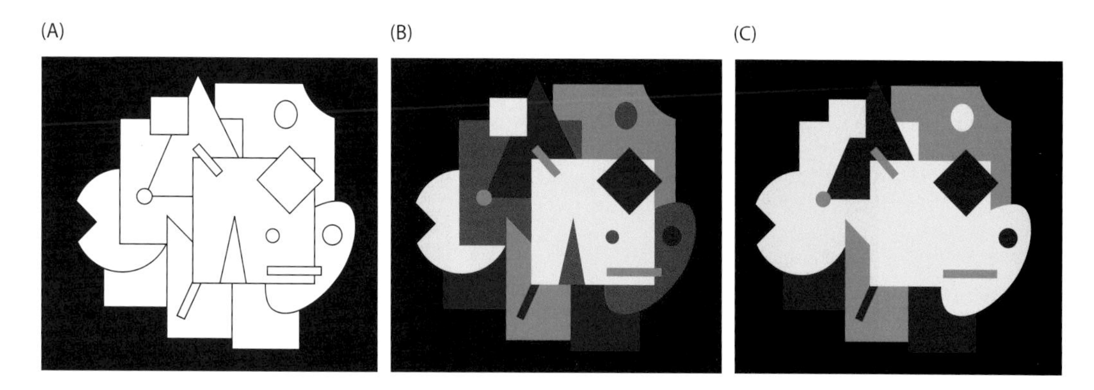

Figure 3.7 The four-color map problem. (A) An arbitrarily complex collection of surfaces in a two-dimensional array. (B) The four-color-map problem alludes to the fact that four differentiating categories are necessary and sufficient to unambiguously distinguish any such collection of surfaces. (C) If fewer than four categories for comparison are available, then some surfaces in the array will be conflated. Resolving this topological problem may explain why humans experience color in terms of four categories, each defined by a unique hue.

If the purpose of color vision is to enhance behavioral success based on the physical similarities and differences of objects in the world, then to be optimally effective, the visual system would presumably want to ensure that no two areas separated by a common boundary in a two-dimensional array such as the retinal image were treated as the same if they are actually different (Purves et al., 2000; Lotto and Purves, 2002).

Color Contrast and Constancy

Like lightness and brightness, a major challenge to understanding color vision is the remarkable influence of context on the colors perceived. These perceptual effects are typically discussed in terms of color contrast and color constancy. Color contrast effects are much the same as the lightness/brightness contrast effects discussed in Chapter 2: When two target patches returning identical spectra to the eye are surrounded by regions that return different spectra, the sensations of color elicited by the targets are no longer the same (Figure 3.8). Color constancy refers to an opposite phenomenon: In some contexts, when two targets return different spectra to the eye, they can elicit similar color sensations (Figure 3.9). Thus, people see a banana as more or less yellow and an apple as more or less red whether the fruits are observed in the "bluish" light of the sky, the "reddish" light of sunset, or the "yellowish" light from incandescent sources such as the tungsten filaments in ordinary lightbulbs. In each of these situations, the illuminant—and, therefore, the spectral composition of light reflected from the relevant surfaces and reaching the eye—is different.

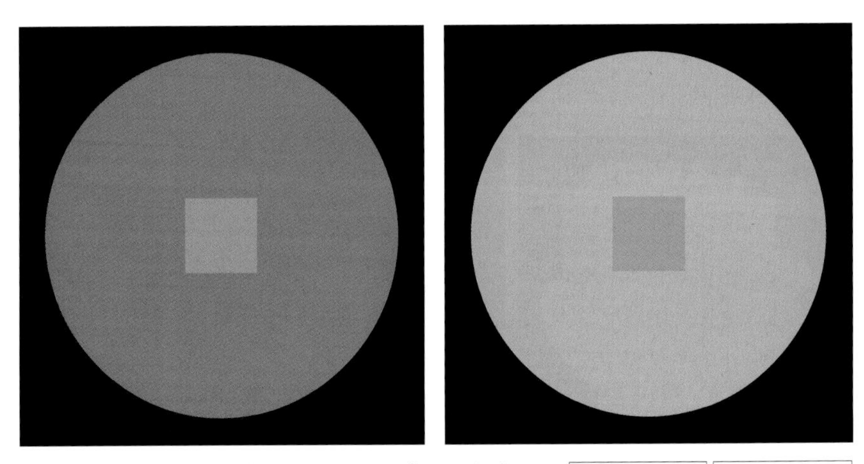

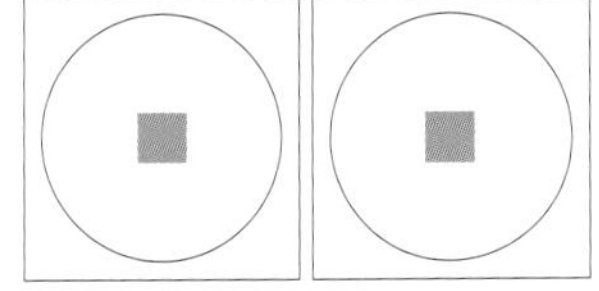

Figure 3.8 A typical color contrast stimulus. As indicated by the inset, the two targets (the central squares) are spectrophotometrically identical and indeed appear so on a neutral background. However, the same target in a reddish surround looks yellowish and, in a yellowish surround, reddish.

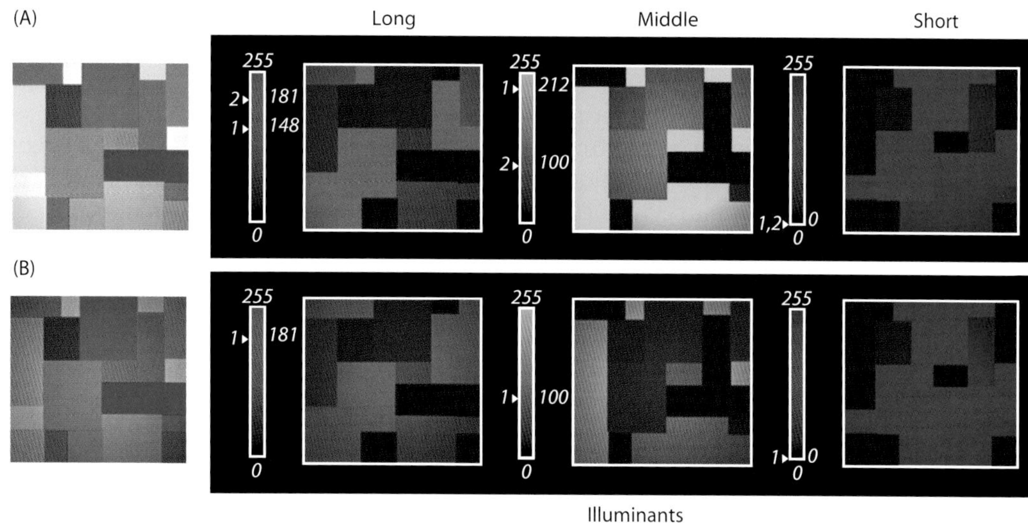

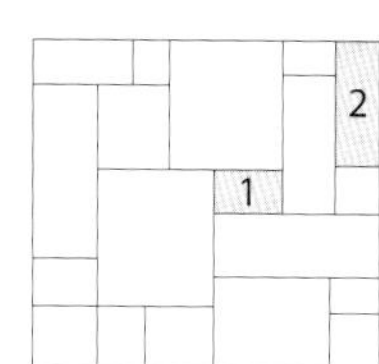

Figure 3.9 Color constancy demonstrated in a printed replication of a "Land Mondrian." (A) The appearance of the collage (left) as it would look when illuminated by a mixture of long-, middle-, and short-wavelength light in the amounts indicated by the three panels on the right; the specific values for the light returned for each of the three sources from the patches marked (1) and (2) in the inset are indicated by arrowheads. (B) The appearance of the collage (left) after readjustment of the three sources so that three values returned from patch (2) in (A) are now the same as those returned from patch (1). Despite the drastic change in the spectral returns, patch (1) continues to look yellowish, and patch (2) continues to look reddish.

These phenomena were well known in the eighteenth century (Mollon, 1995), and recognized as important issues by Helmholtz and Hering in the nineteenth century. Color contrast and constancy were subsequently studied by David Katz (1911/1935), Harry Helson (1938), Deane Judd (1940), and others in the first half of the twentieth century and eventually brought forcefully to the attention of other vision scientists by Edwin Land in the late 1950s (Land, 1959a,b; Land and McCann 1971; Land, 1986).

Land, whose genius had long since been established by his many contributions to photography (preeminently, his invention of the "instant camera"), stimulated a revival of interest in contrast and constancy by a series of extraordinary "demonstrations," as he called them. The most famous of these was a collage of differently reflective matte papers illuminated by three independently controlled light sources that provided long-, middle-, and short-wavelength light, respectively (see Figure 3.9). (Because the collages resemble the work of the Dutch artist Piet Mondrian, these stimuli are often referred to as "Land Mondrians.") Land first adjusted each of the three light sources to some value and then determined the spectral return from one of the surfaces in the array (say, a surface that looked yellowish to observers when it was illuminated by all three lights simultaneously). Under the same illumination, he then showed that another patch in the collage (say, one that looked reddish) provided a substantially different spectral return, as expected from the different reflectance values of the surfaces (Figure 3.9A). Land next readjusted the three illuminators so that the "yellowish" paper now provided exactly the same spectral return as had originally come from the "reddish" paper (Figure 3.9B). The expectation is that the "yellowish" paper should now look like the "reddish" paper in the previous condition of illumination. Like the banana and the apple, however, the yellowish patch continued to look more or less yellow, and the reddish patch (which was also returning a different spectrum of wavelengths to the eye under the new conditions) continued to look more or less red.

Although these results were all predictable from earlier work—Hering, Katz, Helson, Judd, and others even earlier had all shown and discussed similar effects (see Judd, 1960 and Mollon, 1995)—Land's fame, the panache with which he presented his demonstrations, and the provocative conclusions he drew from them, insured renewed attention. The ensuing brouhaha stemmed from Land's challenge to what he portrayed as the conventional wisdom about color vision. "Color in images," Land wrote in 1959, "cannot be described in terms of wavelength and, in so far as color is changed by alteration of wavelength, the change does not follow the rules of color mixing theory" (Land, 1959b, p. 637). "We have come to the conclusion," he went on, "that the classic laws of color mixing conceal great basic laws of color vision" (Land, 1959a, p.115).

The provocative nature of these claims notwithstanding, Land's results, like the less noted observations of his predecessors, do indeed confound the predictions of the classic colorimetry. Why should the same spectral returns from two targets elicit different color sensations (color contrast) and different spectral returns elicit the same sensation of color (color constancy)?

Attempts to Explain Color Contrast

People have wrestled with these perplexing questions for a very long time. Joseph Priestley, Johann Goethe, and Jan Purkinje were all fascinated by one of oldest demonstrations of color contrast, the generation of "color shadows" (see Purkinje, 1823/1969; Goethe, 1840/1967). When a "white" screen is illuminated by two different lights (say, long-wavelength light that looks red and "white" light), the screen—as expected—appears pink. If an object is interposed between the light sources and the screen, two cast shadows result, one lacking light from the "red" illuminant, and one lacking light from the broadband "white" illuminant (Figure 3.10). Since the shadow caused by blocking the white light source allows only the long wavelengths from the "red" source to be returned from the screen, the expectation is that this region should appear as a more saturated red on the pinkish background elicited by the overlapping lights falling on the screen as a whole. By the same token, the shadow caused by blocking the "red" light source should appear white, since only broadband light is being returned to the eye from that region of the screen. The result, however, is

Figure 3.10 Color shadow effects. The picture is didactic only; a physical setup like this is needed to bring out this striking phenomenon. (See text for explanation.)

that whereas the light returned from the shadow caused by blocking the white light is indeed seen as red, the light returned from the shadow caused by blocking the "red" light source appears distinctly greenish.

The phenomenology of color contrast effects was codified in a remarkably complete description published at about the same time. In 1824, Michel Chevreul (1839/1987), a well known chemist and rising star in the firmament of French science, was commissioned by Louis XVIII to investigate the adequacy of the dying methods then in use at the state tapestry works. His appointment as Director of Dyes for the Royal Manufacturers lasted 30 years and led Chevreul to extend his interests from color chemistry to color perception. Chevreul concluded that when a test patch is placed on a background that elicits the perception of a different hue (see Figure 3.8), the appearance of the target is biased toward the color that lies opposite the background color on the color circle (see Figure 3.5). For example, a patch that looks gray on a neutral background appears distinctly greenish when placed on a "red" background (the effect apparent in the color shadow experiment illustrated in Figure 3.10). If, on the other hand, the target is placed on a background of similar hue, then the apparent hue of the target remains the same but appears less saturated (i.e., more grayish, or washed out). This further observation explains, in descriptive terms, why a reddish patch on a red background tends to look duller than the same patch on a neutral background. Color contrast effects have long been appreciated in art and commerce, the domains relevant to the king's appointment of

Chevreul. The discrepancies that ultimately turned out to be perceptual had raised the suspicion that some dyes then in use at the royal tapestry works were substandard. The nineteenth century French painter Delacroix, a contemporary who knew Chevreul's work, endorsed the value of color contrast in art. "Give me the mud of the streets," he exclaimed, "and I will turn it into the luscious flesh of a woman, if you will allow me to surround it as I please" (Signac, 1921/1992, p. 238; see Werner, 1998 for more on color contrast in art).

If only the area immediately adjacent to the target were pertinent to its perceived color, then the receptive field properties of color opponent neurons might, in principle, provide a reasonable account (Daw, 1968; Wyszecki, 1986; Hubel, 1988; Jameson and Hurvich, 1989). But, as Helson, Judd, Land, and others made clear, the entire scene is somehow relevant to the perceived color of any part of it. As in attempts to explain perceptions of lightness or brightness, these remote effects on the perceived color led to more complex ideas about how the surround might generate the perceived color of a target. Many investigators suggested integrative strategies of color vision based on lateral interactions among neurons at various (often higher) levels of the visual system (e.g., von Kries, 1905; Helson, 1938; Judd, 1940; Land, 1986; Jameson and Hurvich, 1989; Creutzfeldt et al., 1990, 1991). The best known of these attempts to elaborate an algorithm that might accomplish this integration of visual scenes is Land's "retinex theory" (see Land and McCann, 1971; Land, 1986). The term "retinex" is a contraction of "retina" and "cortex," reflecting Land's well-justified belief that whatever the explanation of color vision, the retina and cortex must work together as a system. The gist of retinex theory is that contextual effects arise by a calculation of a "ratio of the integrals of the product at each [of three] wavelength[s] of the absorbance of the cone pigment times the reflectance of the colored patch of interest times the illuminant" (Land, 1986, p. 128; the application of retinex theory to explaining the perception of luminance is similar). Although Land's retinex equations have limited explanatory and/or predictive value (see, for example, Marr, 1982, pp. 257ff.; Brainard and Wandell, 1996), retinex theory has been an important point of reference in rationalizing color contrast and constancy.

Other investigators more focused on the physiology of retinal interactions suggested that the basis of contextual effects involves the adaptation of cones (and perhaps subsequent color processing neurons) to the predominant spectral return (and/or spectral contrasts) in a scene (D'Zmura and Lennie, 1986; Creutzfeldt et al., 1990; Creutzfeldt et al., 1991a,b; Chichilnisky and Wandell, 1995; Walsh, 1995; Webster and Mollon, 1995; Hurlbert, 1996; Engel and Furmanski, 2001; see also von Kries, 1905; Helmholtz, 1866/1924–25, vol. II, pp. 181ff.). Like Land's retinex theory, adaptation supposes a discounting of the effect of the illuminant so that the reflectance values of objects can be more accurately represented in perception. Although the perceptual manifestations of chromatic adaptation are less obvious than those of light and dark adaptation, it is a robust and significant aspect of color perception.

Nonetheless, reducing color contrast to incidental effects of integrative mechanisms has not been successful. In the first place, integration cannot account for the full range of color context effects observed (e.g., Brown and MacLeod, 1997; Kraft and Brainard, 1999). As in lightness/brightness, it is not difficult to create stimuli that elicit perceptions inconsistent with those predicted by color adaptation, retinex theory, or other explanations based on integrative algorithms. Consider, for example, the demonstration in Figure 3.11. The central targets (the pairs of circles and crosses) are physically identical, and all the components of the stimuli are equiluminant. These physical realities notwithstanding, the left-hand targets in both the upper and lower panels appear to be about the same shade of yellow, whereas the right-hand targets appear to be about the same shade of red. The similar perception of the upper and lower targets on the left and right thus occurs despite the fact that the chromatic surrounds in the relevant panels are now physical opposites (Figure 3.11B is a chromatic version of a stimulus first created by Dejan Todorovic [1997], which is a variant of White's illusion [Figure 2.2B], which is a variant of von Bezold's nineteenth century "arabesques"). The induction of these opposing sensations of target color by chromatically similar surrounds (and vice versa) is obviously inconsistent with explanations of color contrast and constancy based on adaptation to (or integration of) the predominant spectral return from a scene.

Explaining Color Contrast in Empirical Terms

Given the ability of an empirical framework to rationalize many otherwise puzzling aspects of the relationship between luminance and lightness/brightness described in Chapter 2, it is reasonable to ask whether the same framework can explain the relationship between the distribution of intensities in light stimuli and the color perceptions elicited. The fundamental problem is, after all, the same for the perception of color as for achromatic stimuli. As indicated in Figure 3.12, color stimuli are typically products of the spectrum of the illuminant, the reflectance efficiency function of surfaces, and the transmittance of the intervening medium. Inevitably, the spectral distribution of light intensities in any stimulus conflates the contribution of these factors. If color contrast and constancy are signatures of an empirical strategy that generates color percepts based on successful behaviors in response to light stimuli, then color percepts should accord with—and be predicted by—experience with the way illumination, reflectance, and transmittance combine to generate the spectra humans have always witnessed.

Although the spectral stimuli arising from natural scenes vary widely (e.g., Burton and Moorhead, 1987), the relationship between the

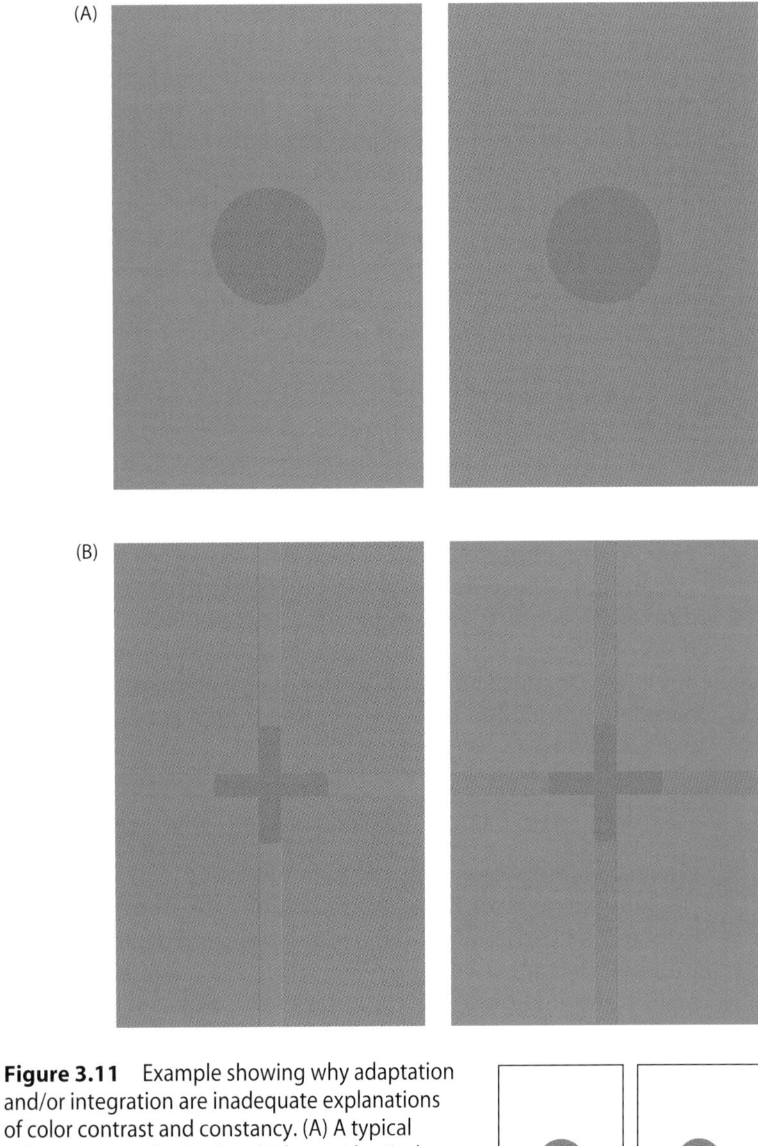

Figure 3.11 Example showing why adaptation and/or integration are inadequate explanations of color contrast and constancy. (A) A typical color contrast stimulus in which two identical central targets embedded in different spectral surrounds appear differently colored. The target on the reddish surround (on the left) appears yellowish, whereas the target on the yellowish surround (on the right) appears reddish, as expected. (B) In this configuration, the target on the left continues to appear yellowish and the target on the right reddish, even though the target on the left is now predominantly surrounded by a yellowish background and the target on the right by a reddish background.

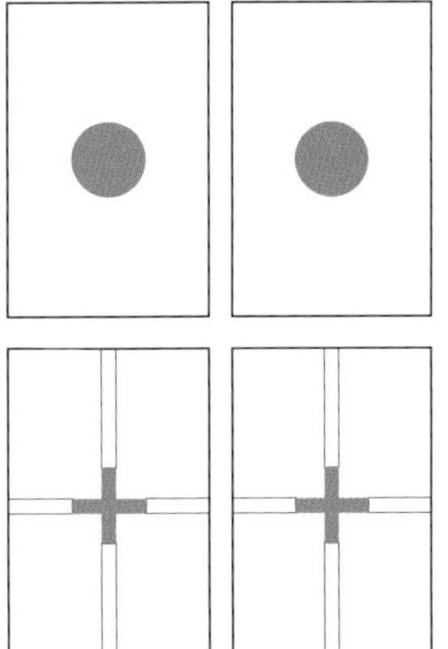

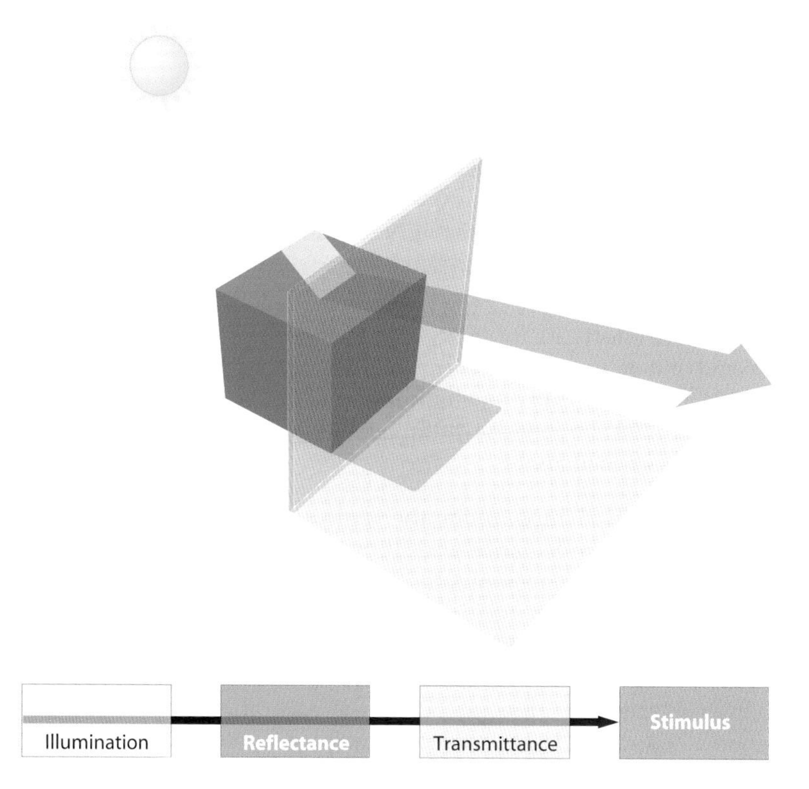

Figure 3.12 As indicated for light intensity in Figure 2.4, the contributions to the stimulus—illumination, reflectance, and transmittance—are conflated in the power spectrum of light reaching the eye. As a result of this entanglement, light stimuli as such cannot specify the nature of the underlying object surfaces, the illuminant, or the effects of transmittance.

reflectance of surfaces, illuminants, the intervening transmittances and the stimuli generated exhibit some fairly obvious relationships (Figure 3.13). These relationships can be summarized as follows (in each case, the effect is considered relative to the same surface under an equally intense illuminant whose power is uniformly distributed):

1. When a surface is illuminated by light that has a distribution of power similar to the surface's reflectance efficiency function (as in Figure 3.13, case 1), the intensity of the return from the surface is increased, whereas the width of the distribution narrows. The location of the peak (or peaks) of the distribution, however, is not much changed.
2. When a surface is illuminated by light that has an opposite distribution of power to the surface's reflectance efficiency function (as in Figure 3.13, case 2), the intensity of the return from the surface is decreased, whereas the width of the distribution broadens and flattens. Again, the location of the peak(s) of the distribution is not much changed.

3. When the same surface is illuminated by light whose distribution of power is neither the same as nor opposite to the surface's reflectance efficiency function (as in Figures 3.13, cases 3 and 4), the power and width of the return are affected to an intermediate degree. In distinction to cases 1 and 2, however, the location of the peak(s) of the distribution of power in the spectral return becomes less correlated with the return from the same surface under white light illumination. In other words, the spectral return shifts along the x axis in the direction of the spectral profile of the illuminant.
4. Finally, all these effects are influenced by the width of the spectral distribution of the illuminant and/or the reflectance efficiency function of the surface: The narrower the distribution of power in the illuminant relative to the reflectance efficiency function of the surface, the greater all the effects described above.

These several relationships define the typical experience of observers with the interaction of reflectance and illumination in spectra. If spectra elicit percepts as a result of accumulated visual experience, then the colors perceived should accord with the predictions made by these empirical relationships, since these relationships would have been incorporated in visual circuitry as a consequence of feedback from successful behavior.

By having subjects adjust the perceived hue, saturation, and brightness of a target on a neutral background until its apparent color matches that of an initially identical target on a chromatic background of the same luminance, one can measure the apparent change in the target color induced by the context (Lotto and Purves, 2000). The results of such matching tests (many studies of this general sort have been undertaken since Chevreul's classical work) can be summarized as follows, all comparisons being relative to the appearance of the target on a neutral surround:

1. When a target is presented on a background whose perceived hue is similar to the hue the target would elicit on a neutral background, the apparent color of the target decreases in saturation and brightness, with little change in hue.
2. When a target is presented on a background whose perceived hue is opposite that of the target on a neutral background, the apparent color of the target increases in saturation and brightness, with little change in hue.
3. When a target is presented on a background whose perceived hue is neither the same as nor complementary to that of the target on a neutral background, the apparent hue of the target shifts away from the apparent hue of the surround, with little change in saturation and brightness.
4. These contextual effects on hue, saturation, and brightness are all enhanced by increasing the apparent saturation of the chromatic surround.

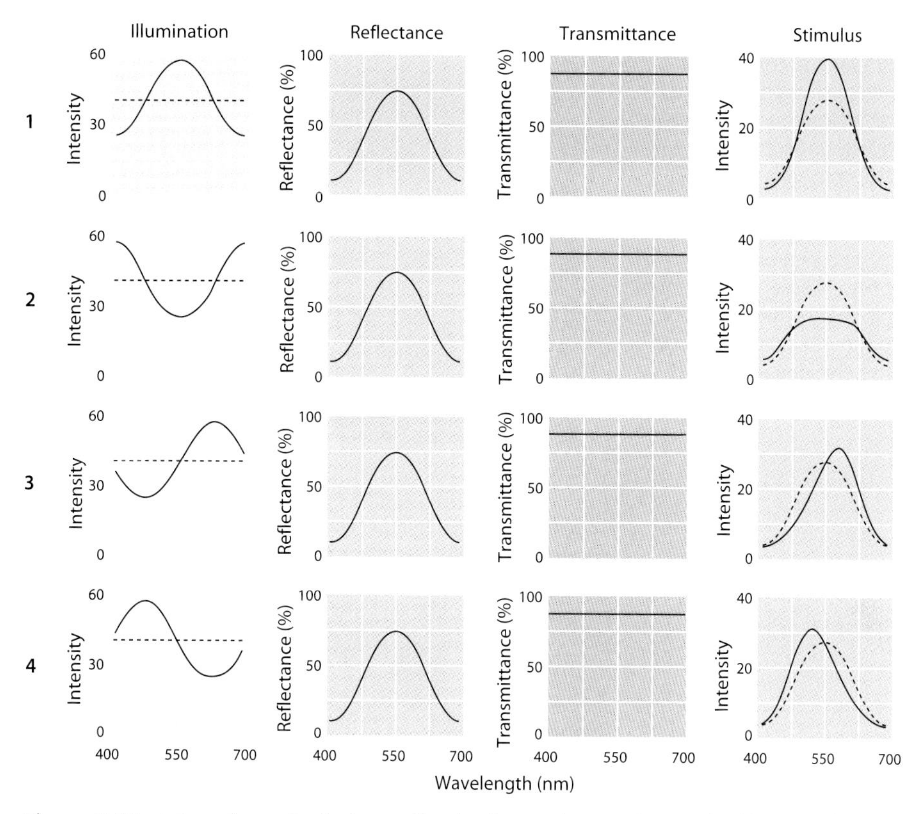

Figure 3.13 Interactions of reflectance, illumination, and transmittance. In this example, a hypothetical surface with a given reflectance efficiency function (illustrated here as a sine wave) is illuminated by spectra of the same overall intensity whose configuration, relative to that of the surface's reflectance, is either the same (case 1), the opposite (case 2), or shifted toward long (case 3) or short (case 4) wavelengths. For comparison, the dashed line represents an illuminant with a uniform distribution of wavelengths (i.e., "white light"); for the sake of simplicity, transmittance is assumed to be uniform. The spectra generated by the product of illumination and reflectance vary systematically compared with the spectra arising from the same surface under white light. (See text for explanation.) (After Lotto and Purves, 2000.)

These changes in the perception of the hue, saturation, and brightness of a target produced by different chromatic surrounds are those that would be needed to cope behaviorally with the routine changes in stimuli arising from the interaction of reflectance efficiency functions and illuminants described in the preceding section. Consider, for example, the stimulus in Figure 3.14A. The light coming from the targets in this standard color contrast stimulus could have been generated by many different combinations of reflectance and illumination (transmittance is again omitted for the sake of simplicity). The visual system must nevertheless generate behavior that works in the world. Achieving this goal means that what an observer

Figure 3.14 An empirical explanation of color contrast. The standard color contrast stimulus in (A) could have been generated by similarly reflective targets on differently reflective backgrounds under the same white light illumination (B) or by differently reflective surfaces on similarly reflective backgrounds under different spectral illuminants (C). As a result of having incorporated the frequency of occurrence of these (and innumerable other) possible images into the visual circuitry that generates the perceptual and behavioral response, the identical targets in (A) appear differently colored. (After Purves et al., 2001.)

(A) Standard color contrast stimulus

(B) Similar surfaces under similar illuminants

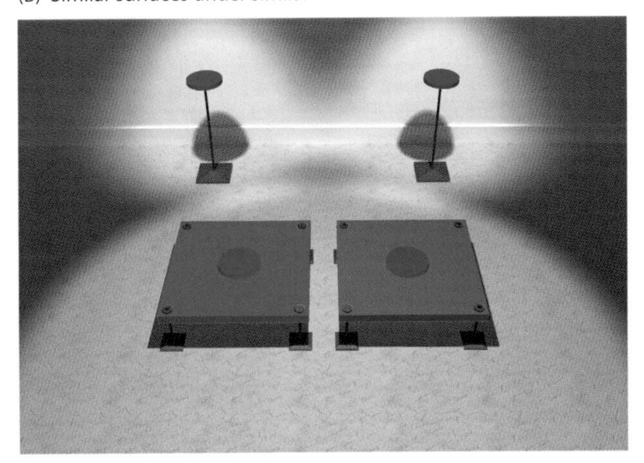

(C) Different surfaces under different illuminants

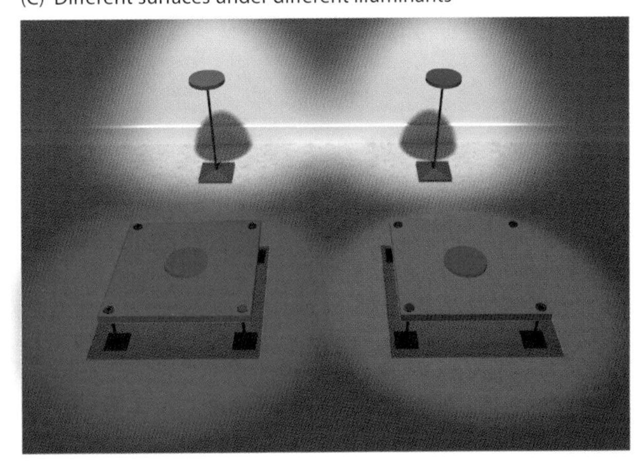

sees should be determined by whether the identical spectral returns from the targets in a matching test are consistent with similar surfaces under the same illuminant (Figure 3.14B), in which case they should appear similarly colored, or with different surfaces under different illuminants (Figure 3.14C), in which case the targets should look differently colored. Accordingly, identical targets on different chromatic backgrounds should give rise to different color percepts, as they do. Since the two target returns in Figure 3.14A (as well as those in Figures 3.8 and 3.11A) will often have been generated by physically different objects under different qualities of illumination, this experience will have been incorporated into visual circuitry and thus the response to the stimulus.

This framework also explains the complementarity of the physical and perceptual relationships enumerated earlier. When different chromatic lights illuminate two targets but nonetheless generate identical spectral returns, the reflectance values of the two targets will always have been different. Given the relationships between reflectance efficiency functions, illuminants, and the resulting spectral returns that observers will have experienced, the spectral distribution of the reflectance of the left target surface in Figure 3.14C would have to occupy a position in perceptual color space that is shifted away from the effect of the "blue" illuminant implied by the surround. It would therefore be a surface that, on a neutral background under white light illumination, would appear relatively purple. Conversely, the reflectance of the right target under "purple" light would, for the same reason, have to have been shifted in the opposite direction and would therefore be a surface that appeared relatively blue on a neutral surround under white light.

Manipulations of Illumination and Reflectance

If these ideas are correct, then the effects of contrast and constancy should be enhanced or diminished by changing the spectral characteristics of a scene. As for lightness and brightness, making a stimulus more (or less) consistent with different combinations of reflectance values and illuminants, or indeed of any other factors pertinent to the behavioral significance of the spectral return, should change perception in a predictable way.

One way of testing this idea is to alter the number of surfaces in a scene while keeping all other qualities the same, as mentioned in Chapter 2. The American psychologist David Katz first described the effects of "articulation" in the 1930s. Because the average intensity of the area surrounding a test target is left unchanged by such manipulations, the altered perceptions Katz described belie explanations based on the idea that the perceptual effects arise from limitations in visual processing and/or adaptation to the average intensity or spectral qualities of a stimulus (see, for example, Gilchrist, 1994; 2008).

These articulation effects are, however, consistent with an empirical strategy of vision. For example, Figure 3.15A presents four brightness

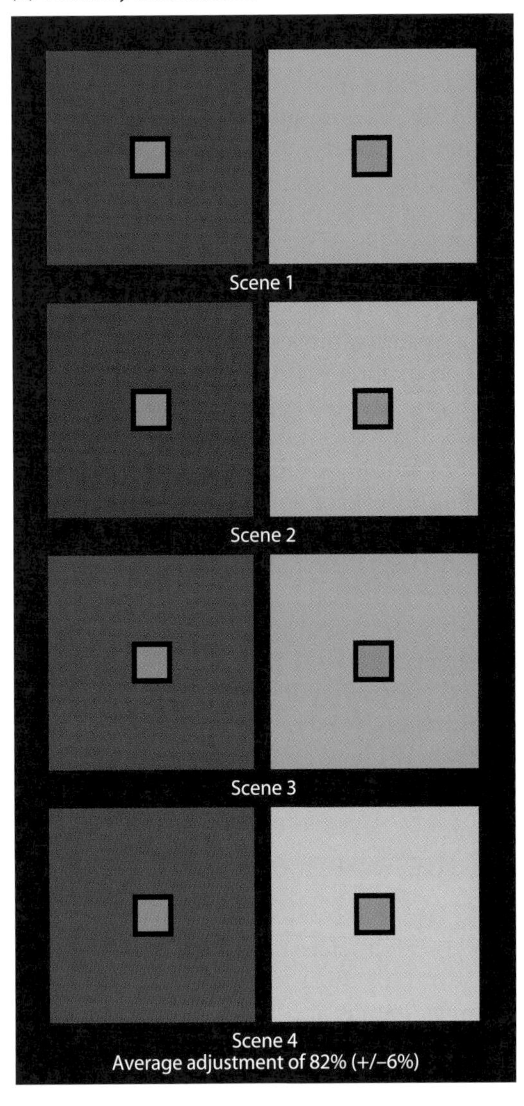

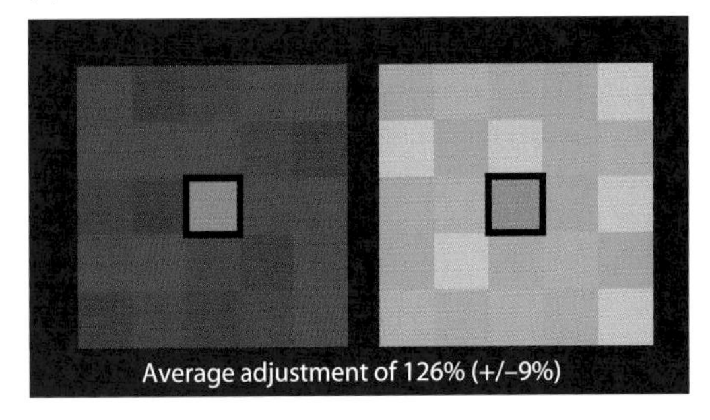

Figure 3.15 Effect of spectral articulation on lightness/brightness. (A) Four scenes comprising equiluminant gray targets embedded in uniformly colored surrounds that differ in luminance (the surround on the right is in each case is 10 times more luminant than the surround on the left). (B) A scene in which each of the four equiluminant color surrounds in (A) is broken up into patches. A greater average adjustment is required to make the two targets look the same in the scenes with multiple equiluminant color patches than in the scenes with uniform colored surrounds. These stimuli only approximate the scenes shown to subjects because some of the colors used are outside the gamut of the printer used to produce this book; the effects are also stronger when the scenes are presented separately, since the empirical significance of one scene affects the others. (From Lotto and Purves, 1999.)

contrast stimuli with identical luminance relationships; all the differently colored dark surrounds are the same intensity, as are all the differently colored light surrounds. Subjects perceive the strength of the effect to be the about same in all four instances. When, however, the surround is broken up into a number of distinct tiles in Figure 3.15B, a stronger effect is seen even though the manipulation changes neither the average spectral content nor the intensity of the stimuli. Thus, the change in perception cannot be attributed to either the greater variance of the surrounds or adaptation, the explanations often offered for why articulation alters what we see. In empirical terms, when a central target is surrounded by "articulated" patches, behavioral responses are more likely to succeed if the perceived intensity of the target is greater compared with the same intensity target and surround in the absence of articulation. Although not included here, color assimilation, in which a neutral target tends to take on the color of its surround, can also be explained in these terms (Long and Purves, 2003).

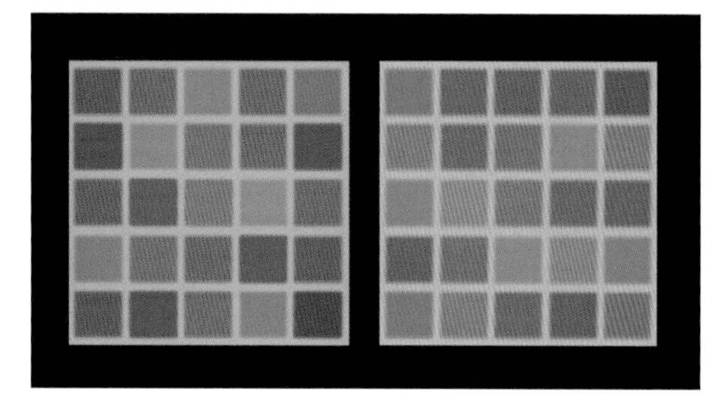

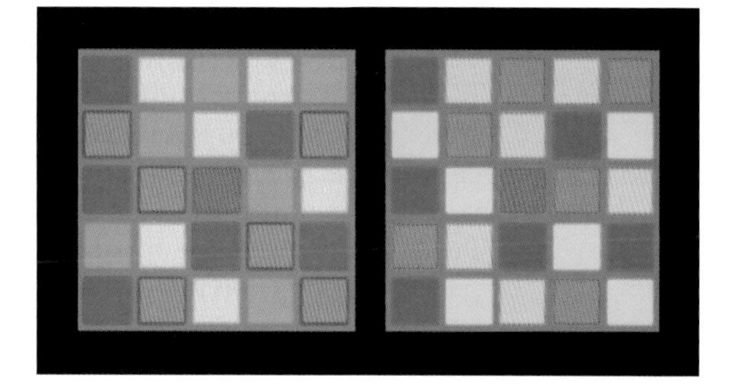

Figure 3.16 Comparison of the perceptions elicited by a standard color contrast stimulus and a more complex scene that changes the empirical significance of the stimulus. (A) Two spectrally identical targets on uniform chromatic surrounds. (B) Stimulus in which the same central squares are presented in the context of different tiles whose spectral returns are in each case consistent with the returns from the targets signifying different objects under different illuminants (the values of each tile are based on the physical interactions of reflectance, illumination, and spectral returns described earlier). Even though the average spectral content in the surrounds in (B) is the same as that of the corresponding surrounds in (A), the perceived color difference of the identical targets is greater in (B). (C) The scene is the same as in (B), but now surrounded by tiles whose spectral returns are inconsistent with the targets being differently reflective surfaces under different illuminants. As a result, the color contrast effect is weaker than in (B) (or the effect in [A]). (After Lotto and Purves, 2000.)

The effects of articulation are equally apparent when the central target is perceived as colored rather than gray. Thus, two spectrally identical targets appear only moderately different in color when presented on chromatically different but otherwise uniform surrounds, as expected from the limited information about the behavioral relevance of the stimulus (Figure 3.16A). However, when, as in Figure 3.16B, the same targets are presented on backgrounds that comprise a variety of chromatic patches having the same average spectral content as the corresponding scenes in Figure 3.16A, the apparent color difference between them is increased. Again, this effect is difficult to explain by adaptation or other input-level mechanisms. In empirical terms, however, the change in color perception makes good sense, since the large number of different spectral returns in Figure 3.16B (like 3.15B) is, by design, mutually consistent with the two identical targets being under different conditions of illumination.

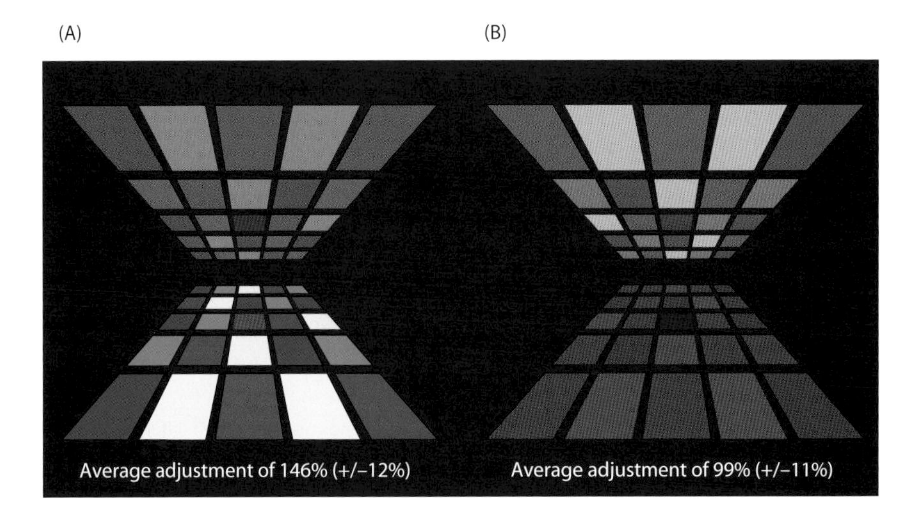

Figure 3.17 Different perceptions elicited by scenes that are spectrally identical. The spectra coming from the scene in (A) are consistent with the lower array being in light and the upper array being in shadow. However, since light typically comes from above, the same stimulus rotated (B) is less consistent with this possibility. As a consequence, the perceptions of the identical gray targets at the center of the lighter and darker arrays differ more in (A) than in (B), as indicated by the adjustments subjects made to equalize their appearance. (From Lotto and Purves, 1999.)

Finally, two identically articulated surrounds can be manipulated to change the strength of the effect without changing the colors themselves. Thus, in Figure 3.17A the physically identical gray tiles in the center of each pattern appear differently light or bright, as expected, since the luminance of the colored surrounds differs. However, the upper surround in (A) is illustrated as pointing upward, consistent with prevailing illumination coming from above. In Figure 3.17B the same stimulus is rotated such that the dark surround is now positioned above the light surround. Rotation causes a 50 percent decrease in the apparent lightness difference of the central gray targets, a phenomenon that is difficult to explain other than in empirical terms.

Color Constancy is Explained in the Same Terms as Color Contrast

Many accounts—Land's writings, for example—have regarded color contrast and constancy as different phenomena, with constancy being the perceptual goal and contrast being the "failure" of constancy: Whatever the circumstances, bananas, apples, or any other object should continue to look their respective "colors" so that objects can be properly identified. Color contrast has been interpreted by many investigators, though certainly not all (see, for instance, Mollon, 1995), as a "low-level illusion" generated by imperfect physiological mechanisms that evolved to serve color constancy.

The evidence indicates that color contrast and constancy are only superficially different consequences of the same empirical strategy of color vision, however. Identical targets in contexts that, through experience, have become associated with behaviors appropriate to physically different surfaces will elicit target perceptions that differ accordingly (color contrast); on the other hand, if spectrally different targets have become associated over time with behaviors that succeed when the targets are perceived as similar, then the targets will look more or less the same (color constancy). The goal of this strategy is not color constancy, but contending in any circumstance with the inverse problem as it pertains to the behavioral significance of the distribution of spectral intensities (see Figure 3.12).

Take, for example, the demonstration in Figure 3.18, in which two target surfaces are presented under conditions in which the variable is the intensity (luminance) of the different components of the scene. Even though the spectral returns from the central squares on the two faces of the cube in

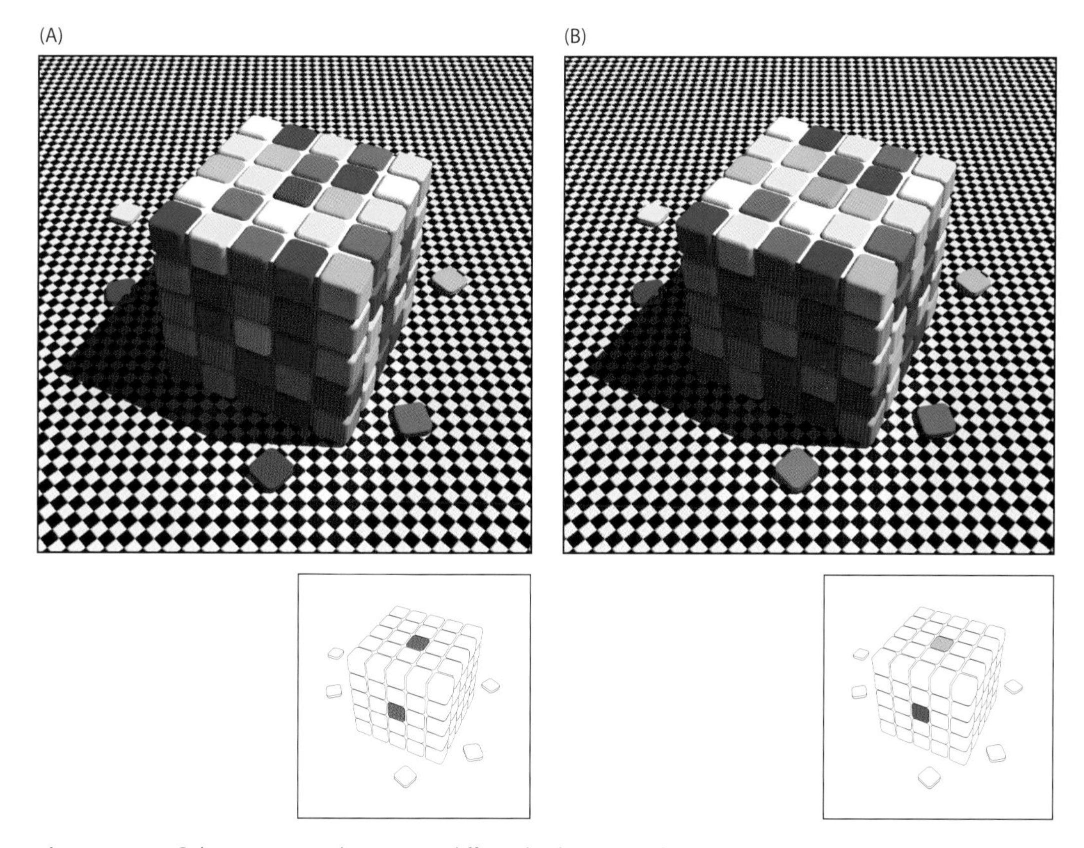

Figure 3.18 Color contrast and constancy differ only phenomenologically. The two panels demonstrate the effects on the color appearance when two similarly reflective target surfaces (A) or two differently reflective target surfaces (B) are presented in the same context consistent with illumination that differs only in intensity. The appearances of the relevant target surfaces in a neutral context are shown in the insets below.

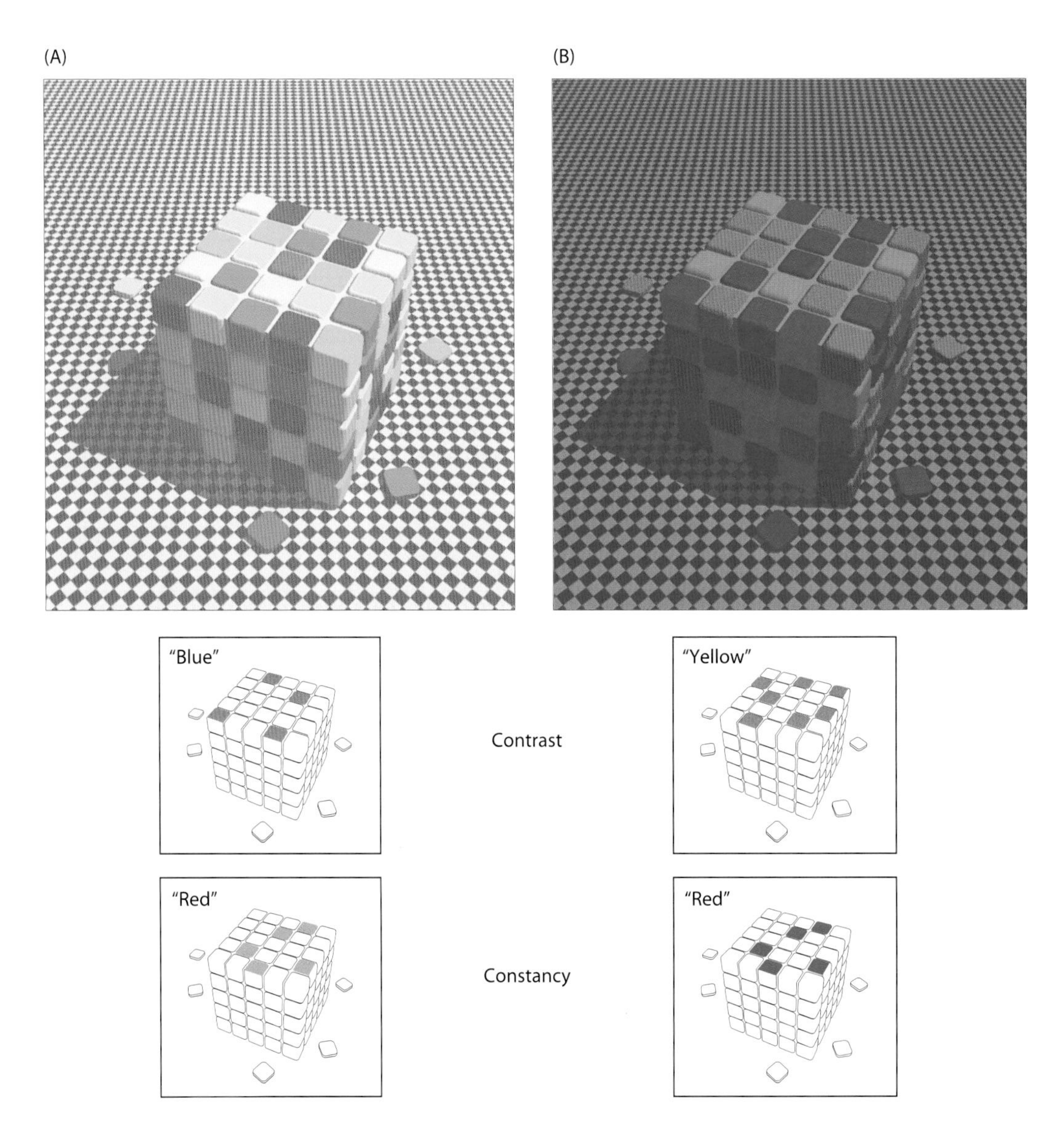

Figure 3.19 The effects on color when the same or differently reflective surfaces are presented in scenes consistent with illumination by spectrally different light sources. The same objects as in Figure 3.18, but now all the information in the scene has been made consistent with "yellowish" illumination (A) or "bluish" illumination (B). As indicated by the insets below, the "blue" tiles on the top of the cube in (A) are physically identical to the "yellow" tiles on the cube in (B). Both sets of tiles appear gray when the contextual information is taken away. This comparison is a far more powerful example of color contrast than the usual demonstrations. Conversely, the "red" tiles on the cube in (A) appear similar to the "red" tiles on the cube in (B), even though their spectral returns are very different (as shown by the masked versions of the scene below). This comparison is a powerful demonstration of color constancy.

Figure 3.18A are identical, their colors appear different because the information in the stimulus increases the probability that the identical returns originate from differently reflective surfaces under different illuminants (color contrast). If, however, the spectral returns from the central targets are different, as in Figure 3.18B, the targets on the two faces of the cube appear similar because the information in the scene is consistent with the spectral returns arising from similarly reflective surfaces under different illuminants (color constancy). Exactly the same context creates contrast or constancy, a result that is difficult to explain other than by visual circuitry that has linked images and behaviors by virtue of experience, and determines what we see.

This framework also explains why color sensations such as browns, navies, and olives can't be elicited by combinations of long-, middle-, and short-wavelength light in standard colorimetry experiments (see Figure 3.3). Such colors will be perceived only when a stimulus pattern is consistent with the intensity of the target illumination being low relative to the rest of the scene. Figure 3.18A, for instance, shows how empirical information about illumination can make two identical targets that look dark tan in a neutral setting appear either orange or brown when embedded in surrounds that have different luminance values.

If additional information in a pair of stimuli is made mutually consistent with different spectral illuminants, as in Figure 3.19, an even more dramatic example of contrast and constancy can be generated.

In summary, color contrast occurs when stimulus patterns are consistent with identical spectra having arisen from physically different combinations of illumination and reflectance; color constancy is apparent when stimulus patterns are consistent with two different spectra having arisen from similar combinations of these factors. To regard color contrast effects as "illusions" arising from the need to achieve color constancy is not only wrong, but misses the significance of these phenomena and the explanatory power of understanding color perceptions in empirical terms.

The Empirical Influence of Transmittance

Whereas these demonstrations illustrate the perceptual consequences of changing the probable contribution of reflectance and illumination to scenes, the effects of transmittance, the other universal determinant of the spectral qualities of light that reaches the eye (see Figure 3.12), have so far been neglected. It is not hard to show, however, that the possible contribution of transmittance to a spectral stimulus has equally predictable effects on perceived color that can be rationalized in terms of the typical physical influence of the medium interposed between the observer and objects in the world.

When surfaces are viewed through a medium that subtracts light more or less evenly from the spectrum of a stimulus, the spectral profile of the stimulus broadens (Figure 3.20A) (D'Zmura et al., 1997; Lotto and Purves,

(A)

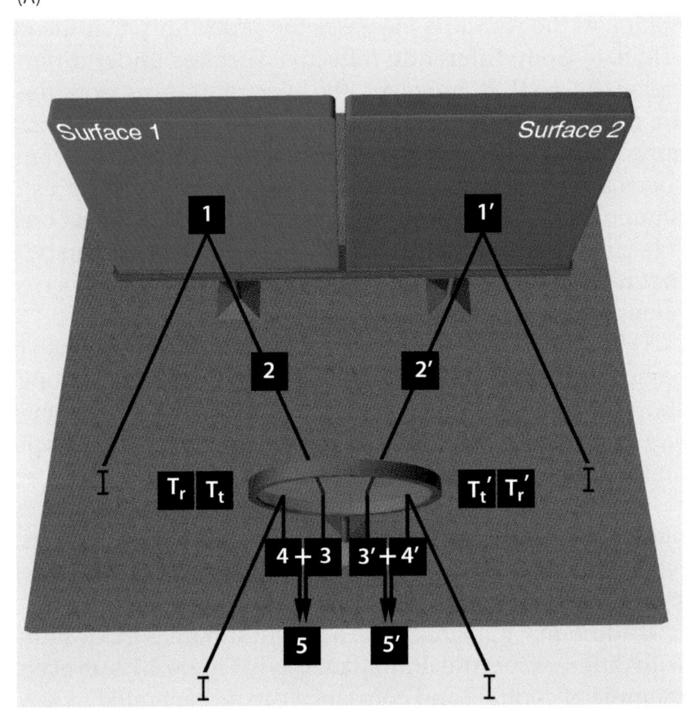
Surface 1
Surface 2
1
1′
2
2′
T_r T_t
T_t' T_r'
I
I
4 + 3
3′ + 4′
5
5′
I
I

(B)

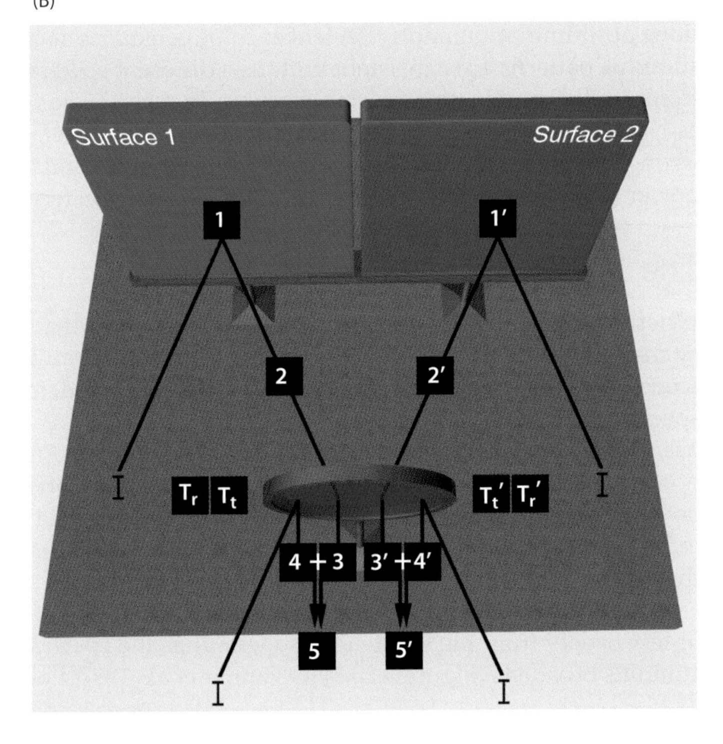
Surface 1
Surface 2
1
1′
2
2′
T_r T_t
T_t' T_r'
I
I
4 + 3
3′ + 4′
5
5′
I
I

◀ **Figure 3.20** The physical effects of imperfectly transmitting media on the spectra returned to the eye from the surfaces of chromatic targets. (A) Effect on the spectra from target surfaces generated by the interposition of a neutral filter, which broadens the spectrum by adding uniformly to both spectra and generating changes in apparent saturation. (B) Effect on the spectra returned from target surfaces arising from the interposition of a chromatic filter, which adds selectively to the two returns, thus generating changes in the perceived hues. (After Lotto and Purves, 2001.)

2001). This physical effect occurs because the light reflected by the medium adds uniformly to the distribution of wavelengths in each of the transmitted returns from the surfaces seen through it, thereby decreasing any spectral differences (consider the spectra at 2 and 2′ compared with those at 5 and 5′ in Figure 3.20A; see also the discussion of the Chubb effect in Chapter 2). By the same reasoning, when surfaces are viewed through a chromatic filter (Figure 3.20B), both returns are shifted toward the spectral characteristics of the medium (consider spectra 2 and 2′ compared with 5 and 5′ in Figure 3.20B). This influence arises because the medium in this instance not only filters out wavelengths from each return, but also adds light characteristic of the reflectance efficiency function of the medium itself. Given this routine experience with the way transmittance affects the quality of light returned to the eye, if color vision is determined empirically, then manipulating the probable contribution of transmittance to spectral stimuli should affect the colors seen.

This prediction can be tested using stimuli similar to those described in Chapter 2 for evaluating the effects of transmittance on perceptions of luminance (i.e., on lightness and brightness). In Figure 3.21, for instance, the average chromatic qualities of the surrounds of all four of the targets are the same. Despite the physical similarity of the surrounds and the identity of the targets in the left and right panels (see insets), the target elements in Figure 3.21B appear less saturated (i.e., the reds and blues in the target look grayer or "duller") than the target elements in Figure 3.21A. The empirical rationale for this effect is that, relative to the uniform surround in Figure 3.21A, the chromatic relationships in Figure 3.21B are consistent with experience of viewing the "red" and "blue" elements of the central target through an achromatic medium of lower transmittance than the medium intervening between the observer and the surround. Similarly, the spectral returns in Figure 3.21D are consistent with viewing the texture of the target through a centrally located "purple" filter, whereas the information in 3.21C is less consistent with this possibility.

Thus, like stimuli that elicit color contrast and constancy effects according to experience pertinent to the contributions of reflectance and illumination and their implications for behavior, the color of identical target stimuli changes according to experience with transmittance.

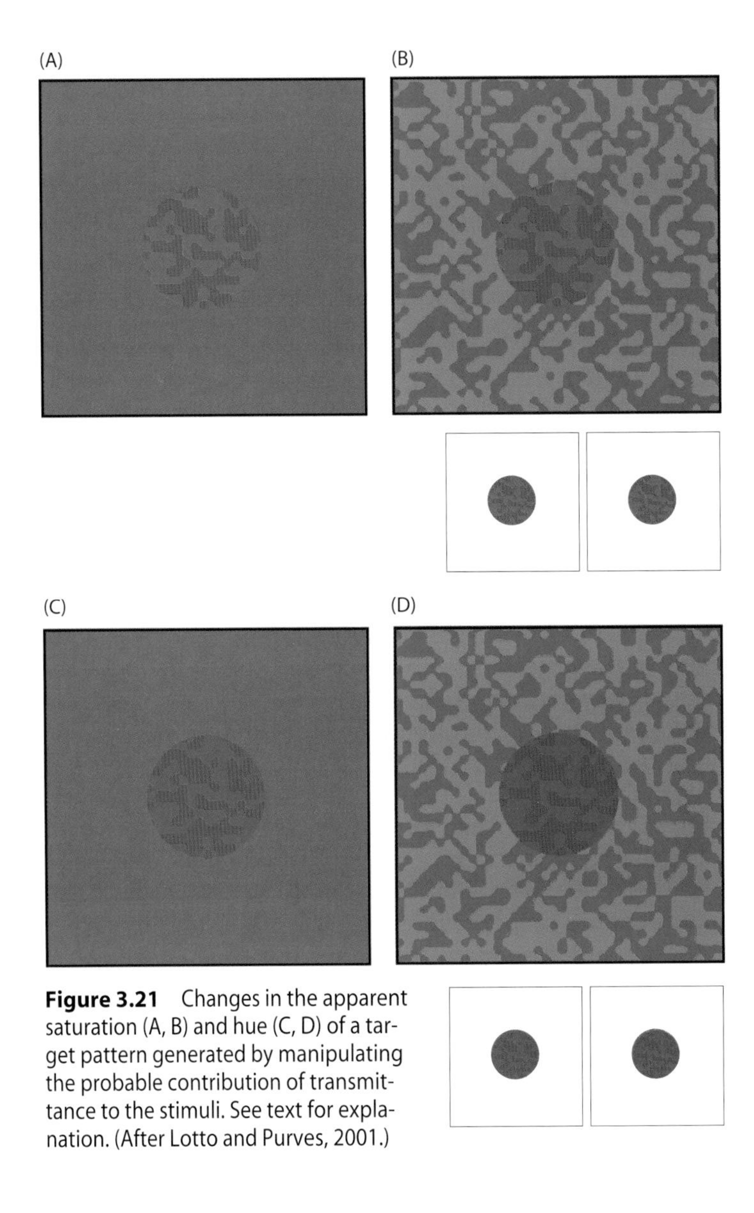

Figure 3.21 Changes in the apparent saturation (A, B) and hue (C, D) of a target pattern generated by manipulating the probable contribution of transmittance to the stimuli. See text for explanation. (After Lotto and Purves, 2001.)

Explaining Colorimetry Effects in Empirical Terms

Another challenge in rationalizing color vision is the results of colorimetry testing, in which targets are presented in a uniform neutral surround with the explicit intent of eliminating contextual variables (see Figure 3.3). The psychophysical functions determined in colorimetry testing can be divided into those derived from color discrimination testing and those derived from color matching paradigms. In color discrimination tests, the ability to distinguish equally noticeable (or "just noticeable") differences in hue or saturation varies in a complex way as a function of the wavelength of a monochromatic stimulus (Bedford and Wyzecki, 1958; Evans and Swenholt, 1968; Wyzecki and Stiles, 1982). The results of other colorimetry tests are even more complicated:

1. Saturation varies as a function of luminance (the Hunt effect; Hunt, 1952).
2. Hue varies as a function of stimulus changes that affect saturation (the Abney effect; Abney, 1910; Roberson, 1970).
3. Hue varies as a function of luminance (the Bezold-Brücke effect; Purdy, 1931; Boynton and Gordon, 1965).
4. Brightness varies as a function of stimulus changes that affect both hue and saturation (the Helmholtz-Kohlrausch effect; Wyzecki and Stiles, 1982).

Explanations of these phenomena have generally been based on assumptions about neuronal interactions early in the visual pathway and have not lead to any consensus (Hurvich and Jameson, 1955; Stiles, 1972; Hurvich, 1981; Wyzecki, 1986; Pokorny and Smith, 2004). The details of these colorimetric functions thus present a daunting set of data that any theory of color vision must be able to explain.

The empirical account of the classic colorimetry functions is predicated on the idea that in the absence of the specific contexts considered so far, the perceptual qualities of hue, saturation, and brightness have nonetheless been determined by trial-and-error behavior in response to the human spectral experience. As further signatures of this visual strategy, colorimetry functions should be predictable on empirical grounds. Much as in assessing the frequency of occurrence of luminance relationships in Chapter 2, a database of natural images is needed, in this case like those in Figure 3.22. Tallying the frequency of occurrence of spectral relationships in many images indicates the experience that would have shaped successful behavior and the associated color percepts—in this case, the classic colorimetry functions (Long et al., 2006; reviewing the dice analogy presented in Chapter 2 may be helpful here).

As an example of this approach, consider the human ability to discriminate hue as a function of wavelength. If people are asked to report when they perceive a change in color as the wavelength of a stimulus is altered in small increments, the result is the complex curve in Figure 3.23A. In empirical terms, this function should be predicted by accumulated human experience with the spectral relationships at each point in retinal images produced by natural illuminants and surface reflectance values (the stand-ins for the points are the pixels in the image database; see also Hendley and Hecht, 1949). Once again, this experience will have been incorporated in visual circuitry based on feedback from successful behavior, and should come to the fore in colorimetry tests that minimize the contextual effects considered earlier in the chapter.

Figure 3.22 Analysis of a spectral database. (A) Examples of spectral images. (B) By examining the spectra in each of millions of pixels, accumulated experience with spectral relationships in the absence of additional context can be determined and used to predict colorimetry functions derived in psychophysical testing. (After Long et al., 2006.)

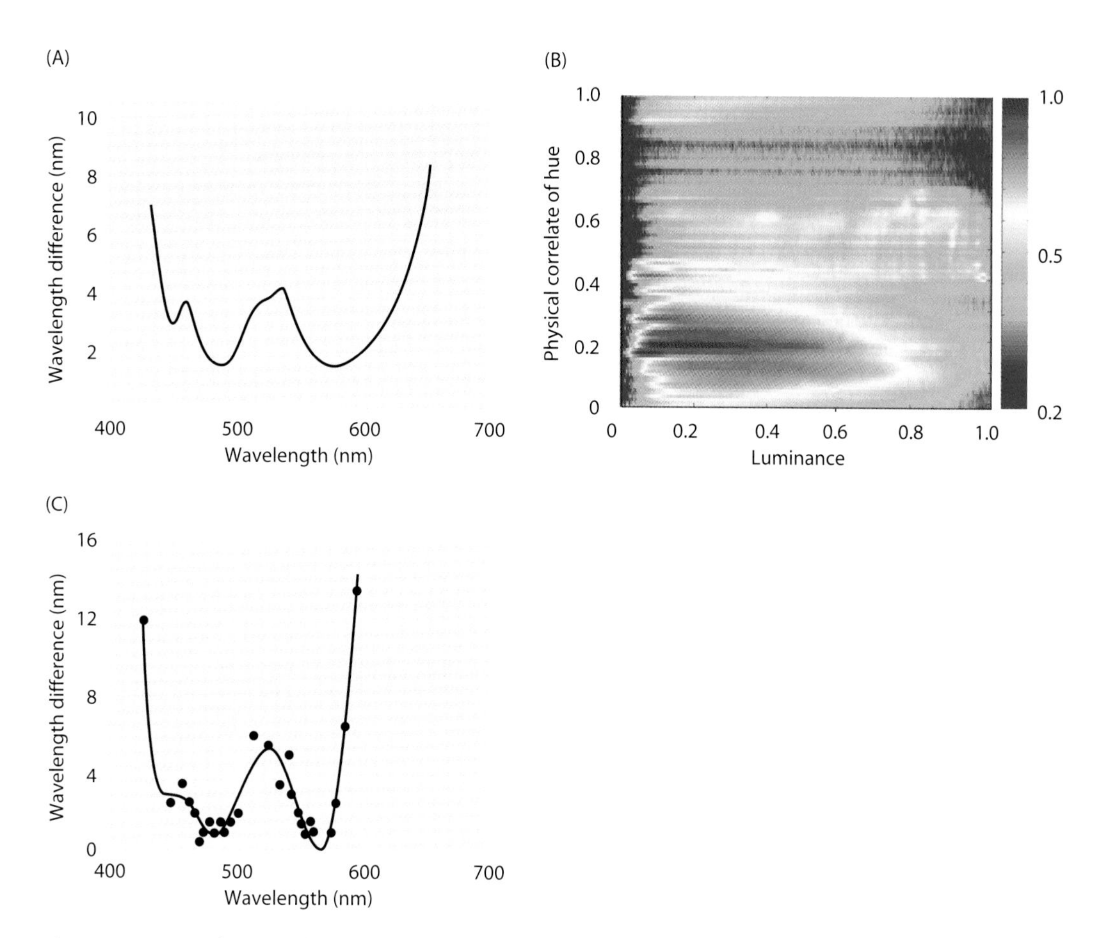

Figure 3.23 Predicting colorimetry functions. (A) Wavelength change in a monochromatic stimulus needed to elicit a just-noticeable difference in hue over approximately 400–700 nm. (B) Probability distribution of the physical correlates of hue (wavelength) and brightness (luminance) derived from scenes such as those is Figure 3.22 (probabilities are indicated by the color-coded bar on the right). (C) The predicted function derived from the data in (B); the curve is a polynomial fit to the points. (After Long et al., 2006; data in [A] from Bedford and Wyszecki, 1958.)

Figure 3.23B shows how the physical correlates of hue (wavelength) and brightness (luminance) co-vary in millions of pixels analyzed in natural scenes such as those in Figure 3.22. These data thus indicate human experience with wavelength variation at given luminance level. In regions where experience with wavelength relationships is richest (yellow/red regions of the graph), the ability to make discriminations among wavelengths should be enhanced; conversely, where experience is relatively sparse (blue regions), this ability should be poorer. By assessing this experience in the database of scenes, the changes in wavelength needed to generate a constant difference in perceived hue over the visible spectrum can predicted (Figure 3.23C). Comparison of Figures 3.23A and C shows general agreement between the ability to discriminate hues as wavelength varies, and the function predicted by analyzing natural scenes.

Other colorimetry effects, including saturation discrimination, the Hunt effect, the Abney effect, the Bezold-Brücke effect, and the Helmholtz-Kohlraush effect, are also predicted by empirical information extracted from a database of spectral scenes (see Long et al., 2006). The basis of these empirical predictions is straightforward in principle. The physical attributes of light reaching the eye co-vary in specific ways, and this experience should be apparent in functions derived from colorimetry testing—as it is.

Differences in Color Brightness Arising from the Human Luminosity Function

Another issue in human color perception also concerns the differential sensitivity of the visual system to wavelengths (Figure 3.24A). Psychophysical tests of the responsiveness to spectral stimuli (threshold detection, or tests of just-noticeable differences) show that humans are less sensitive to long- and short-wavelength light than to light of middle wavelengths (Kaiser and Boynton, 1996; Stockman et al., 2005; see also Figure 3.23A). As a result, luminance, although a physical measurement, is calibrated to reflect the average sensitivity of humans to different wavelengths; it would make little sense to treat all wavelengths of light as having an equal impact on vision when they clearly don't. As a consequence, surfaces that reflect predominantly long or short wavelengths (which tend to appear red or blue, respectively) and surfaces that reflect a narrow range of wavelengths (which thus tend to appear more saturated) will, on average, generate less luminance than surfaces that reflect predominantly middle wavelengths (Figure 3.24B).

If perceptions are determined empirically, then trial-and-error behavioral responses would have taken into account this difference between the radiance of light coming from surfaces and different luminance of retinal stimuli. In empirical terms, "red" and "blue" surfaces should therefore appear somewhat brighter than equiluminant stimuli that evoke sensations of green and yellow. As originally noted by Helmholtz (1863), this is indeed what humans perceive; moreover, the perceptions are correlated with corresponding levels of primary visual cortex activation (Corney et al., 2009). Artificial neural networks trained to recognize surfaces by spectral reflectance also exhibit the same nonlinear relationship between luminance and color brightness if the input of the networks corresponds to the human luminosity function (op cit.).

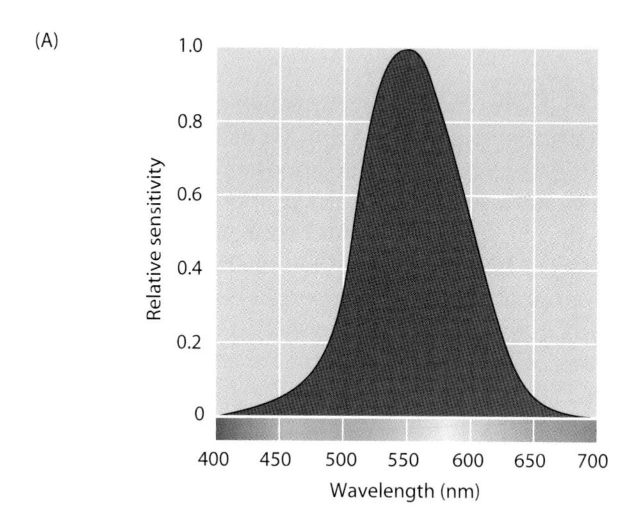

Figure 3.24 The average sensitivity of the human visual system to light of different wavelengths (called the luminosity function). (A) Humans and old-world primates such as rhesus macaques are much less sensitive to long- and short-wavelength light than to light of middle wavelengths, explaining why photometers designed to measure light intensity (luminance) use filters that mimic the sensitivity of the human visual system. (B) Diagram showing the relationship that, as a consequence, will always have been experienced by humans; the width of the colored arrows indicates relative amount or intensity of light. ([A] is redrawn from the 1926 CIE study, as modified by Vos, 1978.)

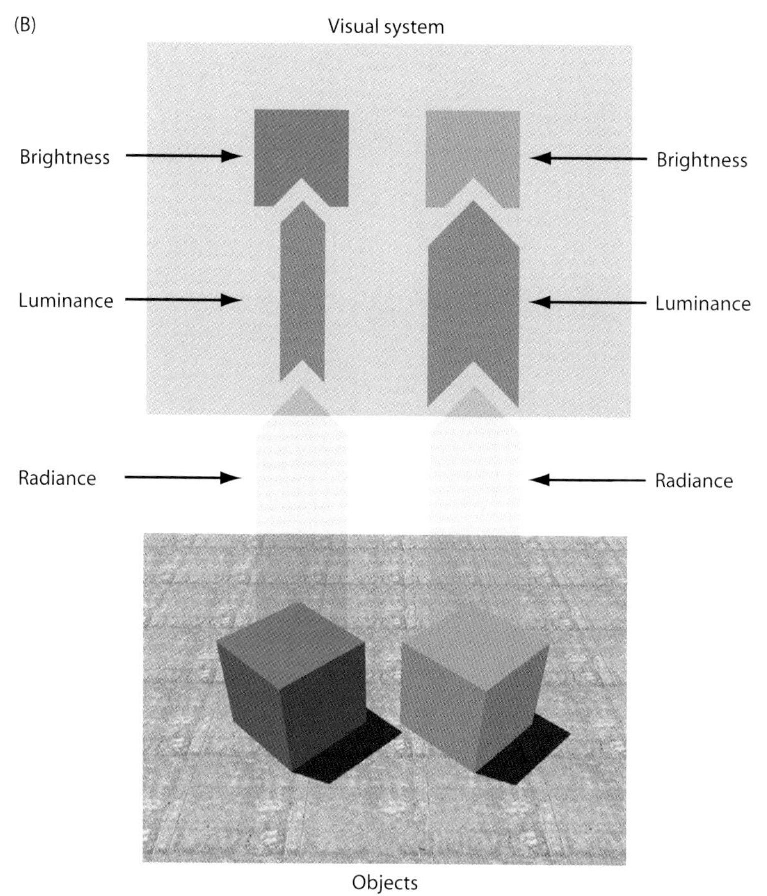

Conclusion

Color vision is initiated by trichromatic mechanisms and subsequently processed by neurons that often have color opponent receptive fields. Nonetheless, understanding the colors seen in response to spectral stimuli is not easily rationalized by knowledge about the properties of spectrally sensitive photoreceptors and higher order visual neurons. The phenomenology of color perception indicates that, like the relationship between luminance and lightness/brightness, the colors perceived are determined by circuitry whose organization has been shaped by feedback from successful behavior as a means of contending with the inverse problem. The evidence for this assertion is the ability to explain color contrast, color constancy, and colorimetry functions in wholly empirical terms. If this conclusion is correct, then trichromacy and color opponency have evolved because they serve this strategy, rather than the other way around.

4

Seeing Intervals, Angles, and Object Sizes

Introduction

In addition to conflating the physical parameters that determine the quantity and quality of light reaching the eyes, the parameters that define the location and arrangement of objects in space—size, distance and orientation—are also inextricably intertwined in the retinal image. As a result, the spatial characteristics of the sources of visual stimuli are, like the sources of luminance and the distribution of spectral power, unknowable by any operation on the retinal image as such. Since successful visually guided behavior depends on the size, distance and orientation of objects in space, the inverse problem is as much an obstacle in determining geometrical relationships as in determining the relative contributions of illumination, reflectance, and transmittance discussed in the preceding chapters. Thus, the perception of geometry should also be understandable in terms of empirical associations made between images and behavioral success. This chapter summarizes evidence that perceptions of spatial intervals, angles, and object sizes are indeed based on this strategy.

The Inherent Uncertainty of Image Geometry

The inability of retinal images to directly specify real-world spatial relationships arises most obviously from the fact that a three-dimensional world is projected onto a two-dimensional sheet of photoreceptors. As a result, objects that have different sizes, orientations, and distances from observers can project the same retinal image (Figure 4.1).

As in other domains of vision, it seems inevitable that this problem be resolved by trial-and-error interactions with objects in the world, the outcome of which is then incorporated into visual circuitry according to the relative success of visually guided behavior over the course of evolution and individual lifetimes. In principle, then, it should be possible to predict perceptions of geometry in these terms.

Discrepancies between Geometrical Stimuli and Their Perception

Observers over the centuries have pointed out that measurements made with rulers or protractors often disagree with the relevant visual perceptions. This seems remarkable since the way we see the geometry of the world obviously works; indeed, it works so well that most people find it difficult to believe that perception isn't representing the physical world as it really is. As a class, these discrepancies are called "geometrical illusions," and constructing them was a cottage industry in the nineteenth and early twentieth centuries. A good "illusion" of this type has provided eponymous immortality to a number of investigators (reviews can be found in Luckiesh, 1922; Rock 1995; Gregory, 1998; Robinson, 1998; the book by Robinson is an especially detailed and authoritative summary of this phenomenology).

Figure 4.2 illustrates some well-known instances. The example attributed to Hering shows two parallel lines (indicated in red) that appear bowed away from each other when presented on the background of

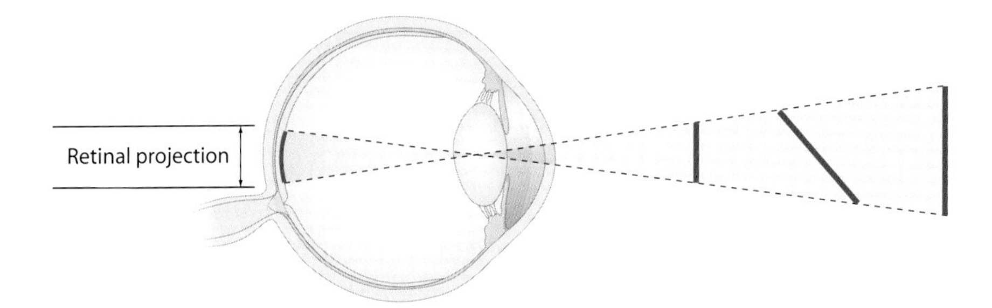

Figure 4.1 The geometrical significance of any object projected onto the retina is uncertain. As indicated in this diagram, the same image can be generated by objects of different sizes, at different distances from the observer, and in different orientations.

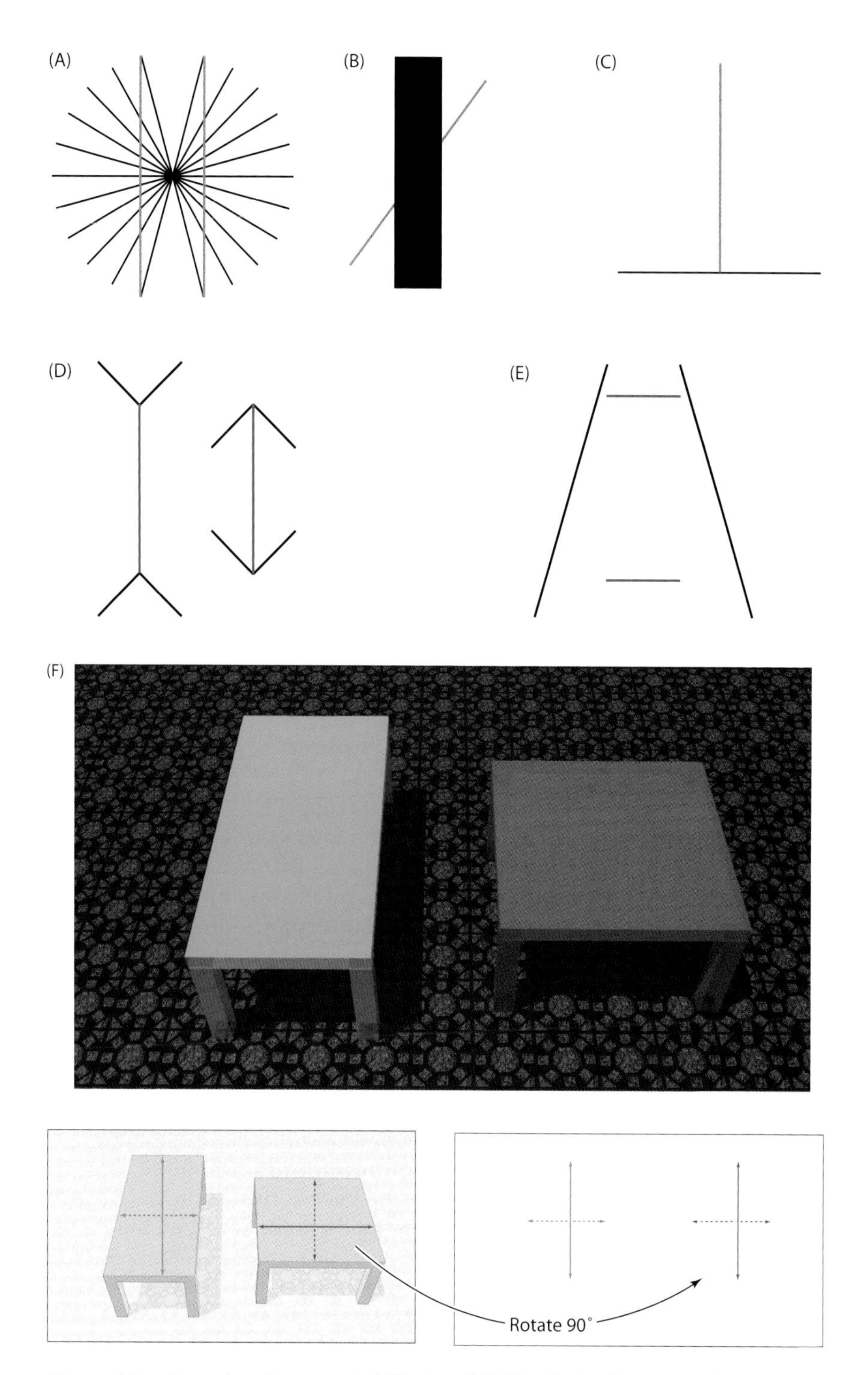

Figure 4.2 Examples of geometrical "illusions." (A) The Hering illusion. (B) The Poggendorff illusion. (C) The T-illusion. (D) The Müller-Lyer illusion. (E) The Ponzo illusion. (F) The "table top" illusion. (See text for explanations.)

converging lines (Figure 4.2A). Another example from the nineteenth century is Poggendorff effect (Figure 4.2B). In this stimulus, a line obscured by a bar appears to be displaced upward to the right of the occluder, even though the line segments are collinear. A number of other geometrical illusions involve the comparison of the length of two lines. The simplest is the "T" illusion, in which the vertical line appears longer than the horizontal line, despite the fact that they are equal in length (Figure 4.2C). In the Müller-Lyer illusion the line terminated by arrowheads looks longer than the same line terminated by arrow tails (Figure 4.2D), and in the Ponzo illusion the upper horizontal line appears longer than the lower one, despite the fact that they are again identical (Figure 4.2E). Finally, Figure 4.2F shows a variation of a tabletop illusion devised by Roger Shepard (1992). The length of the green table looks greater than the breadth of the red table; conversely, the width of the green table looks to be much less than the length of the red table. As shown below, however, when the red table is rotated it is apparent that the dimensions of the two surfaces are identical (although the angles at the corners differ). Many other instances of discrepancies between the geometry of a stimulus and the percept it gives rise to can be found in various popular books on illusions, although most of these stimuli are variations on the basic themes illustrated in Figure 4.2 (Seckel, 2000; 2002).

Understanding Perceived Geometry in Empirical Terms

An empirical explanation of these phenomena follows from the approaches to lightness/brightness and color, the idea being that perceptions elicited by geometrical stimuli are fully determined by the relationships humans have always experienced between the geometry in retinal images and behaviors needed to deal with the geometry of real-world objects. If the provenance of geometrical stimuli is directly unknowable by means of light stimuli and if this uncertainty is resolved the same way as the uncertain provenance of luminance and/or spectral content, then it should be possible to predict how observers perceive any given contour based on the frequency of projected geometries.

Given a database that reasonably represents the physical arrangement of objects in the sorts of circumstances that humans have experienced, one can test whether the perceptions elicited by the geometrical stimuli in Figure 4.2—or any other geometrical stimulus—can be rationalized on this basis. As in the preceding chapters, the frequency of occurrence of light patterns pertinent to a geometrical stimulus indicates the relative rank of any particular stimulus within the totality of experience, and thus the relative rank it should have in perception (e.g., a longer or shorter interval, a bigger or smaller angle, etc.).

Acquiring and Sampling a Geometrical Database

Although acquiring detailed spatial information about the structure of the real world would have been difficult to imagine not many years ago, advances in technology have made at least one aspect of this problem straightforward. In the construction industry, geometrical conformance to architectural plans is routinely monitored using laser range scanning (Figure 4.3). This technique provides accurate measurements of the distances of all the points (pixels) in a digitized scene from the image plane at

Figure 4.3 Determining the physical geometry underlying stimuli generated by typical scenes using laser range scanning. A mirror inside the rotating head of the scanner directs a laser beam in a pattern that scans a scene; the laser beam is pulsed, such that each point (pixel) in the scene is evaluated sequentially. The signal reflected back from object surfaces is detected by a photodiode. The tiny interval between the transmitted pulse and signal returned is determined by a quartz clock; based on the speed of light, the distance from the image plane of the scanner to each surface point is then calculated by a microcomputer. The distances determined in this way are accurate to a few millimeters over a range of ~300m. (A) Digital camera images of a fully natural scene and an outdoor scene that contains human artifacts. (B) The corresponding range images acquired by the laser scanner. Color-coding indicates the distance from the image plane of the scanner; black areas are the points in the scene from which no laser reflection was recorded (the sky). (From Howe and Purves, 2005a.)

the origin of the scanner's laser beam. By setting the height of the scanner at the average eye level of human observers, the distance and direction of each point in an image to its physical source provides a good approximation of the spatial relationships between images and real-world spatial arrangements that humans have typically experienced.

On the face of it, acquiring this information by laser scanning may seem different in kind from the acquisition and use of the databases of luminance or spectral content in Chapters 2 and 3. It is not. In those instances as well, a machine—a photometer or a spectrophotometer—was used to determine how physical properties are related to light patterns in images, thus indicating what accumulated experience would have taught empirically. Whereas a photometer or spectrophotometer measures luminance or spectral content directly in images, the distance and direction of each point in an image must be provided by a method such as laser scanning.

Of course, a database of real-world geometry acquired in this way, like the databases of luminance or spectral content, has serious limitations. First, the range of distances that can be analyzed (from approximately 2 meters to 300 meters) and resolution of the analyses (approximately 0.15°) are quite restricted. Humans obviously see things that are closer than 2 meters and more distant than 300 meters, and can readily resolve points more closely spaced than the resolution of the scanner (normal humans can resolve points as close as 0.005°). Using data with relatively limited range and resolution requires the assumption that the 3-D information obtained is representative of the additional information that would have been gleaned if the analysis included the more highly resolved nearer and farther objects that we see and interact with on a daily basis. Another deficiency is that the database does not represent the variety of landscapes and spatial arrangements found in environments worldwide; all the scenes in the database were acquired in a particular locale (the Duke University campus) in a particular season (summer). Other visual information pertinent to the perception of scene geometry —e.g., stereoscopic disparity (important in binocular depth sensations; see Chapter 5), motion parallax (see Chapter 6), and much more—is not included. Finally, humans and other visual animals do not observe the world in the systematic fashion of a laser scanner, but fixate on objects and parts of objects that contain information particularly pertinent to the species. As the Russian physiologist Alfred Yarbus showed more than 60 years ago, human observers are highly biased in the time they devote to viewing different components of scenes, being particularly interested in human figures, faces, and the most informative parts of faces, such as the eyes and mouth (Yarbus, 1959; see also Parkhurst and Niebur, 2003). Since the scanner samples all portions of a scene uniformly, this deficiency is also inherent in the database of scene geometry.

These limitations notwithstanding, the frequency of occurrence of projected geometries generated by the real world predicts perceived geometry remarkably well, as described in the sections that follow.

Predicting the Perceived Length of Lines on This Basis

A simple test of this scheme is how well it predicts perceptions and psychophysical functions having to do with line length. For ordinal qualities such as length, in which the perception concerns magnitude, predictions can be made on the basis of how often a particular length has occurred in the full range of retinal line lengths experienced. This is same approach that was used in predicting perceptions of lightness and brightness of the same target luminance in different surrounds in Chapter 2 and of colorimetry functions such as the just-noticeable difference in color as a function of wavelength in Chapter 3.

Since there seems no reason not to see lines for what they are, most people suppose that the perceptions of lines drawn on a piece of paper or a computer screen are veridical. Accordingly, if a series of lines of different lengths were shown, observers would expect the apparent lengths to scale in agreement with their measured lengths. As Figure 4.2 illustrates, however, this expectation is not met.

In empirical terms, the perceived length of the line should be determined by how often in past experience a projected line of that length occurred within the entire range of projected lengths on the retina experienced by human observers. For example, if the minimum possible length of projected lines is arbitrarily taken to be 1 and the maximum length 100, then, as indicated in Figure 4.4, a line that measured 25 units in length would rank at the

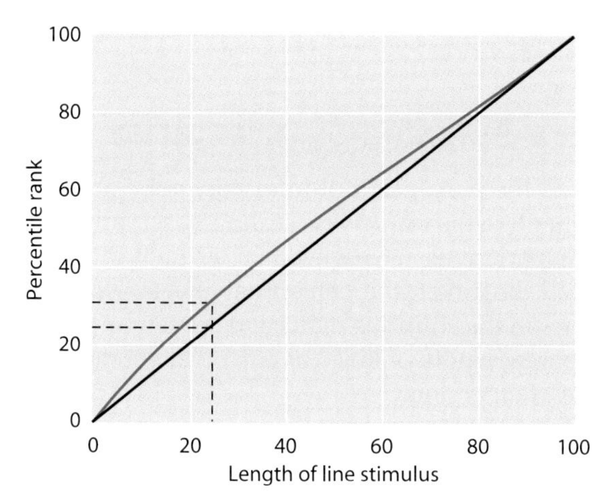

Figure 4.4 Ranking line lengths in images. The red curve indicates the relative frequency of occurrence (expressed as percentiles) of different projected lengths (in arbitrary units), determined from analysis of the laser-scanned database. The black line indicates lengths on a linear scale determined by physical measurements of projected lengths. Because a line of any given length in the retinal image (e.g., 25 units) has a different rank on these two scales (dotted lines), there is always a discrepancy between the perceived and the measured lengths of projected lines. (After Howe and Purves, 2005b.)

25th percentile on a linear scale extending from 1 to 100. The empirical scale of line length, however, is based on how frequently linear stimuli shorter or longer than 25 units have actually been generated in accumulated human experience. If stimulus lines shorter than 25 units have occurred more often than the lines that are longer, then the empirical rank of the 25-unit stimulus line would be higher than the 25th percentile, and conversely. Because many factors determine the frequency of occurrence of even the simplest image features, the rank of any projected line length in accumulated experience will differ from its rank on linear scale. A consequence is routine discrepancies between the linear metrics and the "metrics" of perception.

The advantage of perceiving lengths or intervals on this basis is contending with the inherently uncertain significance for behavior of any particular projected length, as illustrated in Figure 4.1. The empirical rank of a retinal image feature will order the perceptions of that quality in a manner that works behaviorally, the inverse problem notwithstanding. Despite the universal discrepancies between perceptual and measured lengths on the retina or in the environment, seeing this and other visual qualities according to their empirical rank maintains in perceptual space the relative similarities and differences among the generative objects, thus allowing successful behavior. In this way, projected lengths on the retina are related to a correspondingly ordered perceptual length, albeit in a nonlinear manner (see Figure 4.4). This framework is the same as that described in the perceptual "space" defined by the totality of experience with retinal luminance values (explaining the nonlinear lightness and brightness functions discussed in Chapter 2) and spectral distribution values (explaining the nonlinear color functions discussed in Chapter 3). The basis in each case is trial-and-error interactions with the world that link these various aspects of retinal images to perceptions and behaviors that work.

Perceived Length as a Function of Orientation

The previous section concerns the perception of intervals in the absence of any particular context, rather like the brightness scaling observations in Chapter 2 or the colorimetry experiments in Chapter 3. In virtually all of the stimuli illustrated in Figure 4.2, however, the perceptual effects depend a context provided by other lines in the stimulus pattern. If perceived geometry is empirical, then the apparent length of lines (or any other geometrical feature) should be affected by their context.

The simplest "contextual" variable to assess is the orientation of a line, an aspect of the stimulus not yet considered. As indicated in Figure 4.5A (see also Figure 4.2C), people have long noted that a vertical lines looks a little longer than the same lines presented horizontally. Investigators have shown repeatedly over the last 150 years that the perception of line length actually varies continuously as a function of orientation, the maximum length of a line being seen, oddly enough, when it is oriented about 30° from vertical (Wundt, 1862; Shipley et al., 1949; Pollock and Chapanis, 1952;

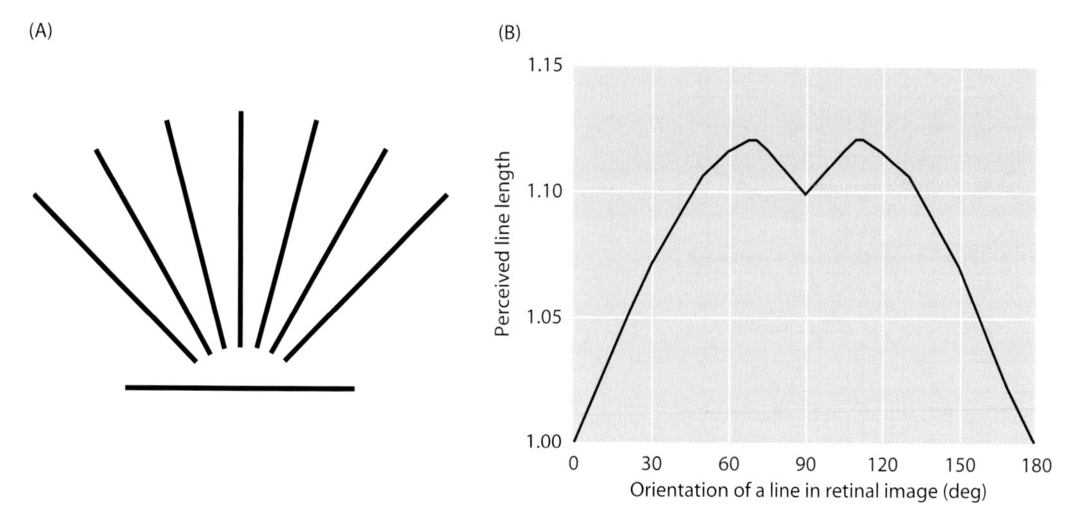

Figure 4.5 Variation in apparent line length as a function of orientation. (A) The horizontal line in this figure looks somewhat shorter than the identical vertical or oblique lines (see also Figure 4.2C). (B) The apparent length of a line reported by subjects as a function of its orientation in the retinal image, expressed in relation to the apparent length of a horizontal line (1.00). The maximum length seen by observers occurs when the line is oriented approximately 30° from vertical, at which point it appears about 10%–15% longer than the minimum length seen when the orientation of the stimulus is horizontal. The graph is a compilation of the psychophysical results reported in the literature. (After Howe and Purves, 2002.)

Cormack and Cormack, 1974; Craven, 1993) (Figure 4.5B). This effect pertains to any spatial interval. Thus, the apparent distance between a pair of dots varies systematically with the orientation of an imaginary line between them, as Wilhelm Wundt first showed in 1862, and a perfect square or circle appears to be slightly elongated along its vertical axis (Sleight and Austin, 1952; McManus, 1968). Despite extensive study of these phenomena, no generally accepted explanation has been forthcoming.

In an empirical framework, the apparent lengths elicited by lines in different orientations on the retina are predicted by the relative rank of the projected length with respect to the frequency of occurrence over the full range of projected lengths in that orientation in the experience of the species and individual. The apparent length of any given line should be determined by how many lines in that orientation have been either shorter or longer than the projected line in question. In consequence, the relative rank will come to correspond to the perceptions and behaviors that have been successful in the past, thus contending with the fact that the physical length and orientation of a projected line cannot be known from the image length as such. The orientation of a line provides a "context" for the projected length, much as the context of luminant or spectral targets improves

(A)

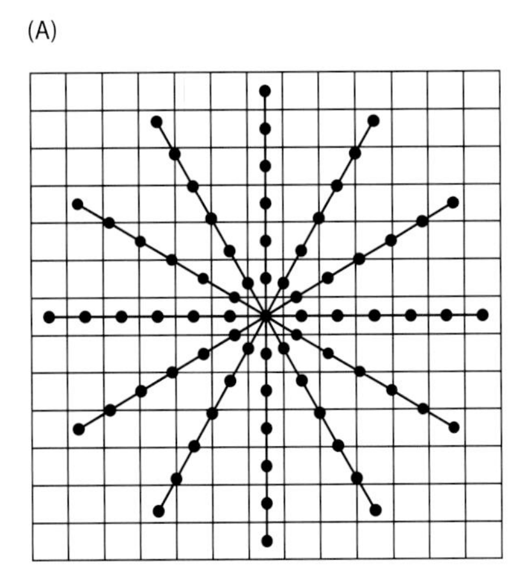

(B)

Figure 4.6 Sampling the frequency of occurrence of straight-line projections in different orientations generated by real-world objects. (A) The pixels in a region of one of the digital images in the database are represented diagrammatically by the grid squares; the connected black dots indicate a series of templates for determining the frequency of occurrence of straight lines projected at different orientations arising from straight lines in the 3-D world in the corresponding range data (see Figure 4.3). (B) Examples of straight-line templates overlain on a typical image. White templates indicate sets of points that corresponded to straight lines in 3-D space, and were thus accepted as valid samples of straight-line projections in that orientation. Red lines indicate sets that failed to meet this criterion and were therefore rejected. (After Howe and Purves, 2005a.)

the operational success of the corresponding lightness, brightness, or color perceptions (or the way the context of a second die can refine behavior in the dice game analogy described in Chapter 1).

This explanation of apparent length as a function of orientation can be evaluated by measuring the relative frequency of occurrence of straight-line projections in different orientations in a database of range images (Figure 4.6). The approach entails tallying up how often a straight-line template applied to the image plane corresponds to a geometrical straight line in the world. Each of the probability distributions derived in this way (Figure 4.7A) can then be expressed in cumulative form to provide an empirical scale of the frequency of occurrence of line lengths projected at any specific orientation (Figure 4.7B). The rank of any line in these distributions indicates the percentage of projected lines in a given orientation that, in the experience of human observers, has been shorter than the line in question, and the percentage that has been longer. As expected from the non-uniform nature of the physical world, the empirical rank of projected lines varies as a function of orientation.

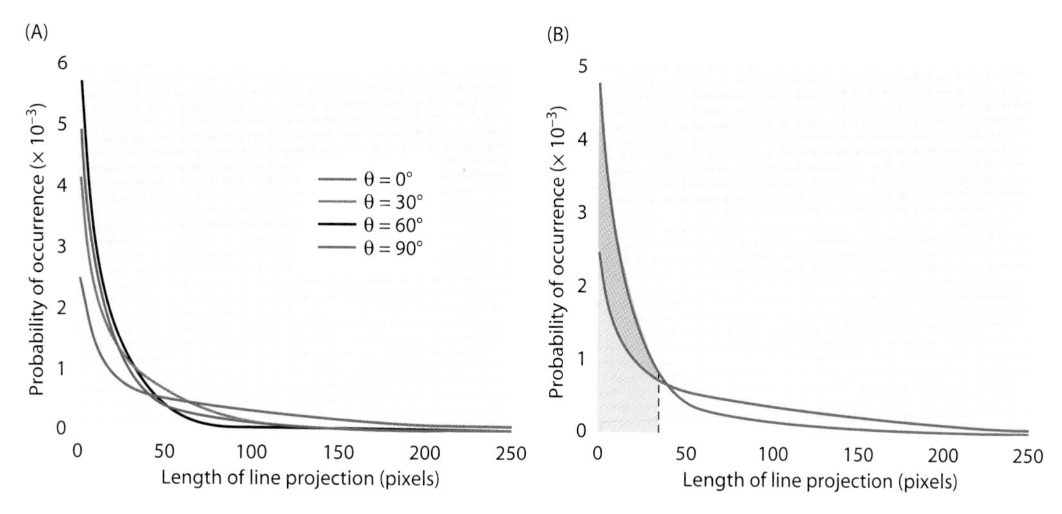

Figure 4.7 The frequency of occurrence of lines in different orientations projected from the real world onto an image plane. (A) Probability distribution of lines of different lengths and orientations (θ) in retinal images. (B) Cumulative probability distributions calculated from the distributions in (A). The cumulative values for any given point on the abscissa are obtained by calculating the area underneath the curves in (A) that lie to the left of a line of that length in the relevant distribution (colored areas). These values indicate how often projected lines in that orientation have been shorter or longer than the projected line being evaluated. (After Howe and Purves, 2005b.)

Given an empirical strategy, the variation in the apparent length of a line as a function of its orientation in the image plane (see Figure 4.5) should be predicted by the variation in empirical rank of a given line as its orientation changes. For instance, a line that is 7 units (pixels) in length oriented at 20° in a retinal image (this length corresponds to approximately 1° of visual angle, a value used in many psychophysical studies) has an percentile rank of about 15. This rank means that about 15% of the physical sources of lines oriented at 20° generated from the laser-scanned database gave rise to projections equal to or less than this length, and 85% gave rise to longer lines. The corresponding ranks of lines 7 pixels in length at orientations ranging from 0° to 180° can be similarly determined. Figure 4.8 shows how these rankings vary as a function of line orientation. A line of this length oriented vertically (90°) in the image plane holds a higher rank than the same line oriented horizontally. Moreover, the lines with the highest rank are those oriented about 20° to 30° from vertical. The shape of the function defined by these empirical rankings for a given line in different orientations follows the psychophysical function of perceived line length for visual stimuli of roughly this size remarkably well.

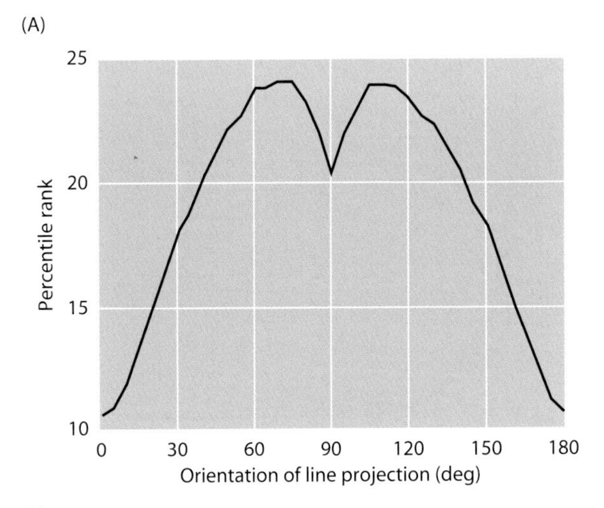

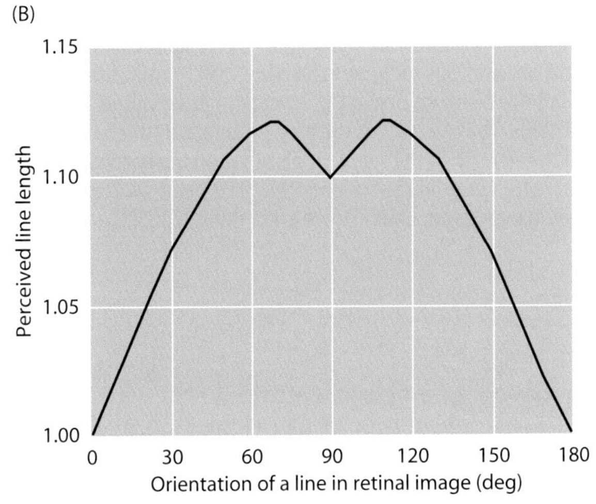

Figure 4.8 The predicted perception of line length as a function of orientation. (A) The percentile ranks for lines 6 pixels in length, derived from the frequency of occurrence of projections in different orientations generated by the real-world sources of linear projections (i.e., of projected lines that corresponded to geometrical straight lines in the world). (B) Figure 4.5B is shown here for comparison. The predicted function in (A) accords with perceived line length as a function of orientation. (After Howe and Purves, 2005b.)

In sum, human experience with projected line lengths at different orientations due to perspective and other real-world biases (see below) can explain the idiosyncratic perception of length as a function of orientation.

Basis of the Biased Frequency of Differently Oriented Line Projections

An obvious question remains: What aspects of the natural world underlie the fact that as the orientation of a projected line becomes more vertical, the probability of finding physical sources for relatively short lines increases (and thus their frequency of occurrence in images) (see Figure 4.7)? Similarly, why does this bias reverse as projected orientations approach 90°, meaning that relatively long vertical line projections are more likely than 60° lines, giving vertical lines a somewhat lower empirical rank than lines oriented 20° to 30° away from vertical?

The answers depend on the fact that straight lines in the physical world are generated by planar surfaces: A one-dimensional line as a physical object, or even a close approximation thereof, rarely occurs in nature. This statement may seem odd since lines in images commonly occur as contrast boundaries (i.e., "edges"). But whereas lines arising from contrast obviously provide useful information about object boundaries and dispositions, surfaces are by far the more frequent source of geometrical straight lines (see Howe and Purves, 2005b, pp. 26–29 for further discussion). Therefore, when considering the physical sources of straight lines at different orientations, the pertinent variable is the extension of flat surfaces in space. As shown in Figure 4.9A, 3-D space is defined by an axis that is horizontal and parallel to the retinal image plane, an axis that is vertical and parallel to the image plane, and a depth axis perpendicular to the image plane. A quick inspection of the world shows that the extension of surfaces in the vertical axis is limited by gravity, whereas extension in the horizontal axis is not. Thus, relatively long vertical lines projected onto the retina are less frequent than relatively long horizontal line projections. As a result, for a line of any given length on the retina, there will always have been more

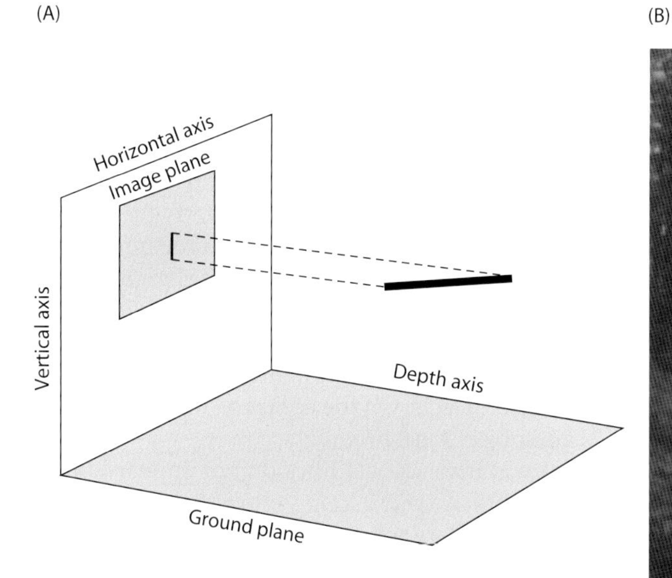

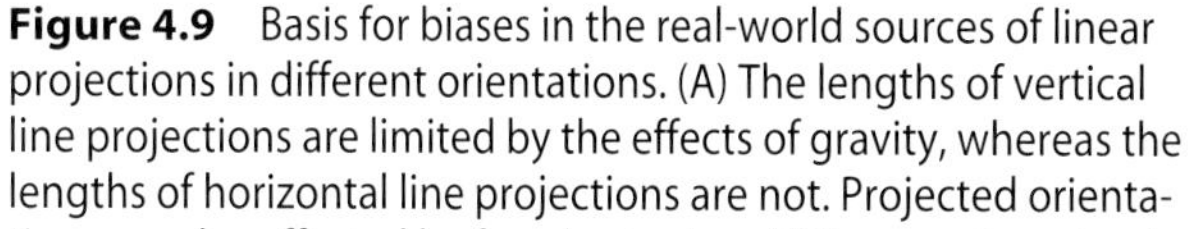

Figure 4.9 Basis for biases in the real-world sources of linear projections in different orientations. (A) The lengths of vertical line projections are limited by the effects of gravity, whereas the lengths of horizontal line projections are not. Projected orientations are also affected by foreshortening. (B) Image of a natural scene with superimposed vertical lines and lines ~25° off the vertical axis. Vertical line projections are more likely to arise from the surfaces of natural objects than are oblique lines, explaining the "McDonald's arches" shape of the functions in Figure 4.5B and 4.8. (After Howe and Purves, 2005a.)

shorter than longer vertical than horizontal retinal projections (see Figure 4.7). This means that the percentile rank of any vertical line will be higher than the percentile rank of a corresponding horizontal line, and should thus be seen as relatively longer.

A different bias accounts for the fact that the fewest relatively long linear projections arising from natural scenes are found at 20° to 30° from vertical (compare the distributions at 60° and 90° in Figure 4.5B). To understand this peculiarity, consider vertical lines and lines about 25° off vertical superimposed on images of the natural scenes (Figure 4.9B). Vertical line projections are more likely to arise from object surfaces than are lines oriented at 25° from vertical because physical objects tend extend vertically rather than obliquely to more efficienctly counter gravity. As a result, the physical sources of relatively long linear projections at 20° or 30° from vertical are less likely than the sources of equally long vertical projections, meaning that their empirical rank will be somewhat higher than the rank of lines at 90°.

The Relationship between the Perceived and Real World Lengths of Objects

Although we humans generally assume that useful behavior requires knowing the qualities of objects and conditions in the world, as already emphasized, that information is precluded by the nature of sensory biology. Nonetheless, it remains extremely difficult to convince people that what they see on a moment-by-moment basis is not a representation of what is "really out there." An effect of this subjective inclination is a tendency to assume that vision must generate perceptions that accord reasonably closely with the relevant characteristics of the underlying objects. Thus, the idea that the line length we see brings the observer into closer agreement with the actual length of the object generating the length on the retina remains attractive. It should be apparent, however, that this interpretation is wrong. The lengths of intervals seen on the basis of empirical rank do not bring perceived lengths into better alignment with the measured lengths of the relevant real-world objects. On the contrary, this way of seeing misaligns perceptions and object measurements.

Another point to emphasize is that vertical lines do not appear longer than horizontal ones because there are more long vertical lines, intervals, or objects in the world; in fact, there are more long horizontal sources, as just discussed. The apparent lengths of intervals we see are a signature of the instantiation in visual circuitry of the neural links between image patterns and behavior. In this counterintuitive concept, perception and reality are in effect parallel domains that are related only indirectly through the aegis of behavior.

The Perception of Angles

Another challenge in geometrical perception is rationalizing the apparent angle made between two lines that meet—explicitly or implicitly—at a point. Like the appreciation of line length, an intuitive expectation is that this basic feature of perceived geometry should scale directly with the dimensions of the angles projected in retinal images. But this is not what people see. Observers tend to overestimate the magnitude of acute angles and underestimate obtuse ones by a few degrees (Figure 4.10A). This subtle yet robust phenomenon was first reported by Wundt (1862) and has been confirmed by several modern studies (Fisher, 1969; Carpenter and Blakemore, 1973; Maclean and Stacey, 1971; Heywood and Chessell, 1977; Greene, 1994; Nundy et al., 2000).

The anomalous perception of angles is easiest to appreciate—and most often demonstrated—using a related series of geometrical stimuli that involve intersecting lines in various configurations. The simplest of these is the so-called tilt illusion, in which a vertical line in the context of an obliquely oriented line appears to be rotated slightly away from vertical in the direction opposite the orientation of the oblique "inducing line" (Figure 4.10B). The direction of the perceived deviation of the vertical line is consistent with the perceptual enlargement of the acute angles in the stimulus and/or a reduction of the obtuse angles. This relatively small effect is enhanced in the Zöllner illusion, which is essentially a more elaborate version of the tilt effect achieved by iterating the oblique stimulus elements (Figure 4.10C). The several parallel vertical lines in this presentation appear to be tilted away from each other, again in directions opposite the oblique orientation of the contextual line segments. A further permutation is the Hering illusion, in which two parallel straight lines appear bowed in the context of intersecting lines whose orientations change progressively (Figure 4.10D). How, then, can accumulated experience with images of angles arising from natural scenes explain these perceptions?

The approach is generally the same as to the perception of line length, and the rationale is identical: Perceiving angles on an empirical basis would allow observers to contend successfully with angles in the world despite the inevitably uncertain significance of projected subtenses. The frequency of occurrence of angle projections can be determined from range images in much the same way that the prevalences of differently orientated straight-line projections can be extracted from a database of laser-scanned scenes. The first step is to identify in a scene a region that contains a valid physical source of one of the two lines that form an angle (the "reference line") by applying to the images a straight-line template at different orientations (see Figure 4.6). If the set of points underlying the reference-line template in the image corresponds to physical points that form a straight line in 3-D space, the physical points are accepted

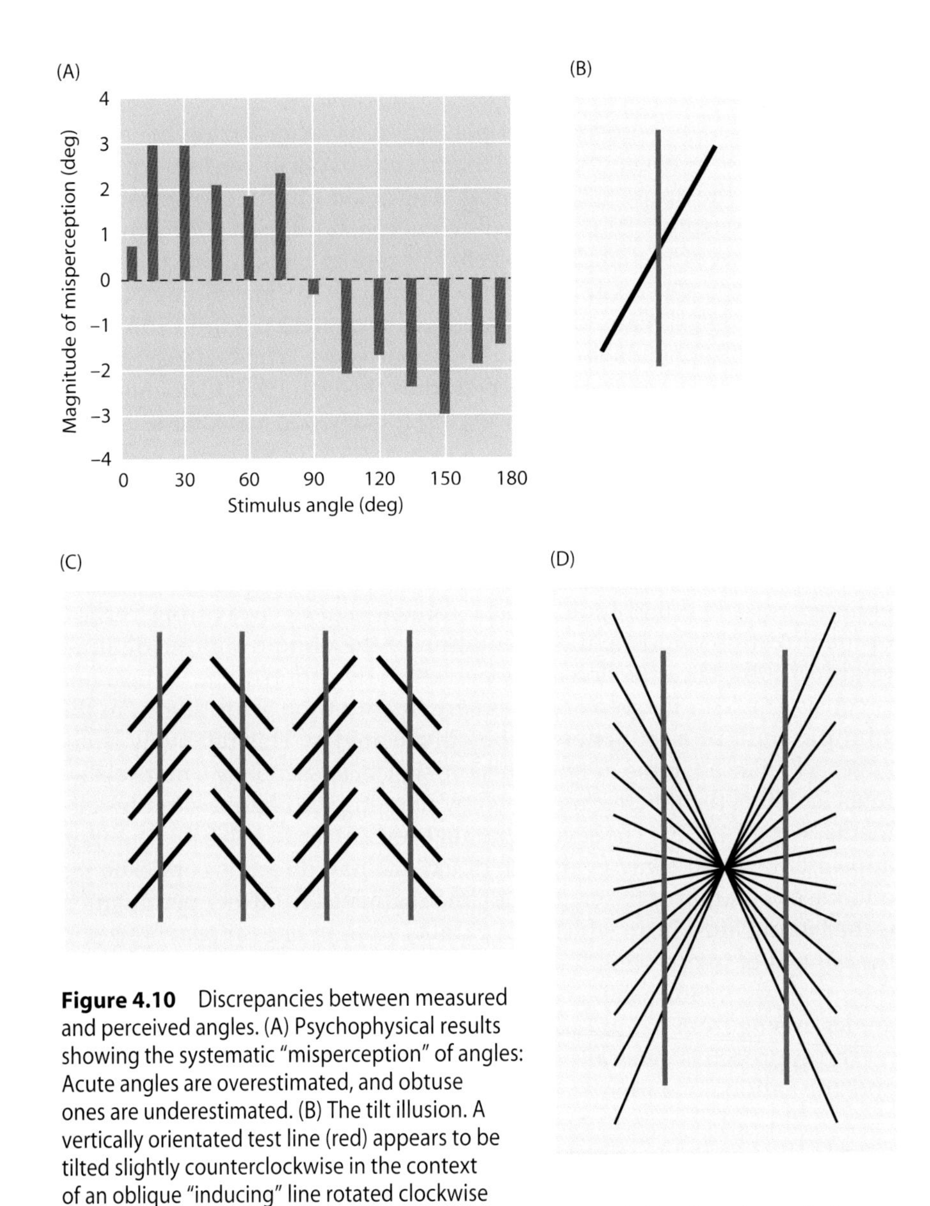

Figure 4.10 Discrepancies between measured and perceived angles. (A) Psychophysical results showing the systematic "misperception" of angles: Acute angles are overestimated, and obtuse ones are underestimated. (B) The tilt illusion. A vertically orientated test line (red) appears to be tilted slightly counterclockwise in the context of an oblique "inducing" line rotated clockwise (black). (C) The Zöllner illusion. In the standard presentation of this effect, the vertical test lines (red) appear tilted in the direction opposite to the orientation of the contextual lines (black). (D) The Hering illusion. The two vertical lines (red) appear bowed when presented in the context of radiating lines. (After Howe and Purves, 2005a; data in [A] are from Nundy et al., 2000.)

as a valid source. Once a physical source of the reference line is found, the probability of occurrence of a line forming an angle with it can be determined. This assessment is made by overlaying a second straight-line template in different orientations on the image, asking whether the points

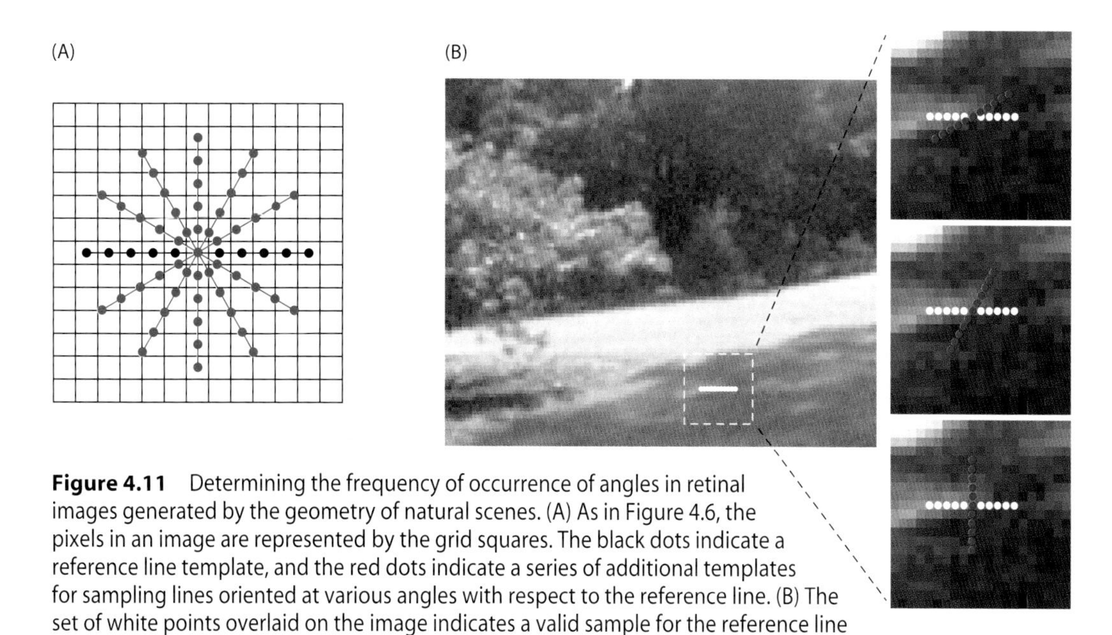

Figure 4.11 Determining the frequency of occurrence of angles in retinal images generated by the geometry of natural scenes. (A) As in Figure 4.6, the pixels in an image are represented by the grid squares. The black dots indicate a reference line template, and the red dots indicate a series of additional templates for sampling lines oriented at various angles with respect to the reference line. (B) The set of white points overlaid on the image indicates a valid sample for the reference line in (A) (the definition of valid is a straight line in the image that corresponds to a geometrical straight line in 3-D space). Blowups of the boxed area show examples of the second template (red) overlaid on the same area of the image to test for the presence of a second straight line in different orientations; in each of the cases shown, the second template also identifies a valid sample (see Figure 4.6). (After Howe and Purves, 2005a.)

underlying the additional line also correspond to a straight line in space (Figure 4.11). By systematically repeating this procedure, the frequency of occurrence of valid projected angles can be tallied.

The results of such sampling are shown in Figure 4.12. Regardless of the orientation of the reference line (indicated by the black line in the icons under the graphs) or the type of real-world scene, the probability distributions of angle projections arising from real-world angle sources form a trough, with lower probability values for angle projections that approach 90°.

This bias can again be understood by considering the provenance of straight lines projected onto the retina. As for single lines, intersecting straight lines in the real world are typically components of planar surfaces. Accordingly, a planar surface that contains two physical lines whose projections intersect at 90° will, on average, be larger than a surface that includes the source of the two lines of the same length that are less orthogonal (Figure 4.13). Since larger surfaces include smaller ones, the probability of finding larger planar surfaces in the world is necessarily lower than the probability of finding smaller ones. Thus, other things being equal, the frequency of occurrence of angles that project at or near 90° is less likely than the occurrence of angles that are nearer 0° or 180°.

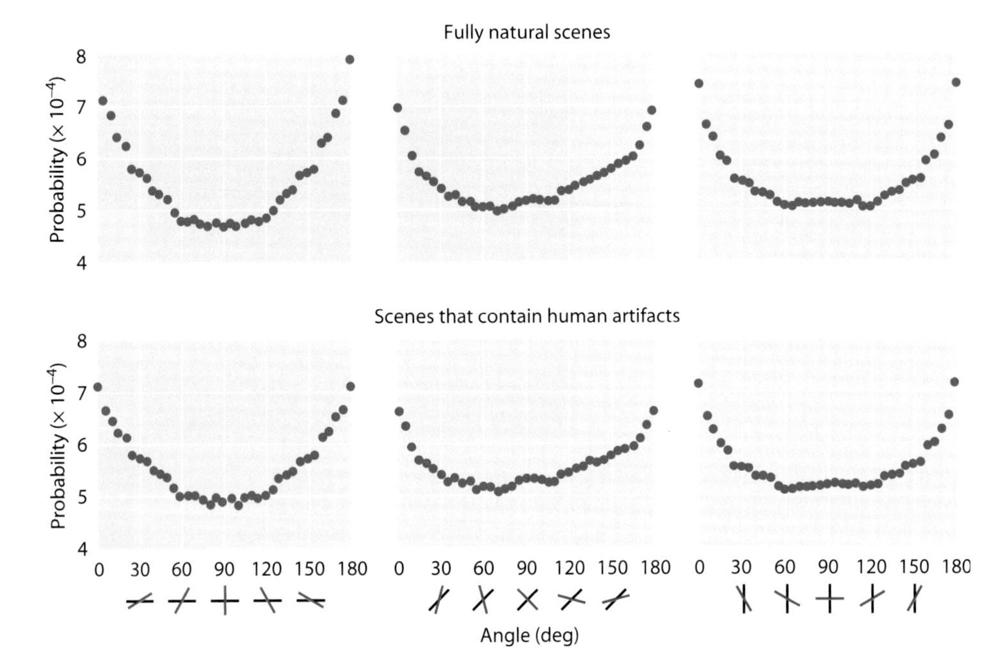

Figure 4.12 The frequency of occurrence of projected angles as a function of their subtense. The three columns represent the results of analyzing images in relation to the database of laser-scanned scenes using a horizontal (left), oblique (middle), or vertical (right) reference line, as indicated by the icons below the graphs. The upper row shows the results obtained from fully natural scenes and the lower row from environments that contained some (or mostly) human artifacts. (After Howe and Purves, 2005a.)

Figure 4.14A shows the cumulative probability distribution of different projected angles generated by physical sources whose geometry was determined by laser scanning. As before, the summed likelihood represents the frequency of occurrence of all the projected angles less than or equal to a given subtense on the retina. Much like its significance in

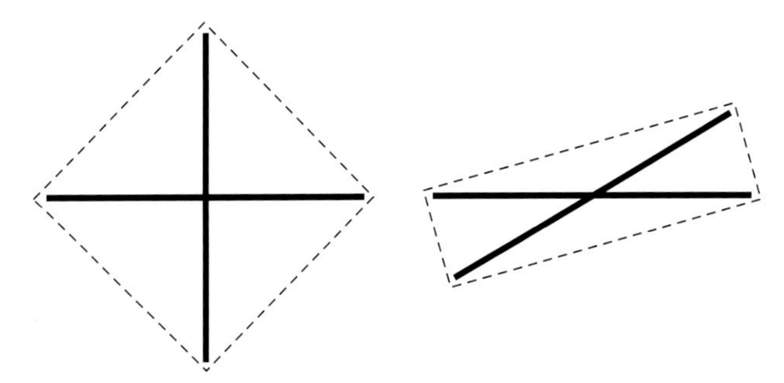

Figure 4.13 The physical source of two lines intersecting at or near 90° is likely to be part of a planar surface (dashed line) that is larger, on average, than a surface that contains the source of the same two lines that are not orthogonal. (After Howe and Purves, 2005a.)

understanding the perception of line length, this function provides an indication of human experience with angle subtenses: For any given subtense, the corresponding cumulative probability value indicates the percentage of all projected angles smaller than the projected subtense in question, and the percentage of larger angles. For example, if on the retina the angle under consideration is 30°, then the corresponding cumulative probability value on the red curve in Figure 4.14A is 0.185; this means that approximately 18.5% of angle projections in human experience have been equal to or less than 30° and that approximately 81.5% of angles have been larger than 30°. When compared to a distribution derived from a hypothetical world in which the probability of experiencing any projected angle is the same (see the inset and the black line in Figure 4.14A), the cumulative distribution derived from laser-scanned scenes gives somewhat

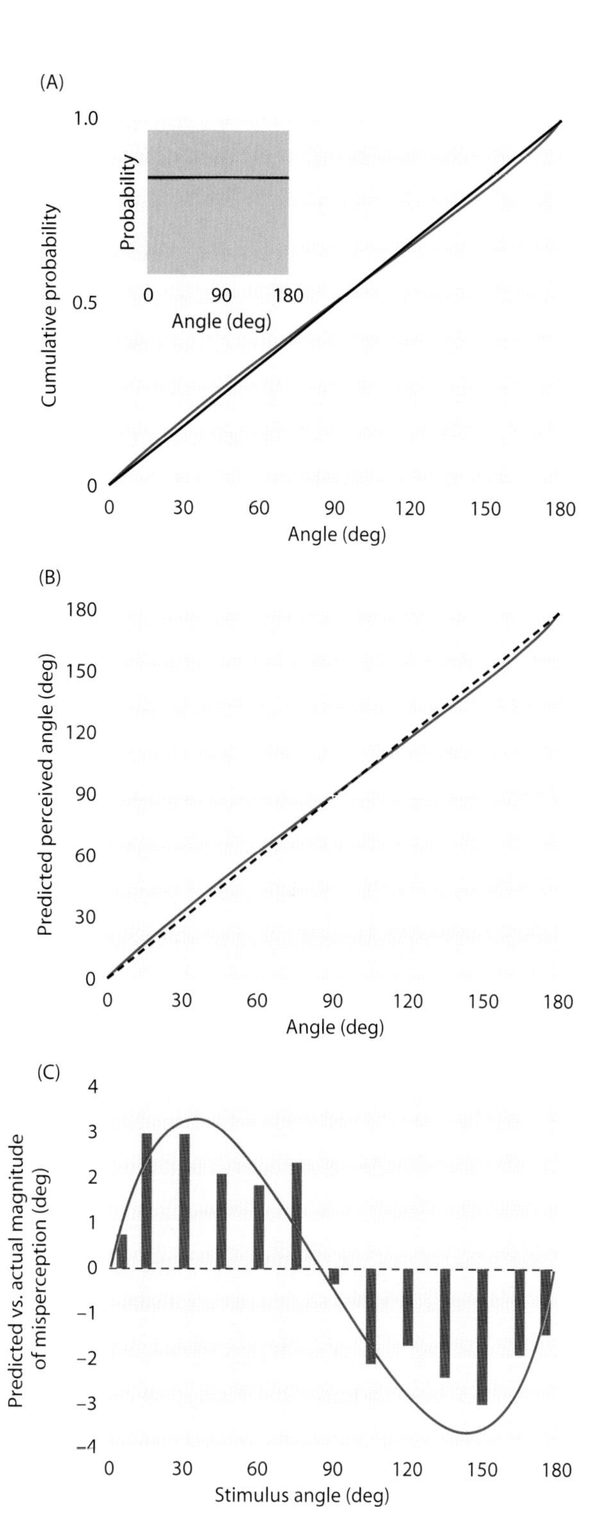

Figure 4.14 Predicting the perceived subtense of angles from the cumulative frequency of occurrence of projected angles generated by real-world sources. (A) The red curve is the cumulative probability distribution derived by pooling the distributions in Figure 4.12. The black line indicates the cumulative probability distribution derived from a hypothetical physical world in which the probability of any given projection is the same (see inset). (B) The predicted perceived angle (red curve) based on in the cumulative probability distribution in (A); the dashed black line indicates the actual angle. (C) The magnitude of angle misperception (red curve) predicted by the difference between the red and the dashed black lines in (B) compared with psychophysical measurements of angle perception (gray bars; see Figure 4.10A). (After Howe and Purves, 2005a.)

higher values for angles less than 90° and somewhat lower values for angles greater than 90°.

If perceptions of angle subtense are generated on the basis of experience with projected angles in relation to the success of behavior directed at their sources, then the angles seen should accord with their relative ranking in the cumulative probability distribution of angle projections. As shown in Figure 4.14A, the cumulative probability for any angle x between 0° and 90° is somewhat greater than $x/180$, meaning that the empirical rank of any acute angle on the scale is shifted slightly in the direction of 180° compared with its position on the geometrical scale of 0° to 180°. The opposite is true for any obtuse angle measuring between 90° and 180°. Figures 4.14B and C show that the predicted perceptions of angle magnitude agree with the psychophysical results described in Figure 4.10A.

The geometrical effects illustrated in Figure 4.10B–D can also be explained in this way. Take, for example, the apparent tilt of a vertical line in the context of an oblique line that intersects it (see Figure 4.10B). The perceptual effect is predicted by the frequency of occurrence of projected angles arising from the physical sources of a second line oriented at various angles, given a reference line oriented at 60° from the horizontal (Figure 4.15A). The dashed line indicates vertical, i.e., a line rotated 30°

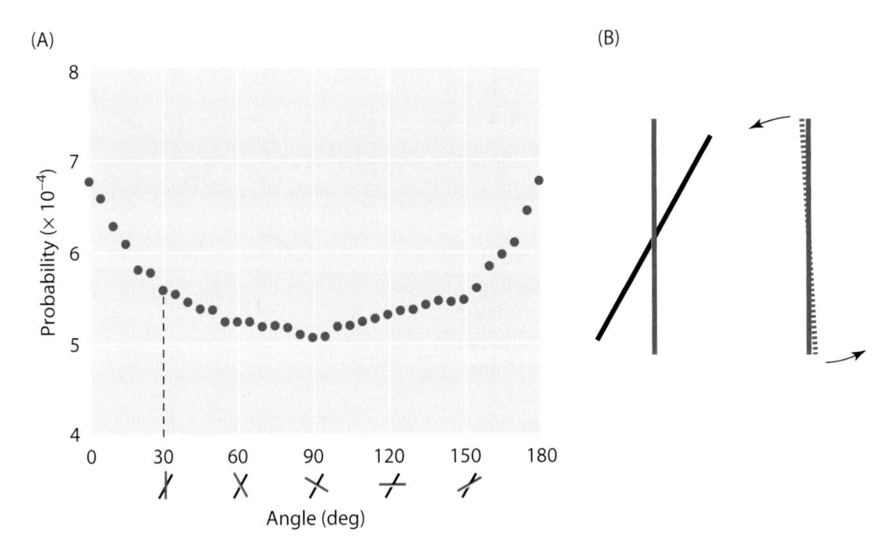

Figure 4.15 Empirical explanation of the tilt illusion. (A) Probability distribution of projected angles arising from the real-world geometry, given a reference line oriented at 60° with respect to horizontal (indicated by the black lines below the graph). (B) Left panel shows the standard stimulus used to demonstrate of the tilt effect; the red line is vertical. Right panel indicates the direction (arrows) and magnitude of the tilt effect predicted by the probability at the dashed line in (A), which indicates the frequency of occurrence of this stimulus pattern; the solid red line is vertical, and the dotted line indicates the predicted shift in the apparent position of the vertical line in the left panel. (After Howe and Purves, 2005a.)

counterclockwise from the reference line. The cumulative probability value associated with a 30° angle projection is 0.184 (see Figure 4.14A). When multiplied by 180, this value predicts that the perceived angle between the reference line and the vertical line should be 33.2°, or 3.2° greater than the actual angle between the two lines (the same argument can be made if the obtuse rather than the acute angle between the two lines is considered). Accordingly, the vertical line in the context of a line oriented at 60° should be perceived as being rotated away from the reference line slightly more than it actually is, thus appearing not quite vertical (Figure 4.15B). This prediction is consistent with what observers see in response to this sort of stimulus (e.g., Bouma and Andriessen, 1970; Wenderoth et al., 1979; Greene, 1994). The Zöllner illusion (see Figure 4.10C) is essentially an iteration of a simple tilt stimulus, producing an overall effect in which the vertical test lines appear to be more markedly tilted away from the contextual lines. Similarly, the parallel lines in the Hering stimulus (Figure 4.10D) appear bowed because of the concatenation of tilt stimuli. In this case, the upper and lower contextual lines tilt in opposite directions, causing the perceived orientation of corresponding components of the test lines to change progressively, resulting in the apparent bowing.

The Perception of Object Size

Although the apparent length of lines and the subtenses of angles are certainly pertinent to size, studies of the perceived dimensions of objects have focused on two other categories of effects referred to as size contrast and size assimilation. These phenomena concern differences in the apparent sizes of two identical targets surrounded by one or more larger or smaller forms, generally of the same type (e.g., a circle surrounded by circles, or a square surrounded by squares) (Figure 4.16). Size contrast and size assimilation are generally similar to the context-dependent contrast and constancy effects in lightness/brightness and color described in Chapters 2 and 3.

The most thoroughly studied size contrast stimulus is the Ebbinghaus figure, in which two identical target circles are surrounded respectively by several larger or smaller circles (Figure 4.16A). The effect of these juxtapositions is that the target surrounded by the larger circles appears a little smaller than the identical target surrounded by the smaller ones (e.g., Zigler, 1960; Pressey, 1967; Massaro and Anderson, 1971). Another aspect of the anomalous perception elicited by the Ebbinghaus stimulus is that when the diameter of the surrounding circles is kept constant, the apparent size of the central target circle decreases as the interval between the central and the surrounding circles increases (Massaro and Anderson, 1971; Girgus et al., 1972; Jaeger and Grasso, 1993) (Figure 4.16B). The same size contrast effects are elicited by a concentric arrangement of circles, in which case the stimulus is referred to as the Delboeuf figure (Luckiesh, 1922; Girgus et al., 1972) (Figure 4.16C).

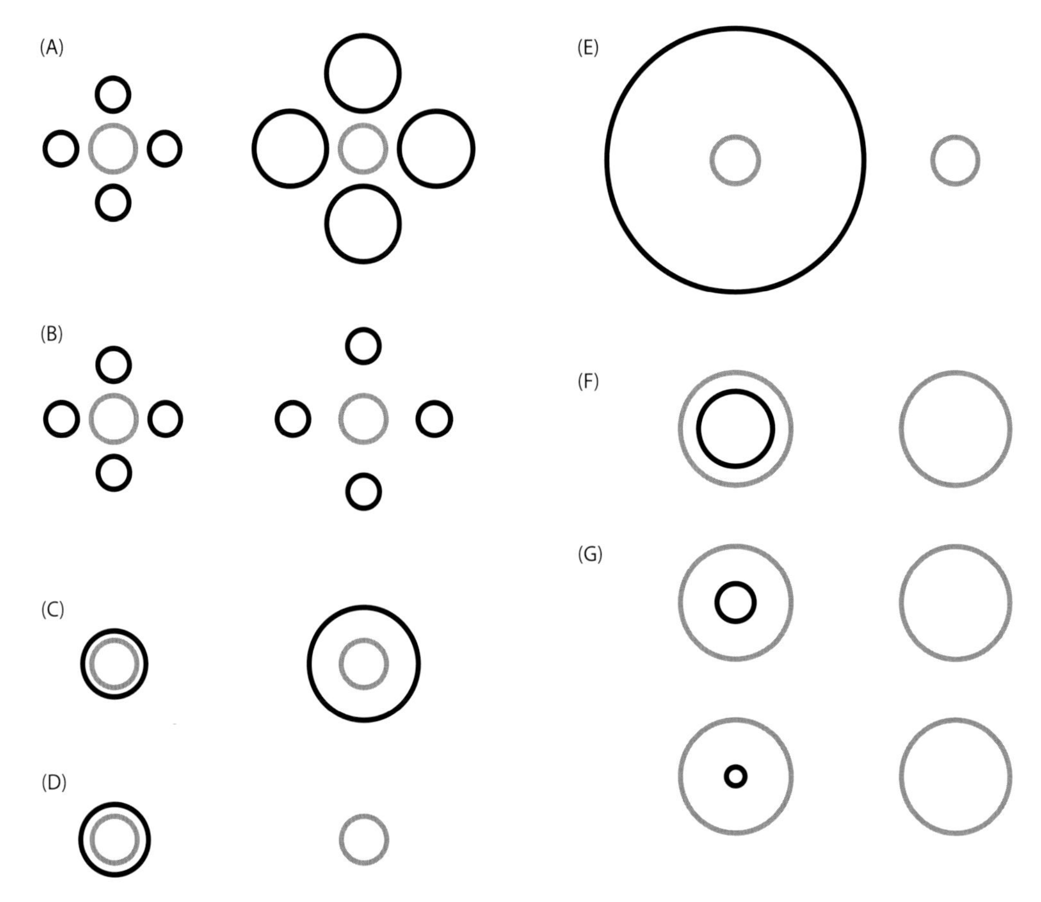

Figure 4.16 Size contrast and assimilation effects (the identical circles to be compared in each stimulus are indicated in gray). (A) Standard presentation of size contrast, often referred to as the "Ebbinghaus illusion." Observers see the central circle surrounded by smaller circles as being somewhat larger than the same circle surrounded by larger circles. (B) The central circle also looks smaller when the interval between the central and the surrounding circles is greater. (C) A similar size contrast effect is generated by a concentric presentation (called the "Delboeuf illusion"). (D) The inner circle of a Delboeuf stimulus also appears larger than a single identical circle if the diameter of the outer circle is not more than about twice that of the inner circle; this effect is referred to as "size assimilation." (E) When, however, the diameter of the outer circle is much larger than that of the inner circle, the inner circle looks the same or even a little smaller than an identical single circle. (F) The outer circle of a Delboeuf figure appears smaller than a single circle of the same size. (G) The effect in (F) diminishes as the inner circle becomes progressively smaller relative to the outer circle. (After Howe and Purves, 2004.)

The Delboeuf figure has a variety of other presentations, some of which elicit a size contrast effect, whereas others give rise to a size assimilation effect. Assimilation here refers to the induction of a similar appearance of object size as a result of the context. For instance, when concentric circles are compared with a single circle identical to the inner target circle of the concentric set, the inner circle appears a little larger than the single circle (Obonai, 1954; Oyama, 1960; Howard et al., 1973; Pressey, 1967) (Figure 4.16D). The effect is called size assimilation because the perceived size of the inner circle appears to be "assimilated" into the size of the surround (Obonai, 1954; Rock, 1995). This phenomenon is diminished, however, when the diameter of the surrounding circle is increased. Indeed, when the diameter of the surrounding circle is sufficiently large, the overestimation of the size of the inner circle changes to a slight underestimation (Obonai, 1954; Oyama, 1960; Pressey, 1967; Jaeger and Lorden, 1980) (Figure 4.16E). Equally puzzling is the observation that the outer circle of the concentric set appears smaller when compared to a single circle of the same size (Figure 4.16F), an effect that decreases and eventually disappears as the difference between the sizes of the outer and the inner circles increases (Ikeda and Obonai, 1955; Oyama, 1960) (Figure 4.16G).

Although various ad hoc explanations of these phenomena have been proposed, there has been no agreement about the basis of size contrast and assimilation (see Robinson, 1998 and below). As with the phenomenology of line and angle perception, the complexity of these effects has made a coherent account difficult.

An Empirical Explanation of Size Effects

If the anomalous perception of size contrast and assimilation effects are a further manifestation of a wholly empirical strategy of vision, then the perceptions generated by the various configurations in Figure 4.16 should be predicted by accumulated experience with the retinal projections of the relevant forms. Identical target objects appear different in size because the frequency of occurrence of the retinal projections generated by natural sources is different. The impact of this experience is, as before, its pertinence to the success of visually guided behavior.

This supposition can be tested by sampling a range image database to determine the frequency of occurrence of retinal projections whose geometrical structure is the same as or similar to a given size contrast or assimilation stimulus (Figure 4.17). The probability distributions of the projections of valid target circles of different sizes in the context of interest indicate the percentage of projections smaller than the size of the target in a particular stimulus pattern, and the percentage that is larger. As in the case of lines and angles, the rank of the target on this empirical scale should predict the apparent size of the target, given the context in question.

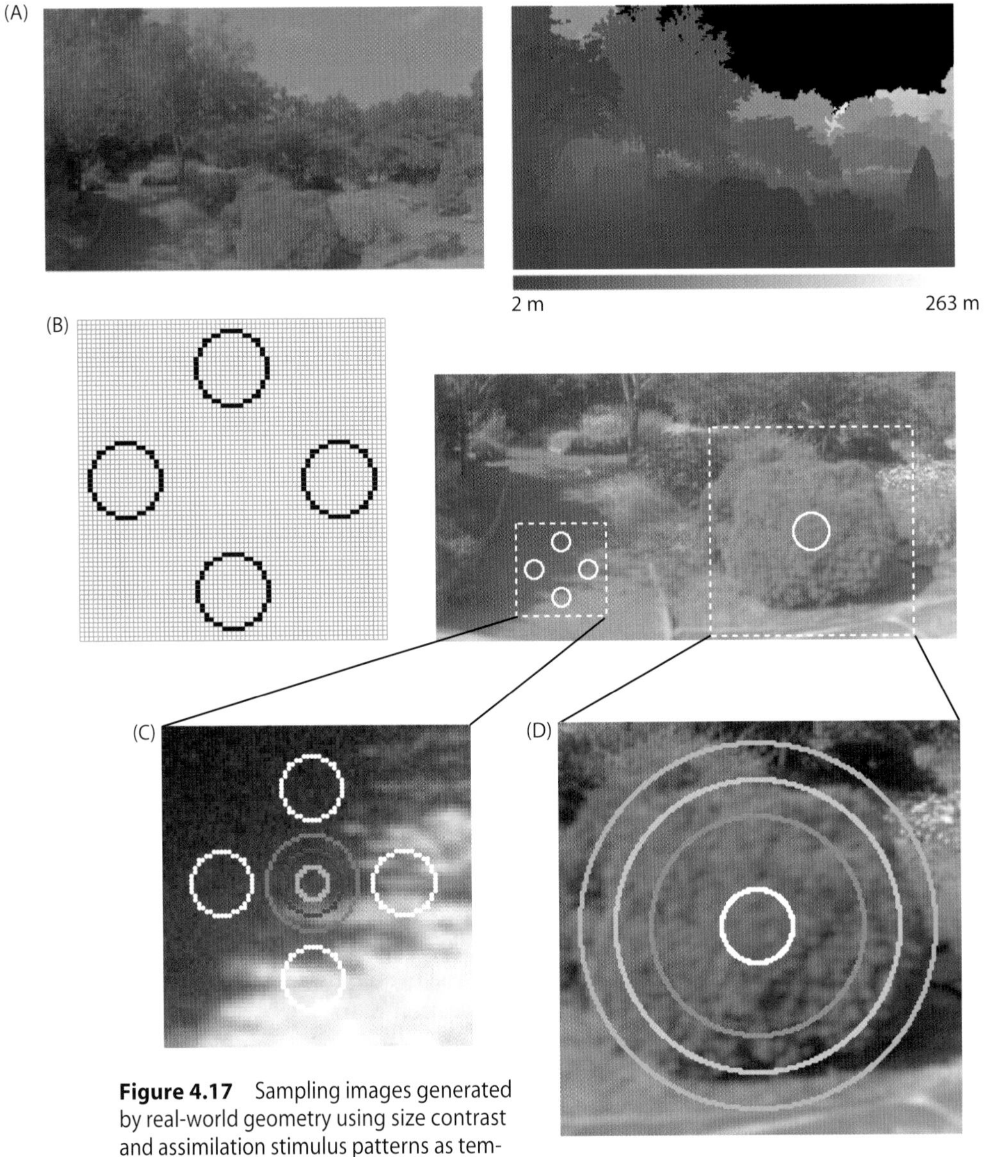

Figure 4.17 Sampling images generated by real-world geometry using size contrast and assimilation stimulus patterns as templates. (A) A typical scene (left) and its corresponding range image (right). (B) As in earlier figures, pixels in a region of the image are represented by grid squares; the black pixels indicate a template of the surrounding circles in an Ebbinghaus stimulus pattern overlain on the image. The pixels covered by the template thus comprise a potential sample of the elements underlying an Ebbinghaus stimulus on the retina. If the set of physical points underlying to the pixels comprising each of the circles corresponded to a plane in 3-D space, the set was accepted as a valid sample for the contextual circles. (C) and (D) are blowups of the boxed areas in a representative image. The pixels underlying the four template circles on the left of the scene are a valid sample of an Ebbinghaus context. The pixels underlying the template on the right are a valid sample of the contextual circle (the inner circle in this case) in a Delboeuf stimulus. The colored circles indicate the templates used to assess the frequency of occurrence of targets with different diameters in the patterns that define Ebbinghaus and Delboeuf stimuli. (After Howe and Purves, 2004.)

Understanding the Ebbinghaus Effect in Empirical Terms

As shown in Figure 4.18, the frequency of occurrence of central circles in images analyzed in this way decreases as the size of the contextual circles increases. This overall decline is more pronounced, however, when the surrounding circles are relatively small than when they are larger (Figure 4.18A; compare the five curves in Figure 4.18B). As a result, the frequency of projections of target circles of different sizes varies systematically as a function of the projected size of the surrounding circles.

Why, then, are two targets identical in size perceived differently when surrounded by contextual circles of different sizes, as in the standard Ebbinghaus stimulus? Consider the target circle 14 pixels in diameter indicated by the dashed line in Figure 4.18B. Since the area to the left of the dashed line compared with the area on the right is greater for the probability distributions derived in the presence of the smaller

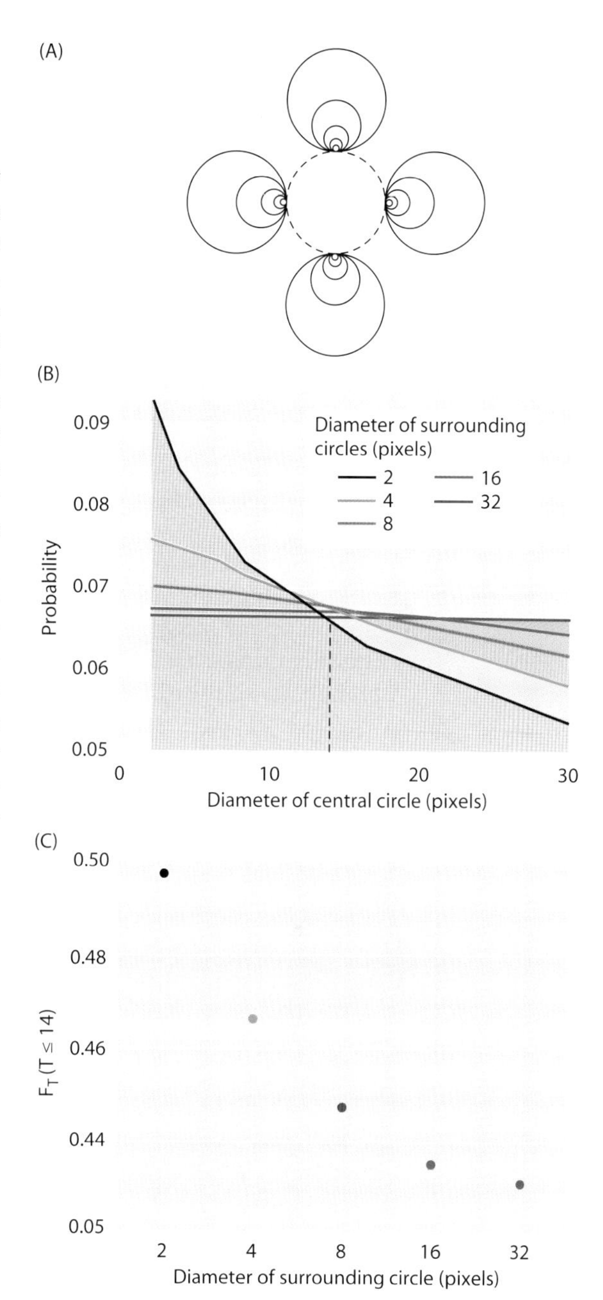

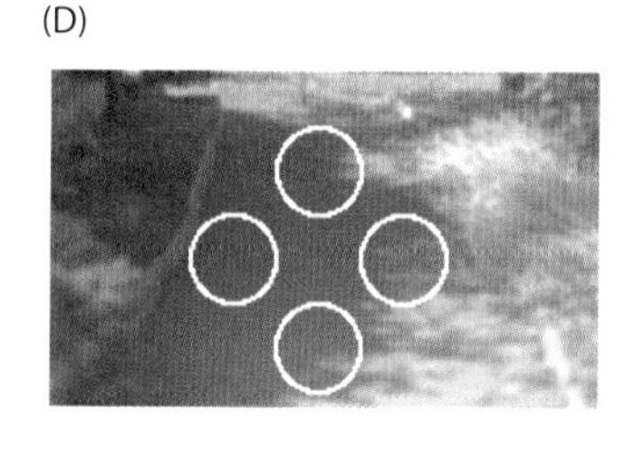

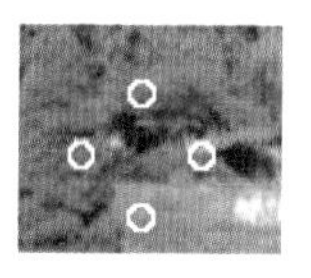

Figure 4.18 Frequency of occurrence of projected circles pertinent to understanding the Ebbinghaus effect. (A) The four circles surrounding a central circle (dotted line; see Figure 4.16A) were systematically varied and the range image database sampled for valid 3-D sources. The procedure was repeated for target circles of different sizes. (B) Frequency of occurrence of valid target circle sources as a function of target and surrounding circle diameters. The dashed line indicates, as an example, a target circle 14 pixels in diameter. (C) The cumulative probabilities for this particular target size (14 pixels in diameter) derived from the probability distributions in (B). (D) Examples of regions in natural scenes from which a valid sample of large contextual circles (left) vs. small contextual circles (right) were more likely to be obtained. (After Howe and Purves, 2004.)

contextual circles than the larger ones, the target circle occupies a different relative rank in each of these empirically determined distributions. If T stands for target size and x the diameter of a circle in pixels, then F_T $(T \leq x)$ denotes the cumulative probability of occurrence of the physical sources of target circles smaller than or equal to a specific target diameter x (Figure 4.18B and C). A greater value of F_T $(T \leq x)$ means a larger percentage of the sources of a projected target circle in a given set of contextual circles generated a target smaller than x pixels in diameter. Thus, the greater the value of F_T $(T \leq x)$, the higher the rank of a given target size x on the empirical scale of target sizes associated with the set of contextual circles, and the larger it should appear. As shown in the figure, the cumulative probability decreases progressively as a function of the size of the surrounding circles associated with each probability distribution. This relationship means that when the surrounding circles are small, a target 14 pixels in diameter ranks relatively high on this empirical scale; conversely, when the surrounding circles are large, the same target size ranks relatively low.

If the apparent size of the identical targets is determined by these empirical ranks, a given target circle presented in the context of small circles should appear larger than the same target in the context of large surrounding circles. This is, of course, what observers see in response to the Ebbinghaus stimulus (see Figure 4.16A).

Understanding the Delboeuf Effect in Empirical Terms

A different perceptual effect is elicited when a stimulus comprises the two sets of concentric circles, as in the Delboeuf stimulus in Figure 4.16C. The inner target circle within the relatively small contextual circle appears larger than the identical target in the relatively large outer circle.

The frequency of occurrence of the physical sources of an inner target circle, given the presence of an outer concentric circle, is shown in Figure 4.19A. Since the largest possible target size is limited by the size of the outer circle, the rank of any specific target size in the distributions varies according to the size of the outer circle. As before, the cumulative probabilities show that a target circle of a given diameter ranks higher on an empirical scale when associated with a smaller outer circle than when associated with a larger one (Figure 4.19B). In accord with what people see, this relationship predicts that the target within the smaller outer circle should appear somewhat larger than the same target in the larger outer circle.

The other effects elicited by the Delboeuf stimuli (Figure 4.16D–G) can also be accounted for in this way. For instance, when the inner circle is compared with a single circle of identical size, the inner circle looks larger than the single circle if the outer circle is less than twice the diameter of the inner circle (see Figure 4.16D). Conversely, it appears smaller than the single circle when the diameter of the outer circle is more than 4 or 5 times that

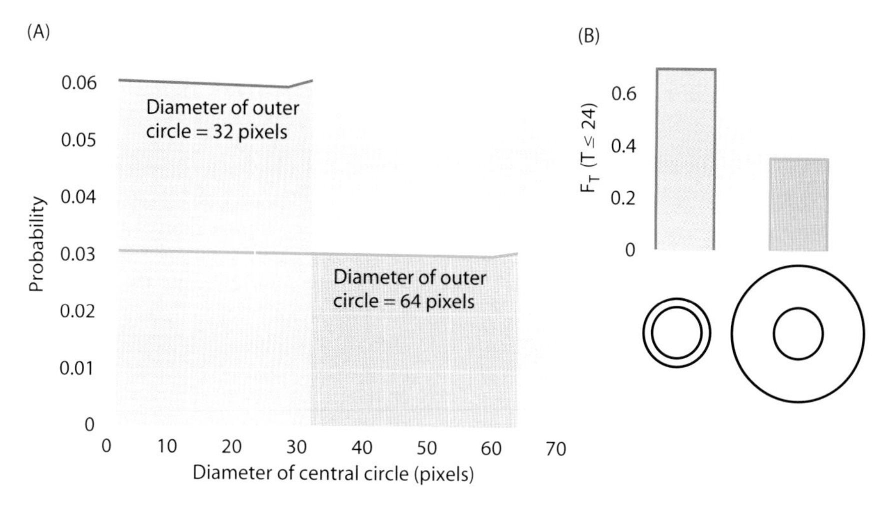

Figure 4.19 Frequency of occurrence of the components of a Delboeuf stimulus (i.e., two sets of concentric circles). (A) The frequency of occurrence of inner (target) circles of different diameters associated with contextual circles of two different sizes. The dashed line indicates, as an example, the position in the two distributions of a target circle 24 pixels in diameter. (B) Cumulative probabilities in images of a target circle 24 pixels in diameter (i.e., F_T $(T \leq 24)$) derived from the two distributions in (A). (See text for explanation.) (After Howe and Purves, 2004.)

of the inner circle (see Figure 4.16E). To understand how these paradoxical phenomena are explained empirically, consider the probability distribution of the physical sources of a single circle in the absence of a contextual circle (Figure 4.20A). For the reasons already given, the probability of encountering a valid real-world source of a projected circle decreases as the projected size of the circle increases; thus, the frequency of projections from such sources will be relatively less. Now, compare the probability distribution for a single circle with the distribution of the physical sources of inner target circles in the presence of an outer contextual circle that is either relatively small (Figure 4.20B) or relatively large (Figure 4.20C). Figure 4.20B shows the position of a target that is 24 pixels in diameter (dashed line) in relation to the probability distribution of valid sources of single circles, and in relation to the distribution of valid sources of inner target circles given the presence of an outer circle that is 32 pixels in diameter. The cumulative probabilities associated with the target that is 24 pixels in diameter derived from these two distributions (Figure 4.20B, inset) show that the same target circle ranks higher on an empirical scale of target size associated with the outer circle than on the empirical scale for single circles. This relationship predicts that when the ratio of the diameter of the outer circle to the target circle is less than 2 (it is 1.33 in the example shown here), the target within the outer circle should appear larger than a single circle of the same size, as it does.

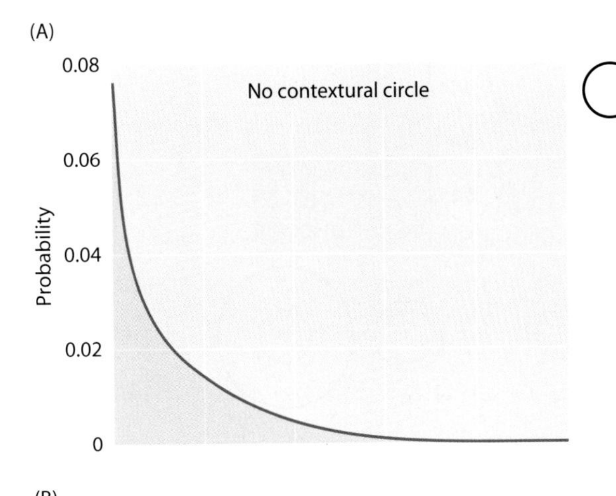

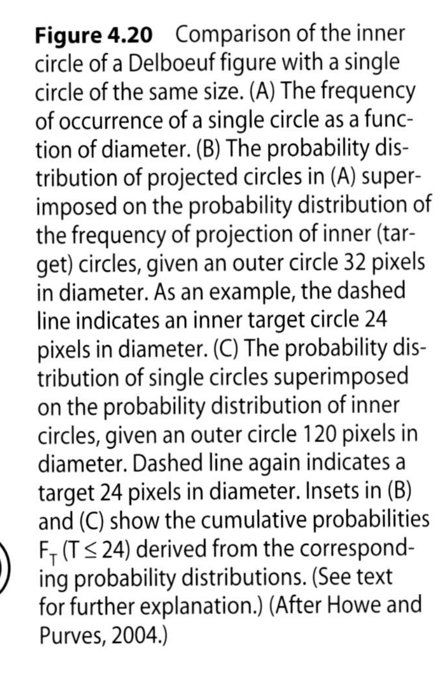

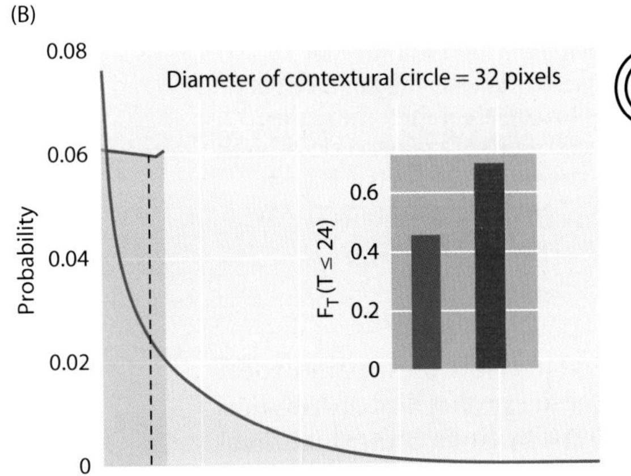

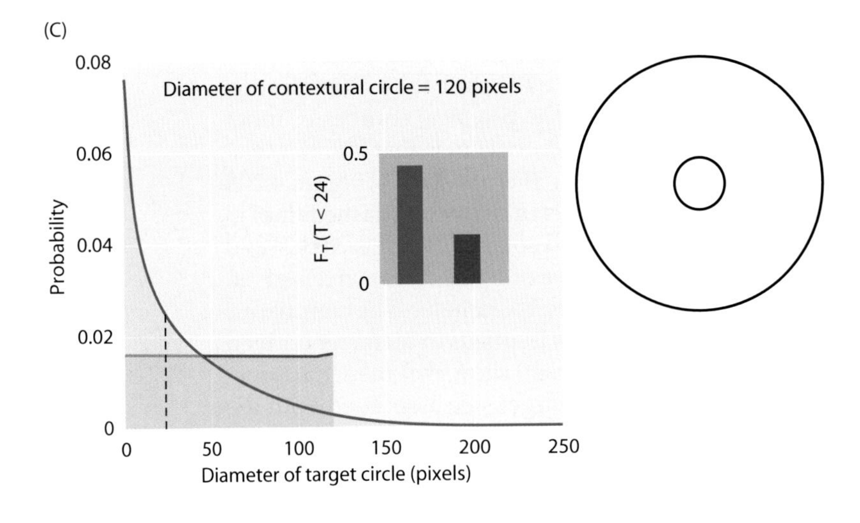

Figure 4.20 Comparison of the inner circle of a Delboeuf figure with a single circle of the same size. (A) The frequency of occurrence of a single circle as a function of diameter. (B) The probability distribution of projected circles in (A) superimposed on the probability distribution of the frequency of projection of inner (target) circles, given an outer circle 32 pixels in diameter. As an example, the dashed line indicates an inner target circle 24 pixels in diameter. (C) The probability distribution of single circles superimposed on the probability distribution of inner circles, given an outer circle 120 pixels in diameter. Dashed line again indicates a target 24 pixels in diameter. Insets in (B) and (C) show the cumulative probabilities F_T (T ≤ 24) derived from the corresponding probability distributions. (See text for further explanation.) (After Howe and Purves, 2004.)

When the contextual circle is significantly larger, however, this relationship no longer holds. Figure 4.20C shows the rank of the same target circle (dashed line) in the probability distribution of single circle physical sources, and in the distribution of inner target circles in the presence of a much larger outer circle. The cumulative probabilities in this case indicate that the empirical rank of the target is now lower in the context of the outer circle than for single circles (Figure 4.20C, inset). This empirical relationship predicts that when the ratio of the diameter of the contextual circle to the target circle is large, the target in the contextual circle should look smaller than an identical single target circle, as it does.

Basis of the Biased Projections Underlying Size Contrast and Assimilation Stimuli

As with lines or angles, an important question remains: What underlying aspects of the real world give rise to the different frequencies of occurrence of circular projections?

A circle in the retinal image can, of course, be generated by projections of an infinite variety of circular or elliptical objects in 3-D space. But, as before, the configurations of Ebbinghaus and Delboeuf stimuli will typically have arisen from planes in the 3-D world. It follows that the probability of encountering the physical source of the target circles generating experience decreases as the size of a target circle increases (see Figure 4.18B), simply because the larger planes will always encompass smaller ones (which is why any shapes will work in these size contrast and assimilation stimuli). Likewise, the presence of contextual circles of various sizes will differentially affect the occurrence of the sources of target circles. Since larger surrounding circles necessarily arise from larger physical planes in the world, the relevant region of the scene is more likely to comprise surfaces that make up a "smoother" whole, as can be appreciated in Figures 4.17 and 4.18. A region with fewer physical discontinuities is more likely to contain larger planar areas capable of giving rise to the projection of larger central target circles. Thus, the presence of larger contextual circles increases the probability of occurrence of projections from sources able to generate larger target circles. As a result, the probability distribution of projected target circles varies according to the size of contextual circles.

Other Explanations of Interval, Angle, and Object Size Perceptions

Many theories have been proposed over the last century or more to explain the range of perceptual anomalies elicited by geometrical stimuli. With respect to the apparent lengths of lines, attempts to rationalize the discrepancies between the measured lengths and their perception have taken several different directions. Older explanations included asymmetries in the anatomy of eye (Kuennapas, 1957; Pearce and Matin, 1969; Prinzmetal and Gettleman, 1993), the ergonomics of eye movements (Wundt, 1862; Luckiesh, 1922), and cognitive compensation for the foreshortening of vertical lines (Gregory, 1974; Schiffman and Thompson, 1965; Girgus and Coren, 1975; von Collani, 1985). More recently, investigators have supposed that these effects arise from inferences made by observers as a result of the rules (heuristics) derived from the appearance of objects in the world. For example, in the 1960s, Richard Gregory argued that the Müller-Lyer illusion (see Figure 4.2D) is a result of seeing the arrow-tails and arrowheads figures as "concave" and "convex" corners, respectively, in the 3-D world. In this scenario, the anomalous percept is taken to be an unconscious result of previous experience with the different distances from the eye implied by such real-world corners (Gregory, 1963; 1968). Intuitively, it indeed seems that convex corners should have been generally closer than concave ones. Gregory (1968) also used this type of explanation to rationalize the apparently different length of the two lines in the Ponzo illusion (see Figure 4.2E). The "inducing lines" in the Ponzo stimulus converge, having much the appearance of railroad tracks extending into the distance. As a result of such experience, the theory goes, observers unconsciously infer that the upper horizontal line is farther away than the lower one. Given this inference, the upper line would have to be a physically longer object, and its projection might therefore be expected to look longer. In a similar vein, Barbara Gillam (1998) suggested that several other geometrical illusions are based on the observer's familiarity with the scale and size of objects, and the way these parameters are typically affected by the perspective generated in projected images.

Although the explanations proposed by Gregory and Gillam are broadly empirical, the idea that the brain uses such rules fails to explain the range of phenomenology outlined in this chapter. The Müller-Lyer effect (see Figure 4.2D) is seen even if circles or squares rather than arrowheads and arrow tails terminate the ends of the two lines (Figure 4.21A). Conversely, the Ponzo illusion (Figure 4.2E), which is often described in terms of a heuristic based on perspective, is diminished when the two lines to be compared are presented as vertical attachments to one of the inducing lines (Rock, 1995) (Figure 4.21B). Moreover, most of the effects elicited by the stimuli in Figure 4.2 do not express any obvious rule, and analyses of real-world geometry often fail to confirm intuitions that seem obvious. For instance, an analysis of laser-scanned scenes shows no significant difference between the average distance in depth of convex and concave corners, the explanation that Gregory suggested for the Müller-Lyer illusion (Howe and Purves, 2005b).

Some psychologists have endorsed yet another explanation of geometrical illusions known as the theory of "contrast and confluence" (Obonai, 1954; Rock 1995; these concepts were initially described by Helmholtz [1924, vol. III, pp. 236–240]). In this interpretation, observers are supposed to perceive an object's properties based on a comparison with other nearby

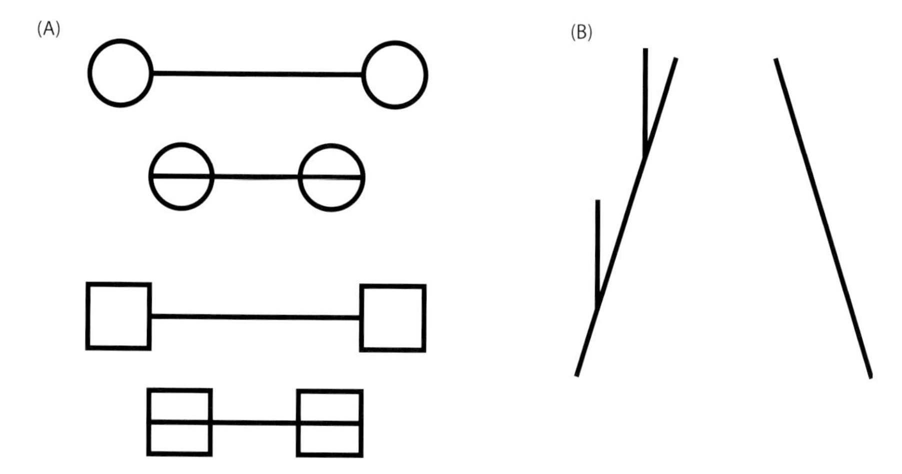

Figure 4.21 Examples that confound rule-based explanations of classical geometrical "illusions." (A) Persistence of the Müller-Lyer effect in the absence of possible differences in depth implied by arrowheads and arrow tails. (B) Diminishment of the Ponzo effect by an altered arrangement that nonetheless maintains the depth information conveyed by the converging lines.

features. Thus, any differences observed between the object and the context would tend to be exaggerated, leading to size contrast; conversely, "confluence" would lead to size assimilation. For example, the lower line in the Ponzo illusion would tend to look smaller in contrast to the relatively large empty space at either end of it, whereas the upper line would tend to look larger in contrast to the relatively small space at its ends. A related approach has been to imagine that "contrast" and "assimilation" signify actual cognitive operations that perform these comparisons. In this conception, size contrast effects are taken to be the consequence of comparing the properties of an object with its context, whereas assimilation is considered to be the result of incorporating the properties of contextual elements into the percept of the target (Coren, 1971; Coren and Enns, 1993; Rock, 1995).

Another explanation often alluded to in psychology is the "adaptation level" theory, which suggests that the overall "level" of a stimulus ("size level," in this instance) provides a reference or "anchor point" that the visual system then uses to make judgments about the relative magnitude of specific stimulus elements (Helson, 1964; Green and Stacey, 1966; Restle and Merryman, 1968; Jordan and Uhlarik, 1986). For a size contrast stimulus such as the Ebbinghaus figure, the different "levels" of adaptation generated by the different sizes of the surrounding circles would, in this interpretation, make the two identical central targets appear different in size. This argument, however, cannot explain the effect of altering the interval between the central and surrounding circles of the Ebbinghaus stimulus (see Figure 4.16B) or many of the other phenomena elicited by the stimuli in Figure 4.16. In particular, the theory is contradicted by size assimilation effects. The adaptation-level theory predicts that the inner target circle in Figure 4.16D should appear smaller than the single circle because the overall size of the concentric circles is larger, thus providing a higher "adaptation level" than the single circle. This prediction is opposite the perception elicited. A general problem with all these ad hoc theories is the lack of any biological reason why such processes would be necessary, or even useful.

More sophisticated explanations of some of these effects angle misperceptions, for instance have turned to electrophysiological data. The key finding that motivated this approach was the discovery by David Hubel and Torsten Wiesel (1959, 1962, 1968, 1974) of cells in visual cortex that respond selectively to stimulus features. Although such studies do not explicitly address the neural basis of perception (much less geometrical illusions), an implicit assumption has been that the stimulus features encoded in the response properties of higher-order cells in the primary and extrastriate visual cortices underlie the relevant perceptions. Putting this evidence together with the well-documented existence of lateral inhibition in the visual system (e.g., Kuffler, 1953; Hartline, 1969), a natural suggestion is thus that lateral inhibitory effects among orientation-selective cells in the visual cortex explain some of the effects illustrated in Figure 4.2 (e.g., Bouma and Andriessen, 1970; Carpenter and Blakemore, 1973; Andrews, 1967; Blakemore et al., 1970; Hotopf and Robertson, 1975). In this interpretation, the cortical response to an angle might differ from a simple summation of the neural activity elicited by each angle arm alone because the orientation domains co-activated by the two arms inhibit each other. If so, the effect might be to shift the distribution of cortical activity toward orientation domains whose selectivity is more orthogonal than would otherwise be the case.

Although this idea rationalizes the perceptual enlargement of acute angles, a difficulty is that interactions among visual cortical neurons are far more complex than the simple inhibitory effects originally envisioned (see, for example, Gilbert and Wiesel, 1990; Li and Li, 1994; Dragoi and Sur, 2000). As it turns out, the effect of a contextual line on the response to a target line can be inhibition, facilitation, or some combination thereof, depending on a host of factors (e.g., the orientation and the length of the stimuli, and the other response properties of the neurons involved).

Conclusion

The long history of attempts to understand the peculiar perceptual consequences of simple geometrical stimuli emphasizes how difficult it has been to provide a comprehensive, biologically plausible explanation of the perception of space and form. The point of this chapter has been to show that the same empirical strategy that explains many aspects of lightness/brightness and color perceptions works equally well in explaining perceived geometry. Knowing the distance and direction of every point in a database of natural scenes allows determination of the frequency of occurrence of lines, angles, and the sizes of simple figures projected onto the retina. By generating percepts and actions that accord with the relative rank of geometrical projections in accumulated past experience, behavioral responses to inherently uncertain meaning of geometrical image features circumvent the inverse problem. The strategy is once again linking images to useful behavior despite the "hidden" status of the physical sources. The ability to explain much otherwise-puzzling perceptual phenomenology using data that reflects human experience with line lengths, orientations, angles, and the sizes of simple forms further supports a wholly empirical interpretation of vision and visual processing.

5

Seeing Distance and Depth

Introduction

The apparent distance of objects, although a component of perceived geometry, is sufficiently important—and in at least one way special—to be considered in its own right. Humans use a variety of cues to generate perceptions of distance and depth (these terms are roughly synonymous, although "distance" is more often used to describe the locations of things that are relatively far away, and "depth" to locations that are closer). Much of this information is available when looking at the world with only one eye rather than two. However, for humans and many other animals with frontally placed eyes, additional information about the disposition of objects in space arises in binocular view. This binocular information generates a special sense of depth called stereopsis. The purpose of this chapter is to consider evidence that the perception of distance and depth are, like the perceptions of lightness/brightness, color, and geometrical form, generated by the accumulated effects of trial-and-error experience during evolution and postnatal development on the organization of visual circuitry.

Approaches to Rationalizing Visual Space

The perceived geometry described in Chapter 4 shows that the subjective experience of space does not correspond to physical space gauged with measuring sticks, protractors and the like. For some people who have thought about the discrepancies between physical space and visual space, an appealing intuition has been that the properties of visual space arise as the result of a systematic neural transformation of the Euclidean properties of physical space (Indow, 1991; Hershenson, 1999). Theoretical studies of how and why we see distance and depth in the way we do have proposed, for example, that visual space might represent a Riemannian space with constant curvature, or an affine space (defined as a vector space without an origin) (Luneburg, 1947; Wagner, 1985; Indow, 1991; Todd et al., 2001). More practically oriented psychologists have suggested that visual space is a computed composite, based on more or less independent information derived from cues such as perspective, texture gradients, binocular disparity, and motion parallax (Gillam, 1995; Loomis et al., 1996; Hershenson, 1999). The most influential advocate of this approach was James Gibson (see Chapter 1), who argued that since human beings are terrestrial, the ground determines a frame of reference built on the Earth's surface (Gibson 1950a,b; 1979).

Whereas each of these approaches is interesting to consider, given the present argument they miss a deeper point. Perceptions of distance and depth, much as perceptions of lightness/brightness, color, and form, must contend with and be understood in terms of the fundamental obstacle presented by the inverse problem. Since the distances and depths of the sources of retinal projections are inevitably uncertain (see Figure 4.1), explaining the relevant perceptual phenomenology seems likely to also depend on accumulated information about the link between images and the relative success of behavior over the course of human evolution and individual development. Absent this empirical information, it is hard to imagine how a useful appreciation of distance and depth could arise.

Monocular Perception of Distance

Many categories of monocular information about distance and depth are available to observers (Figure 5.1). Perhaps the most obvious indicator of the arrangement of objects in space is occlusion: When a part of one object is obscured by another, it will always have been the case that the obstructing object is closer to the observer than the obstructed object. A subtler source of monocular information about distance is the fainter and fuzzier appearance of distant objects. Because the earth has a substantial atmosphere, the farther objects are from the observer, the more interposition of matter. Since the molecules and particles in the atmosphere cause it to transmit light imperfectly, the number of photons that reaches the eye is progressively reduced as a function of distance. Moreover, the farther away an object surface, the more the photons reflected from it that do reach the eye are scattered, which makes increasingly distant objects look

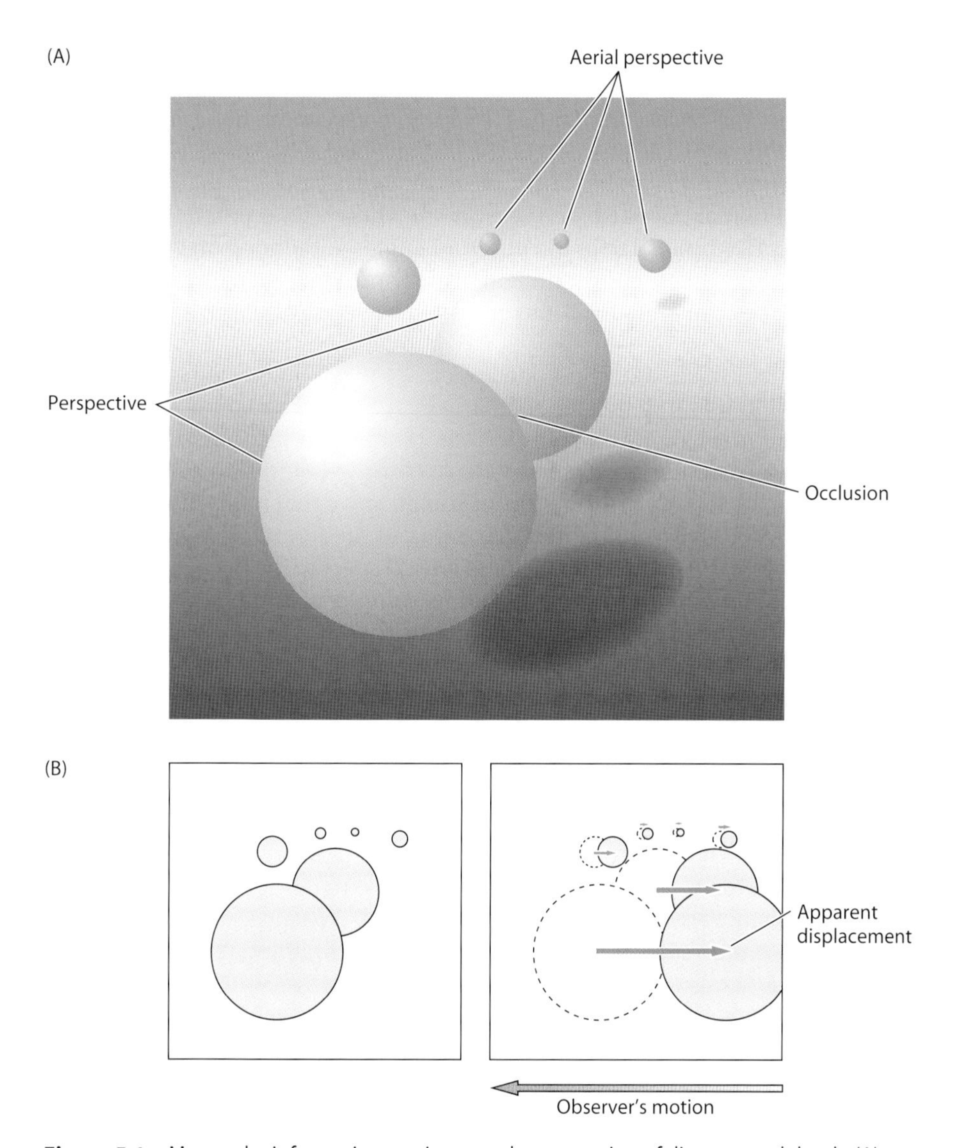

Figure 5.1 Monocular information pertinent to the perception of distance and depth. (A) Illustration of occlusion, effects of the atmosphere ("aerial perspective"), and the diminution of size with distance (perspective). (B) Motion parallax (indicated by the dotted duplications of near and far objects). (See text for further explanation.)

progressively fuzzier (Figure 5.1A). And, because the atmosphere absorbs more long- than short-wavelength light (the interposed medium is effectively "sky"), distant objects also look bluer compared to their appearance nearby, as landscape artists have long recognized (for more information about the color changes induced by the atmosphere see Minnaert, 1937 and the description of Rayleigh scattering in the Glossary).

Another example of monocular information that contributes to the perception of distance and depth is motion parallax (Figure 5.1B). When the position of the observer changes (by movement of the head from side to side, for instance), the position of the background with respect to an object in the foreground changes much more for nearby objects than for distant ones. Monocular cues also derive from the known sizes of familiar objects. For example, knowledge of the different angles subtended on the retina by human beings, trees, buildings, and other commonly seen objects at different distances presumably contributes to the sense of depth and distance in scenes.

It seems obvious—and no one appears to have suggested otherwise—that much if not all of this monocularly available information about depth must be learned (i.e., incorporated into neural circuitry by virtue of the trial-and-error experience of individuals during their lifetime). In one way or another, observers discover through experience that more-distant objects are often occluded; look fainter, fuzzier, and bluer; are smaller in appearance; and tend to change position less with respect to the background when the head is moved. Individual visual learning is of course commonplace; recognizing faces, learning the letters of the alphabet, and identifying neighborhood landmarks are simple examples.

Monocular Distance Percepts Elicited by Stimuli with Minimal Context

Other aspects of monocular distance perception can be examined quantitatively. When subjects are asked to make monocular judgments with little or no contextual information (e.g., to estimate the distance to a luminous but otherwise featureless object in a darkened room), the distances reported differ in several ways from the corresponding physical distances (Figure 5.2) (Sedgwick, 1986; Gilliam, 1995; Loomis et al., 1996; Hershenson, 1999). First, objects in these circumstances are typically perceived to be 2 to 4 meters away, a phenomenon referred to as the "specific distance tendency" (Gogel, 1965; Owens and Leibowitz, 1976) (Figure 5.2A). Second, objects that are relatively near each other in the retinal image appear to be about the same distance from the observer, a phenomenon called the "equidistance tendency" (Gogel, 1965) (Figure 5.2B). Third, when presented at or near eye level, the distance to objects relatively near the observer tends to be overestimated, whereas the distance to objects that are farther away tends to be underestimated (Epstein and Landauer, 1969; Gogel and Tietz, 1979; Morrison and Whiteside, 1984; Foley, 1985; Philbeck and Loomis, 1997) (Figure 5.2C). Finally, the apparent distance to objects on the ground varies with the declination angle of the line of sight (Wallach and O'Leary, 1982): Objects on the ground that are at least several meters away appear closer than they really are, but with increasing distance they are judged to be progressively more elevated than warranted by their physical position (Ooi et al., 2001) (Figure 5.2D). Although a variety of explanations

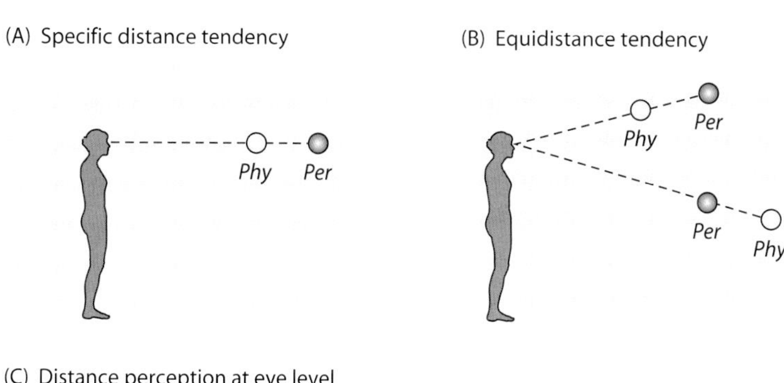

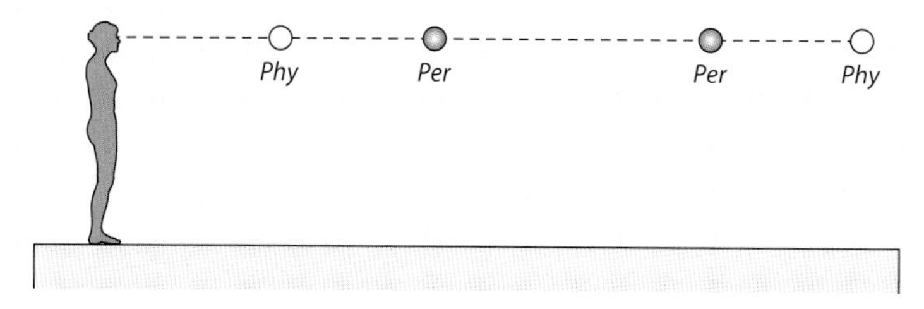

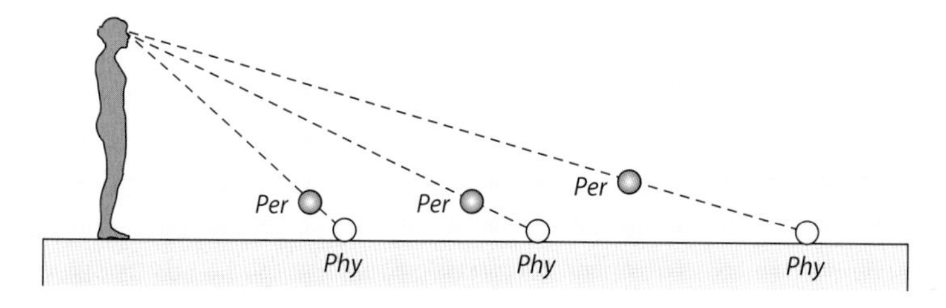

Figure 5.2 Anomalies of perceived distance. In these diagrams, which are not to scale, "Phy" indicates the physical position of the object and "Per," the perceived position. (A) The specific distance tendency. (B) The equidistance tendency. (C) The perceived distance of objects at eye level. (D) The perceived distance of objects on the ground. (See text for explanation.) (From Yang and Purves, 2003.)

have been proposed in these various studies, there has been little or no agreement about the basis of these unusual perceptions of distance. More often than not, the several tendencies illustrated in Figure 5.2 have simply been accepted as facts that are then used to rationalize other aspects of visual space. As with perceptions of light intensity, spectral distribution, and geometry, these peculiar perceptions of egocentric distance present another challenge for any explanation of vision.

In an empirical framework, the question is: Can the frequency of occurrence of visual stimuli explain the odd perceptions of egocentric distance illustrated in Figure 5.2? A first step in this assessment is to use

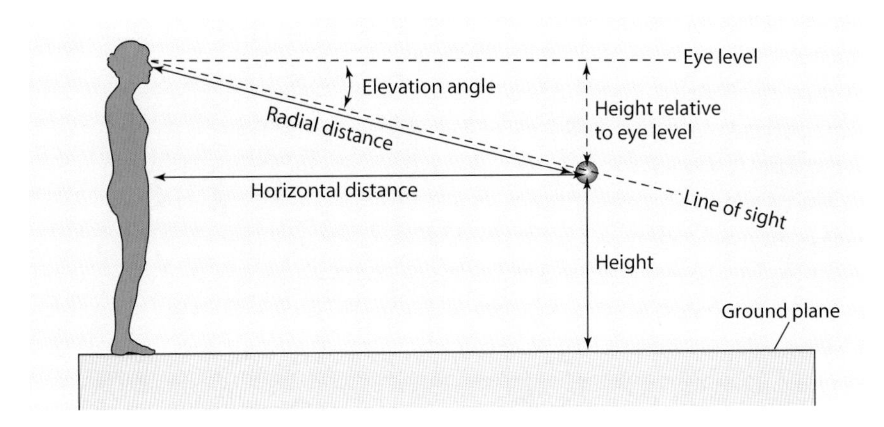

Figure 5.3 Diagram defining some of the terms used to describe distance and viewpoint. (From Yang and Purves, 2003.)

the information about surface distance derived from the range database described in Chapter 4 to compute the distribution of object locations from the image plane in typical physical scenes (Yang and Purves, 2003). As before, the image plane of the laser scanner substitutes for the retinal image plane of an observer, and the distance data is a proxy for the behavioral utility. The relatively special terms required in what follows—i.e., the ground plane, radial distance and elevation angle—are defined in Figure 5.3.

One aspect of projections onto the image plane of the scanner (or the retina) is that the distribution of the distances to the locations of the surfaces in natural scenes is strongly biased toward relatively near values. As shown in Figure 5.4A, the distribution has a maximum at about 3 meters, declining approximately exponentially over greater distances. This decay occurs because the farther away an object is: (1) the more likely it is to be occluded by other intervening objects and (2) the less the area the same physical surface spans in the image plane as a result of transformation by projection ("perspective"). In physical terms, there are of course just as many near objects as distant ones, on average. The falloff at especially near distances arises in large part because in acquiring the data the range scanner was never placed directly in front of large objects that would have prevented the beam from scanning a scene (see Chapter 4); as a result, there is very little data for distances of less than a meter. Although this bias is largely an artifact of the way the environment was sampled, human observers in natural settings generally avoid nearby objects that block a good view of their surroundings.

A second feature emerging from the analysis of images in relation to the physical distance of sources concerns how different locations in natural scenes are related to each other in terms of their radial distances. The distribution of the distances from the scanner to any two physical locations (whether separated horizontally or vertically) is also highly skewed, having a maximum near 0 and a long tail (Figure 5.4B; the maximum difference is close to 0 for horizontally separated locations and about 10 to 15 centimeters for vertically separated locations). Even for physical separations as large as 30° of visual angle, the most frequent difference between the distances from the image plane to two surface locations is relatively slight.

A third statistical feature about images and typical surface locations is that distances from the image plane to surfaces in the horizontal plane change relatively little with the height of locations in the scene (Figure 5.4C). Thus, the distribution of the horizontal distances to surfaces at different heights above and below eye level (see Figure 5.3) has roughly the same configuration as the distribution at eye level. All have a maximum at about 4 meters, with a gradual falloff over greater distances.

Given object presentations with minimal context (e.g., a luminous blob in an otherwise darkened room), an empirical framework predicts that the perceptions of distance illustrated in Figure 5.2 should accord with the accumulated past experience linking stimuli with behavior. Much as in the context-free brightness scaling studies described in Chapter 2, overall experience should explain the subjective

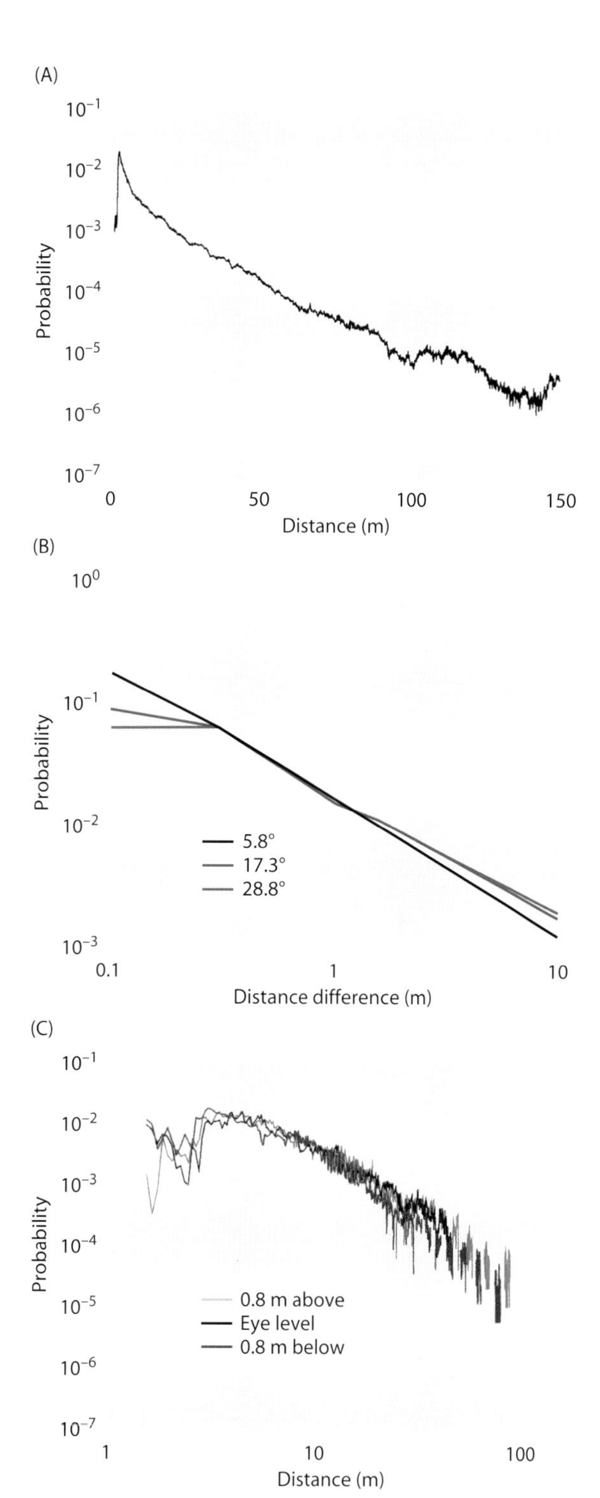

Figure 5.4 Distributions of the physical distances from the image plane of the laser range scanner to the surfaces of objects in natural environments. (A) The distribution of distances from the center of the image plane to all surface locations. (B) The distribution of the differences in the distances of two surface locations in physical space separated by three different visual angles in the horizontal plane (vertical separations, which are not shown here, had a generally similar distribution). (C) Distributions of horizontal distances of surface locations at different heights with respect to eye level (see Figure 5.3). (After Yang and Purves, 2003.)

sense of distance elicited by relatively impoverished stimuli. Although such stripped down stimuli rarely occur in the normal course of events, the way they are perceived can nonetheless speak to the underlying strategy of vision.

As indicated in Figure 5.4A, the overall distribution of the distances to natural object surfaces, estimated by human observers, shows that more surface points are about 3 meters away than other distances. In empirical terms, this distribution implies that in the absence of more specific information in a visual stimulus, the apparent distance in the subjective space that defines this aspect of perception should be about 3 meters. This prediction is in agreement with the evidence in psychophysical studies that observers tend to perceive objects as being 2 to 4 meters away under the experimental conditions just described, thus accounting for the "specific distance tendency" illustrated in Figure 5.2A.

The similar distance to neighboring points perceived in the absence of additional information (the "equidistance tendency"; see Figure 5.2B) also accords with the distributions of the surface distances that would have determined the success or failure of behavior. Since the distribution of distances from the image plane to two locations with relatively small angular separation (the black line in Figure 5.4B) has a maximum near 0, any two neighboring objects should, if subjective distance in response to impoverished stimuli is determined empirically, be perceived to be at about the same distance from the observer—as they do. At large angular separations (the colored lines in Figure 5.4B), however, the absolute difference in the distances to two points tends to increase, and the distribution becomes progressively flatter. Accordingly, the tendency to see neighboring points at the same distance from the observer in the absence of other image information would be expected to decrease as a function of increasing angular separation. This bias has also been observed in psychophysical studies (Gogel, 1965).

The distribution of physical distances at eye level (the black line in Figure 5.3C) can, in much the same way, account for the perceptual anomalies generated by near and far objects presented at this height. Based on the probability distributions illustrated in Figure 5.4, the perceived distance to an object located at eye level should, in conditions with little contextual information, be perceived to be about 2 to 4 meters away and resist change. Thus, the distance to an object located closer than this distance should be overestimated, and the distance to an object farther away should be underestimated (see Figure 5.2C). These predictions also fit the psychophysical data reasonably well. For instance, Philbeck and Loomis (1997) showed that the apparent distance to a dimly luminous object presented at eye level in an otherwise dark environment tends to remain at about 4 meters as the actual distance is varied between 2 and 5 meters (subjects reported apparent distance in this case by walking blindfolded to the place they thought the object was, explaining the relatively small range of distances tested).

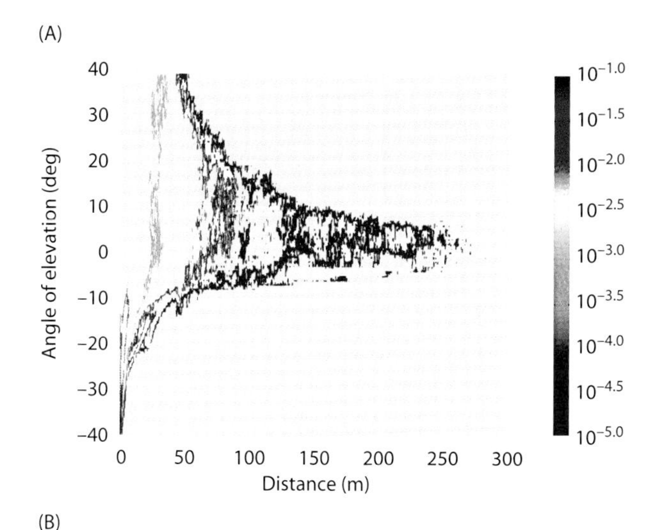

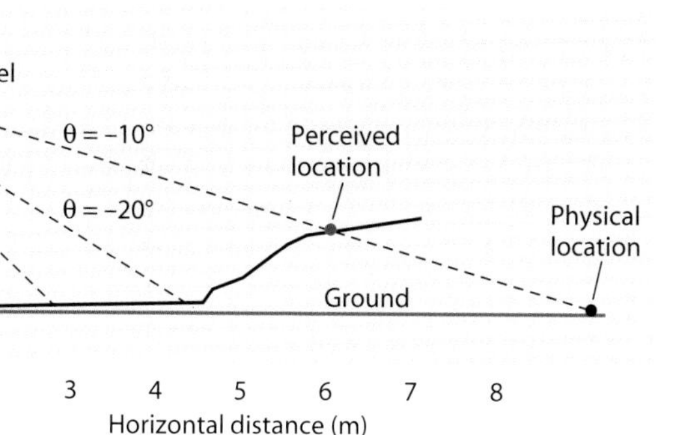

Figure 5.5 Distributions of physical distances at different elevation angles. (A) Contour plot of the distributions of distances from the image plane as a function of elevation; probabilities are indicated by color coding (bar on the right). An elevation angle of 0° is eye level; positive elevation angles correspond to lines of sight above eye level, and negative values correspond to those below. (B) The typical distance from the image plane as a function of the elevation angle (θ) of the location, derived from the data in (A). The vertical axis is the height relative to eye level; the horizontal axis is horizontal distance from the image plane. The blue line indicates the position of the ground plane at ~1.65 meters below eye level. These data predict that the apparent distance of an object on the ground more than a few meters away in a darkened environment should appear closer and higher than the actual location of the object. (From Yang and Purves, 2003.)

The apparent distance to objects on the ground illustrated in Figure 5.2D can also be accounted for empirically. The relevant distribution of surface locations in this instance is distances at different elevation angles relative to a horizontal plane at eye level (i.e., along different lines of sight; see Figure 5.3). As shown in Figure 5.5A, this distribution is more dispersed when the line of sight is directed above rather than below eye level. Moreover, the distribution shifts toward nearer distances as the line of sight deviates increasingly from eye level, a tendency that is more pronounced below than above eye level.

The differences in surface distances as a function of elevation are more obvious in Figure 5.5B, which shows the most frequently occurring physical distances (solid black line) to locations below eye level in laser-scanned scenes as a function of the elevation angle. The most likely distances from an observer to surface locations at various elevations form a curve that is relatively near the ground at closer distances; the curve increases in height, however, as the horizontal distance from the observer increases. The physical basis of this variation is presumably that objects below eye level, which are progressively farther from the observer, have to be increasingly higher in the visual field to be visible; i.e., more distant low-lying objects tend to be occluded by objects that are closer to the observer.

In summary, all these anomalies of perceived egocentric distance in the absence of much context can be rationalized in empirical terms.

Depth Perception Arising from Binocular Information

Up to this point, these considerations of how we see distance and depth concern percepts that are much the same whether an observer views the world with one eye or two. Another aspect of depth perception, however, depends specifically on viewing the world with both eyes. The special sense of depth that arises when viewing relatively nearby objects binocularly is called stereopsis.

Stereoscopic depth is of special interest because it has been thoroughly studied behaviorally, anatomically, and electrophysiologically. Moreover, in the course of this work no one has paid much if any attention to the idea that accumulated experience might be a way of understanding stereopsis. On the contrary, stereoscopic perception has usually been taken as a good example of how perceptions are generated by on-line neural computations based on retinal image features—in this case, retinal disparity.

Retinal disparity refers to the slightly different images on the two retinas that arise when the eyes converge to inspect nearby objects (Figure 5.6A). Because human eyes are separated horizontally across the face by an average of 6.5 centimeters, when both eyes fix on a nearby point, each one has a slightly different viewpoint. The sensory and behavioral significance of this fact can be appreciated by comparing the accuracy of bringing the points of two pencils together in the frontal plane using both eyes versus only one. Making the tips of the pencils touch (or performing other more consequential tasks) is much easier in binocular view. The reason is that perception of nearby objects in depth in binocular viewing is enhanced by a compelling sense of the third dimension of space that is not otherwise appreciated. This added sense of depth can be dramatically demonstrated by comparing the monocular and binocular views of scenes presented in a stereoscope (Figure 5.6B).

As nineteenth century polymath Charles Wheatstone showed in the 1830s by his invention of the stereoscope (the physicist David Brewster independently devised a different type of stereoscope at about the same time; see Wade, 1987), the greater behavioral success with both eyes open in tasks involving relatively fine manipulation (and the richer sensation of depth) arises from the different viewpoints of the two eyes. Wheatstone also pointed out that stereoscopic information is limited to objects relatively near the observer. The difference in the viewpoints of the right and the left eyes decreases progressively as the lines of sight of the two eyes

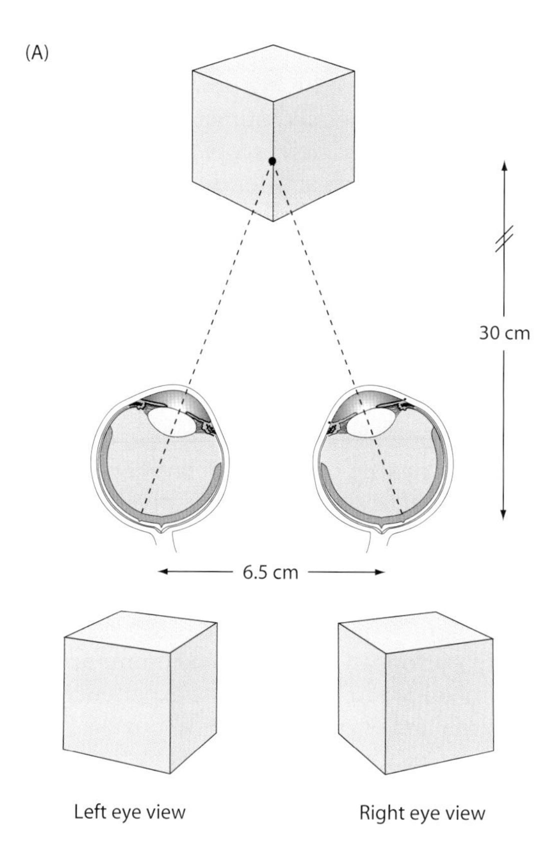

Figure 5.6 The different viewpoints of the two eyes. (A) Viewing any nearby object with one eye and then the other—a cube at reading distance, in this example—makes this difference apparent. (B) If two pictures (or photographic transparencies this modern version of a nineteenth century stereoscope) are depicted or taken from slightly different angles, looking at the 2-D images binocularly in a stereoscope produces a strong sensation of depth not present when the same images are looked at with one eye or the other. This simple demonstration shows that sensations of stereoscopic depth are due to the slightly different vantage points of the two eyes.

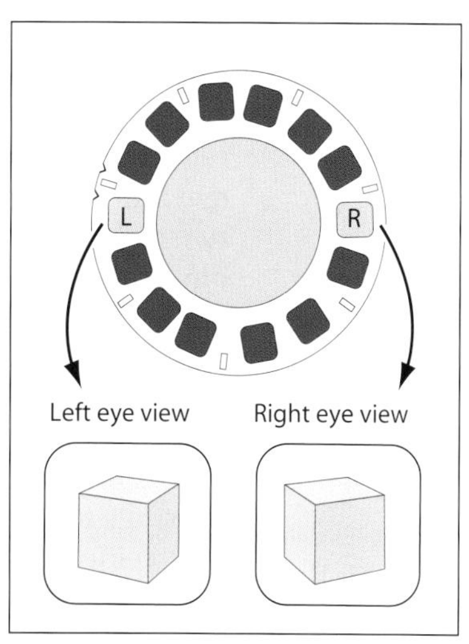

become increasingly parallel, causing the binocular disparity of objects in the image plane to eventually fall below the resolving power of the visual system. The so-called "stereo horizon" is generally estimated to be several hundred meters, although the contribution of stereoscopic information to depth perception becomes negligible at much shorter distances (see, for example, Ogle, 1964). Other animals with frontally located eyes enjoy the same advantage, and even wall-eyed animals have some stereoscopic ability within a small region of binocular overlap.

Binocular Fusion and Rivalry

In addition to stereopsis, two other phenomena critical to understanding binocular vision are binocular fusion and binocular rivalry.

Although humans routinely view nearby objects with both eyes open (and thus have two different retinal images as input from relatively nearby scenes; see Figure 5.6A), the perceived image of the world is different from the perceptions generated by either eye alone; what we see seems to have been generated by a single eye in the middle of the face, a subjective experience referred to as "cyclopean vision." This fusion of two quite different monocular views into a coherent cyclopean percept is so commonplace that we take it for granted. Yet, embedded in this everyday experience is a deep psychological, neurobiological, and even philosophical puzzle: How are the two independent views of any nearby scene conjoined by the visual system to create a single percept having qualities not present in the view of either eye alone?

Conventional explanations of how the two monocular views are fused in the brains of humans and other animals that have substantially overlapping fields of view in the two eyes are predicated on the fact that inputs from the eyes converge on the same cortical neurons in the primary visual cortex (V1) (Figure 5.7). Although right and left eye inputs are kept separate in the dorsal thalamus and in cortical layer IV of the primary visual cortex (which receives the thalamocortical afferents), many neurons in the deeper and more superficial cortical layers in V1 of cats and macaque monkeys (the two animals that have been most intensively studied in this respect) are binocularly driven (Barlow and Pettigrew, 1967; Hubel and Wiesel, 1977; Hubel et al., 1977). The prevalence of binocular cells in the primate visual cortex (and the perceptual quality of stereopsis) suggests that cyclopean vision arises from this demonstrable conjunction of right and left eye inputs at the level of common target cells in V1 and beyond.

Despite this apparently straightforward anatomical and physiological substrate for a perceptual union of the two monocular streams in visual cortex, seeing a cyclopean world as a result of convergence of inputs onto binocular neurons in V1 does not easily align with other evidence. In addition to showing that stereopsis arises from the different views of the two eyes, Wheatstone (1838) found that if a stimulus pattern (e.g., vertical stripes) is presented to one eye and another, strongly discordant pattern (e.g., horizontal stripes) to the other, the observer experiences perceptual rivalry: The

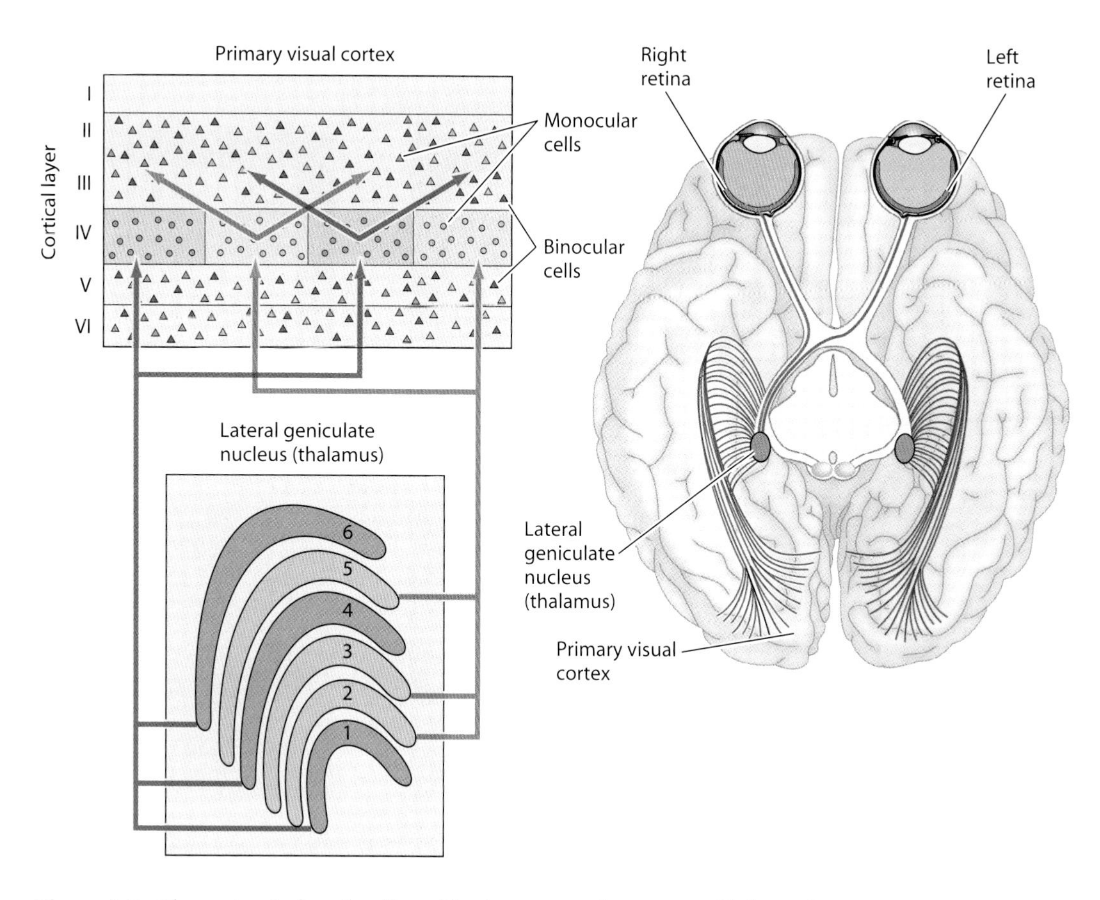

Figure 5.7 The anatomical conjunction of the two monocular streams of information in the visual cortex. Although these studies were carried out mainly in macaque monkeys, based on less direct evidence the same organization describes the human visual system. Inputs related to the right and left eyes first come together in the primary visual cortex, where half or more of the neurons can be activated by the same stimulus presented in the same location in visual space to either the left or right eye. The afferents related to the two eyes remain segregated at the level of the cortical neurons in layer IV of the primary visual cortex; binocularly driven cells are found only above and below this input layer.

same region of visual space is alternately seen to be occupied by vertical stripes or horizontal stripes, but only transiently and incompletely by both (Figure 5.8A). If information from the two eyes were simply united in the visual cortex (as implied by the prevalence of binocular neurons), observers would presumably see a stable integration of vertical and horizontal stripes (i.e., a grid in the most simplistic interpretation) in response to the stimuli in Figure 5.8A.

Binocular rivalry also occurs when substantially different color stimuli are presented to each eye (Figure 5.8B). Using such stimuli, it is easy to show that whether the information available to the left and right eye rivals

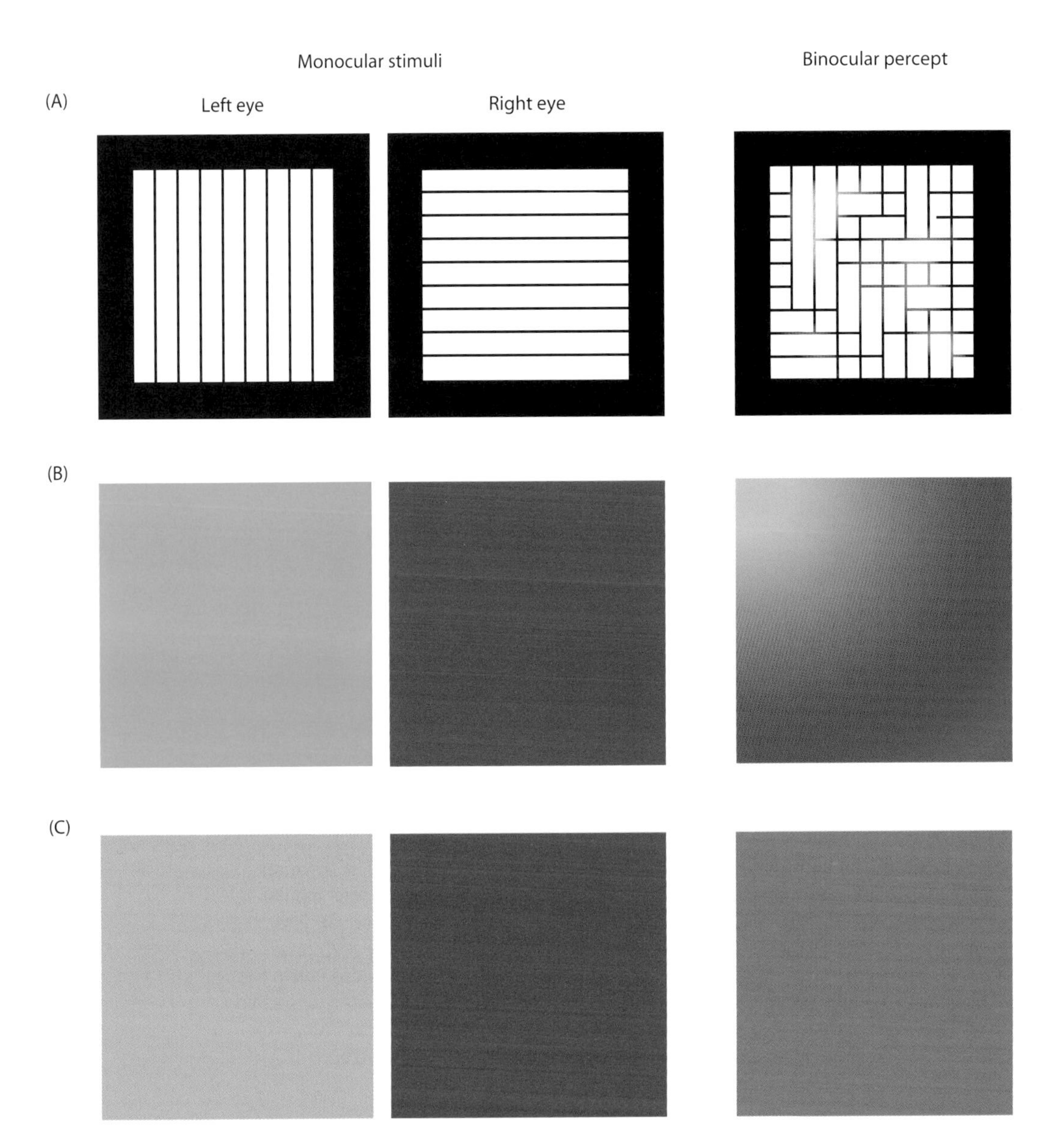

Figure 5.8 Binocular rivalry occurs when substantially different stimuli are presented to the two eyes using a stereogram (as here) or a stereoscope (see Figure 5.6B). (A) In this example, the vertical stripes presented to one eye and the horizontal stripes presented to the other eye are not stably integrated when the views of the two eyes are fused, but elicit a dynamically changing perception whose pattern corresponds (in whole or in part) to the view of one eye or the other. (B) The same phenomenon occurs when markedly different spectral stimuli are binocularly fused. (C) If, however, the spectral differences are not great, as in this example, then fusion gives rise to a more stable sensation that is usually (but not always) approximately the color that would be seen in response to adding the two monocular spectral returns. (For individuals who have trouble with "free fusion," simple stereo viewers can be obtained via the Internet for a few dollars.)

or fuses to generate a stable color percept depends on the degree of difference between the two monocular stimuli (see Figures 5.8B and C). Whereas stimuli whose spectra differ by relatively little in one part of the light spectrum may fuse, stimuli separated by the same physical wavelength distance in another part of the spectrum may rival (Ikeda and Nakashima, 1981; Ikeda et al., 1982; Sagawa, 1982). Moreover, when the effects of different spectra do fuse in perception, the resulting color sensation is not always that predicted by their average (Hoffman, 1962; Erkelens and van Ee, 2002). These several observations again suggest that something more complex than simply bringing together the views of the two eyes must be going on in binocular fusion.

Various explanations of binocular rivalry have been suggested, ranging from the suppression of one or the other monocular view (Ascher, 1953) to an alternation between the two monocular views that occurs normally (Verhoeff, 1935; Walls, 1948). Such propositions have not found wide acceptance and are in varying degrees incompatible with the evidence. Furthermore, work dating back to at least the 1920s indicates that it is not always the images on the two retinas that rival but, at least sometimes, the percepts they trigger (see Diaz-Caneja, 1928; Logothetis et al. 1996; Alais et al., 2000; Miller et al., 2000). These further observations imply that rivalry is a "high-level" phenomenon occurring in the visual association cortices (and perhaps involving interhemispheric switching in some interpretations; e.g., Pettigrew and Miller, 1998; Miller et al., 2000; Miller, 2001). The arguments for rivalry as an extrastriate phenomenon have been disputed, however, based on the observation that fluctuations in V1 activity correlate with perceptions of binocular rivalry (Polonsky et al., 2000; Tong and Engel, 2001; see Blake, 2000 for a review of this debate).

That the process underlying rivalry may not be a competition between the two retinal images is also suggested by the observation that monocular stimuli are capable of eliciting perceptual experiences much like binocular rivalry when the possible interpretations of a stimulus sources are bimodal or multimodal (Breese, 1899; Hodgkinson and O'Shea, 1994; Andrews and Purves, 1997). For example, a stimulus comprising 12 lines in the arrangement shown in Figure 5.9A generates alternating percepts of the two most probable sources (this familiar stimulus is called a "Necker cube" after Louis Necker, the nineteenth century Swiss naturalist who called attention to the "bistable" percepts he experienced during microscopic studies of crystal structure). When the same 12 lines are rearranged to correspond to a cube seen from the particular vantage point at which the retinal projection of the cube is symmetrical, as in Figure 5.9B, it is less easy to see the stimulus as a cube; most people perceive a two-dimensional design resembling a snowflake, presumably because the retinal projection of a real-world cube in this particular way is unlikely. Even the elements of quite ordinary monocular stimuli, such as the red and blue lines in Figure 5.9C, begin to "rival" if one stares at them long enough (Breese, 1899; Andrews and Purves, 1997).

(A)

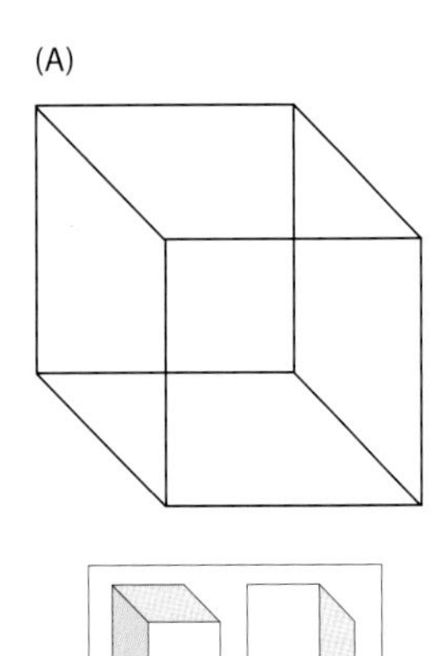

Figure 5.9 Monocular stimuli can generate a variety of experiences, some of which are perceptually similar to binocular rivalry. (A) A wire frame cube depicted in three-quarter view (the Necker cube) alternates between the two different perceptions of orientation shown in the solid depictions. (B) The same 12 lines as in (A), but arranged symmetrically; in this case, the depiction is often seen as a two-dimensional figure. (C) When observers stare at the stimulus shown here for 30 seconds or more, portions of the red or blue lines begin to disappear and reappear in a manner similar to the sensations of binocular rivalry elicited by the stimuli in Figure 5.8A. (C is after Andrews and Purves, 1997.)

(B)

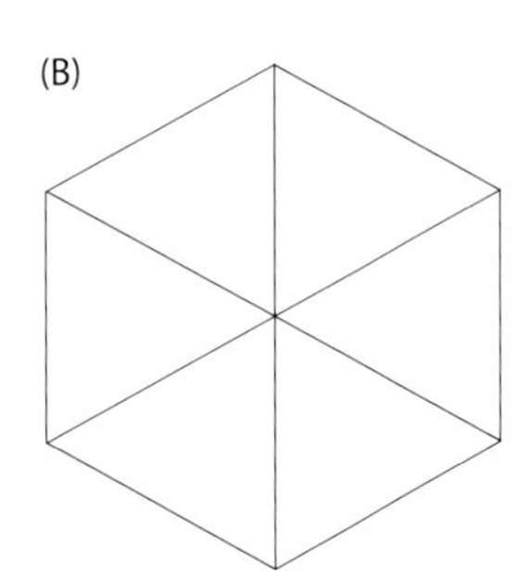

(C)

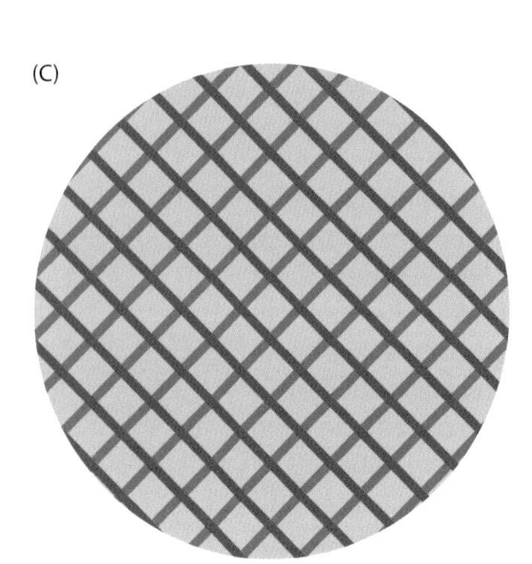

In short, there is no consensus about the physiological or psychological basis of rivalry, and even some aspects of the phenomenology are in dispute.

Sherrington's Experiment

Another example of the puzzle underlying binocular fusion and rivalry is a little-known study carried out more than a century ago by Charles Sherrington (1904; 1906a). Sherrington, whose work on vision has been all but forgotten, wanted to know whether there is a "final common pathway" in sensory systems, similar in principle to the organization that he had discovered in the motor pathways in the spinal cord (see Sherrington, 1906b). To this end, he examined how the monocular views come together in perception. The gist of his ingenious experiment was to present independently to the two eyes a light that alternated between on and off at a frequency he could control (Figure 5.10A). When looking with both eyes at a light going on and off, observers typically cease to detect flickering when the alternation of light and dark reaches about 60 Hz, a point called the "critical flicker-fusion frequency." Sherrington surmised that the critical flicker-fusion frequency when the two eyes are synchronously stimulated ought to be different from the frequency observed when the two eyes are stimulated alternately (i.e., one eye experiencing light when the other eye experiences dark). If the information from the two eyes are brought together in a final, common pathway in the visual brain, then the combined asynchronous left and right eye stimulation should be perceived as continuous light at roughly half the rate of presentation as the normal flicker-fusion frequency (Figure 5.10B). The result, however, was that the perceived frequency was virtually identical in the two circumstances (Sherrington, 1904; 1906a; see

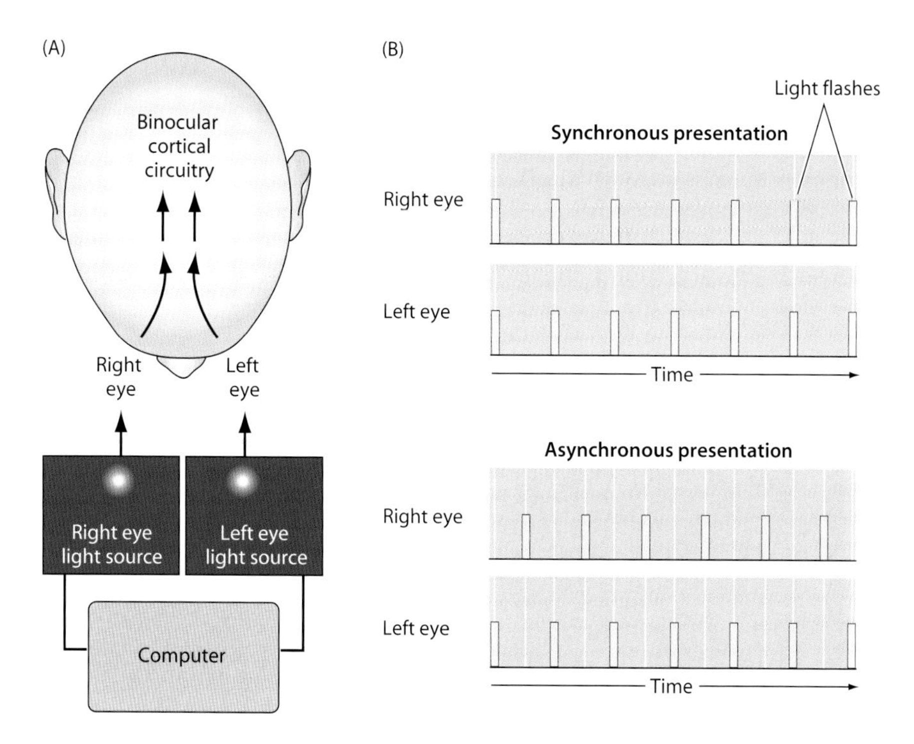

Figure 5.10 A modern version of Sherrington's flicker-fusion experiment, in which synchronous and asynchronous stimulation of the two eyes are compared. (A) A computer program triggers independent stroboscopic light sources whose relationship can be precisely controlled; one series of flashes is seen only by the left eye and the other only by the right. (B) Diagram of synchronous versus asynchronous stimulation of the left and right eyes. (After Andrews et al., 1996.)

also Andrews et al., 1996). Based on this observation, Sherrington concluded, rather despairingly, that the two retinal images must be processed independently by the brain and that their union in perception must be "psychical" rather than physiological, thus lying beyond his ability (or interest) to pursue.

Although this aspect of Sherrington's work is rarely referred to, his conclusion that cyclopean vision must arise in a way that is fundamentally different from the concept of a final common pathway is close to the mark. The phenomenology of cyclopean vision and rivalry implies that, binocular neurons notwithstanding, vision operates, as Sherrington suggested, without ever fusing the views of the two eyes at some level in the visual pathway.

Whatever the ultimate explanation of these several effects, they argue against a simple confluence of monocular information at the level of binocular neurons in V1 (or anywhere else) as the explanation of fusion and rivalry.

An Empirical Explanation of Fusion and Rivalry

An interpretation that can accommodate this perceptual strangeness is again empirical. When the diverse binocular information in the two retinal images is consistent with similar views of the same object(s), the result is cyclopean perception. When, however, this information is inconsistent with that experience, then the views from the two eyes rival. Experience with such stimuli is not in the least unusual: Whenever an object occludes

Figure 5.11 Whether a stimulus elicits fusion or rivalry may be determined by its behavioral significance for successful behavior. (A) As indicated in Figure 5.8B, when two surfaces with significantly different spectra are independently presented to the right and left eyes, the two views rival in perception. (B) However, when the right and left eyes are presented with stimuli that have about the same average spectral differences, but include information that increases the likelihood that the images in the two eyes have arisen from different views of the same object at the same location in space, the two views tend to fuse in perception.

the view of one eye, the two retinal images are rivalrous. We don't usually experience rivalry because our eyes normally move 3 or 4 times a second to examine different locations in a scene.

The rationale for fusion or rivalry, as before, is the need to solve the inverse problem, but now in the context of binocularity. As explained below, the left and right retinal images are formally uninterpretable with respect to stereo depth (many points in 3-D space could underlie corresponding image points on the two retinas). Nonetheless, by linking the left and right eye images with successful behaviors via accumulated experience instantiated in visual system circuitry, the problem can be overcome, leading to a fused perception of the world and stereo depth.

In this conception, cyclopean vision and binocular rivalry, like contrast and constancy in previous chapters, are not fundamentally different, but perceptual poles along the same continuum, the position along it being set by accumulated past experience. The subjective experience of fusion occurs whenever the differences between the two monocular retinal images are consistent with past experience of the binocular views associated with the same object(s) at a given location in 3-D space. If, on the other hand, the images on the two retinas are inconsistent with this possibility, then the two eyes continue to trigger independent patterns of central neural activity that compete for perceptual "attention," giving rise to the sensory experience of rivalry.

Consider, for instance, the stimuli in Figure 5.11. Normally, spectrally different "targets" that occupy the same position in the right and left eye views generate rivalry, as they do when the stereogram in Figure 5.11A is fused. If, however, the empirical information in images generated by a scene is consistent with objects in the same location in space, as in Figure 5.11B, then images with approximately the same spectral qualities might be expected to fuse in a stable way. As can be seen by fusing these stereograms, this is indeed what happens.

Conventional Explanations of Binocular Depth Perception

Ever since Wheatstone's original demonstration that stereopsis arises from differences in the view of the two eyes, most students of vision have accepted the idea that this special sensation of depth is based on neural computations of the magnitude and direction of the retinal disparity (Figure 5.12). In this interpretation of stereoscopy, the visual system "compares" the loci on the two retinas that are stimulated by light rays arising from the same point in visual space. As a result, stereopsis has seemed an especially pertinent domain in which to assess algorithms assumed to underlie visual processing and perception (see, for example, Marr and Poggio, 1979; Grimson, 1981; Marr, 1982; Poggio and Poggio, 1984; Prazdny, 1985; Ullman, 1987; Howard and Rogers, 1995). Indeed, many studies have shown that some neurons in the visual cortex of cats and monkeys are "tuned" to specific

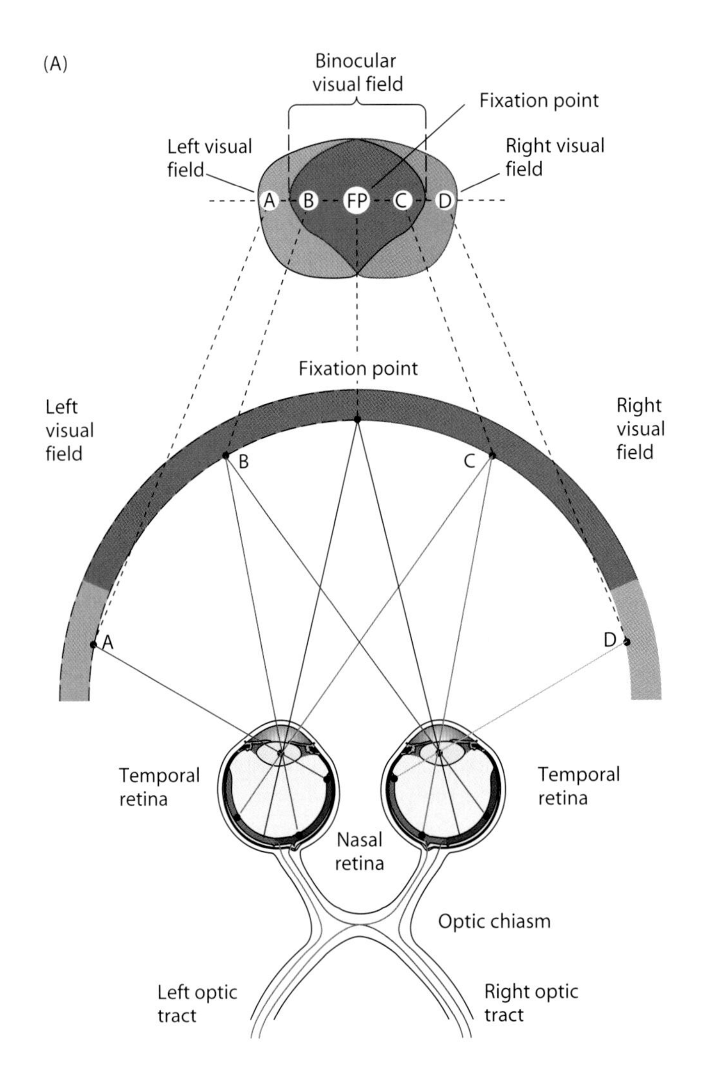

disparities (Barlow et al., 1967; Hubel and Wiesel, 1977; Poggio and Poggio, 1984; Poggio, 1995). These electrophysiological observations, together with the knowledge that stereopsis can be elicited by random dot stereograms (see below), seem to confirm that stereopsis is generated by computations of retinal disparity (Marr, 1982; Julesz, 1995; Ullman, 1995; see Howard and Rogers, 1995 for an extensive review).

Although this way of explaining stereoscopy is eminently logical, understanding how the nervous system implements the geometrical comparison of a stereo pair has been a difficult challenge. There is no agreement about how such a feat could be accomplished (see, for example, Marr and Poggio, 1979; Mayhew and Frisby, 1981; Howard and Rogers, 1995; Julesz, 1995).

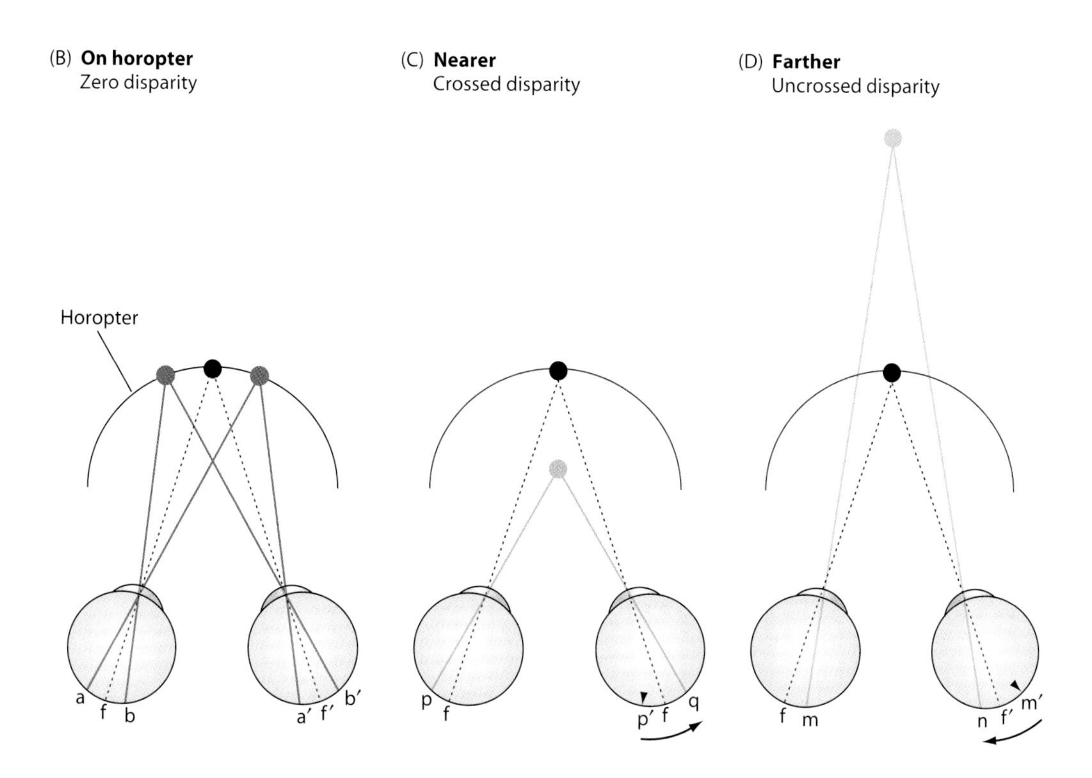

Figure 5.12 Retinal disparity. (A) Diagram of the overlapping visual fields in humans. Some degree of binocular overlap is a prerequisite for stereo vision. (B) Two pairs of geometrically corresponding retinal points—i.e., points located the same distance and direction from the fovea in each eye—are indicated by a, a′ and b, b′; f and f′ indicate the two foveas. All points in visual space that simultaneously stimulate "corresponding" retinal points (indicated by the matching colors of the rays) are perceived as being at the same stereoscopic depth as the point of fixation (the black dot at the intersection of the two foveal lines of sight). The set of these points in visual space defines the horopter (shown here for simplicity as a segment of a circle; the actual configuration of the horopter is more complex). Light rays emanating from all other loci in visual space fall on non-corresponding retinal points; the direction and magnitude of such deviations from geometrical correspondence define retinal disparity. (C) Light rays arising from points nearer than the point of fixation (e.g., the greenish dot) stimulate a retinal point in the left eye (p) that no longer matches the geometrically corresponding point in the right eye (p′). As indicated by the arrow, the stimulated point in the right eye (q) is displaced temporally from the geometrically corresponding retinal point (p′). Such temporal displacement is called crossed disparity and is associated with the perception of objects being nearer the observer than the point of fixation. (D) Light rays arising from points more distant than the point of fixation (bluish dot) also fall on non-corresponding retinal points on the two receptor surfaces (m and n), but in this case the displacement from the geometrically corresponding point (m′) is toward the nose (as indicated by the arrow). Such displacement in the direction of the nose is called uncrossed disparity and is associated with the perception of objects being farther away than the point of fixation. Only horizontal disparity is considered here because this is the direction that is most important in stereopsis; vertical disparity has also been studied and plays a role in vision, but it arises from more subtle geometrical considerations.

Problems Explaining Stereoscopy in Terms of Retinal Disparity

A problem explaining binocular depth perception terms of disparity was already noted in middle of the nineteenth century, namely, that sensations of stereoscopic depth can be elicited by stimuli in which disparity is only implied (Panum, 1858; Helmholtz, 1924. vol. III, Plate V). In Figure 5.13A, for example, disparity is not explicitly represented for one of the black bars (this type of stimulus is sometimes referred to as "Panum's limiting case"). A circumstance that gives rise to this sort of binocular stimulus is when all three objects indicated by the bars diagrammed in Figure 5.13B are visible to the right eye but not the left; one of the more distant objects is made invisible by the interposition of one of the nearer objects. Despite the absence of disparity, stereo depth is accorded the object seen with only one eye.

Another problem apparent from the outset of stereoscopic studies is the failure to reverse stereoscopic depth when all the disparities in a scene are reversed. As Wheatstone described in 1852, a reversal of the right and left eye views can be achieved by viewing the world through a pair of base-in, right angle prisms, a device he called the "pseudoscope" (Figure 5.14). A similar reversal can be achieved by putting the photographic images on the "wrong side" of an ordinary stereoscope such as the one shown in Figure 5.6B (i.e., by flipping the cards over). The pseudoscope switches the line of sight for the eyes so that the left eye now sees objects from the angle seen by the right eye and vice versa, thus reversing the "sign" of the normal retinal disparity. If a computation of disparity were the determinant of stereoscopic depth perception, then reversing the views of the two eyes should reverse the perception: What was closer should appear farther away, and what was farther away should appear closer. Wheatstone observed that

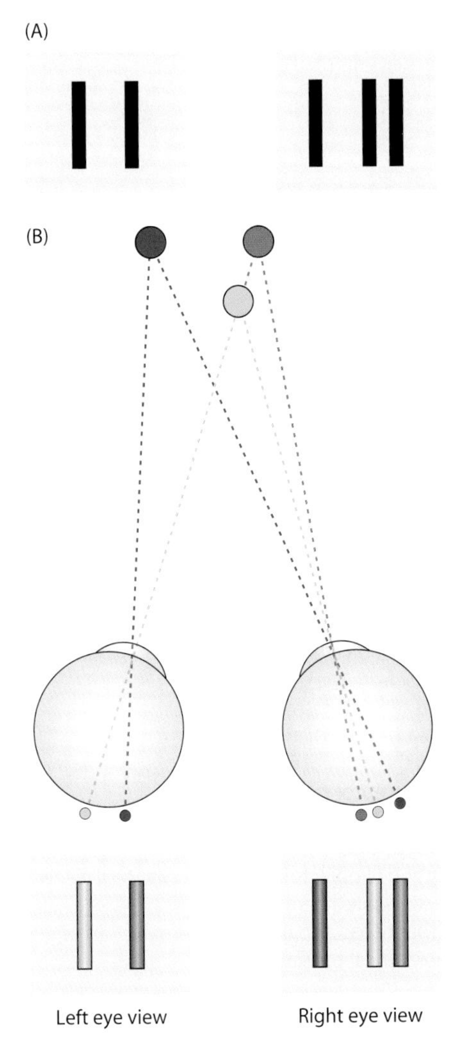

Figure 5.13 An example of the independence of stereoscopic depth sensation and retinal disparity. (A) Stereogram in which two objects are presented in the view of the left eye, and three in the right. Despite the absence of explicit disparity, the stimulus elicits a perception of stereoscopic depth when the two images are fused (one of the bars appears to be at a distance different from that of the other two). (B) Diagram of an arrangement of objects that could have given rise to the two monocular components of the stereogram in (A).

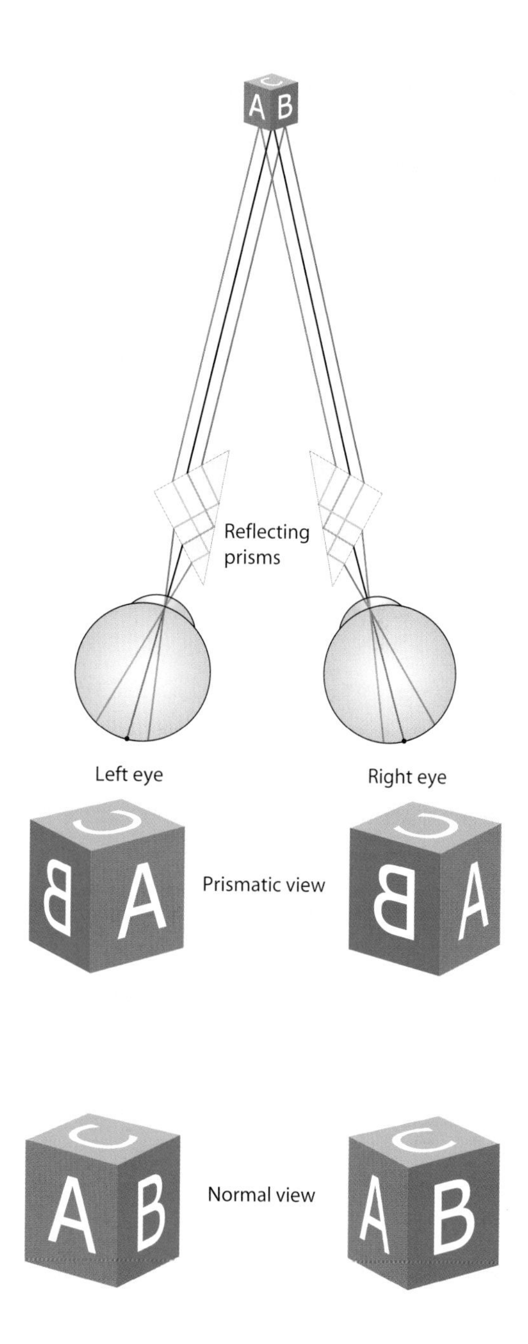

Figure 5.14 "Pseudoscopic" viewing through base-in right angle prisms. In this circumstance, the left eye sees objects from the point of view of the right eye, and vice versa. Completely reversing retinal disparity has relatively little effect on the appearance of distance and depth.

for some sorts of objects stereoscopic depth perception in this circumstance is indeed reversed (see also Shimojo and Nakajima, 1981; Ichikawan and Egusa, 1993). Thus, he reported the bottoms of cups as appearing to come toward the observer, medals cast as bas relief appearing as intaglios, and etched glass cylinders in reversed perspective. But most objects, even nearby ones, look quite normal when observed through reversing prisms. This failure of the pseudoscope to simply reverse binocular depth perception has generally been attributed to the wealth of monocular cues about depth that are not changed by prismatic viewing (e.g., perspective, occlusion, and so on; see earlier). Nonetheless, the nature of pseudoscopic effects raises further doubts about stereoscopic percepts arising simply from computations of retinal disparity.

A further indication that stereoscopic percepts might not represent neuronal computations of retinal disparity as such is the inherent ambiguity of this metric as a result of the inverse problem (Figure 5.15). In much the same way that a given monocular projection can be generated by many possible objects at different distances and in different orientations (see Figure 4.1), the projection of the same point in visual space onto the two retinas (which defines the disparity of that point; see Figure 5.12) can be generated by many pairs of points at different distances along the same lines of sight. This puzzle is referred to as the "false target

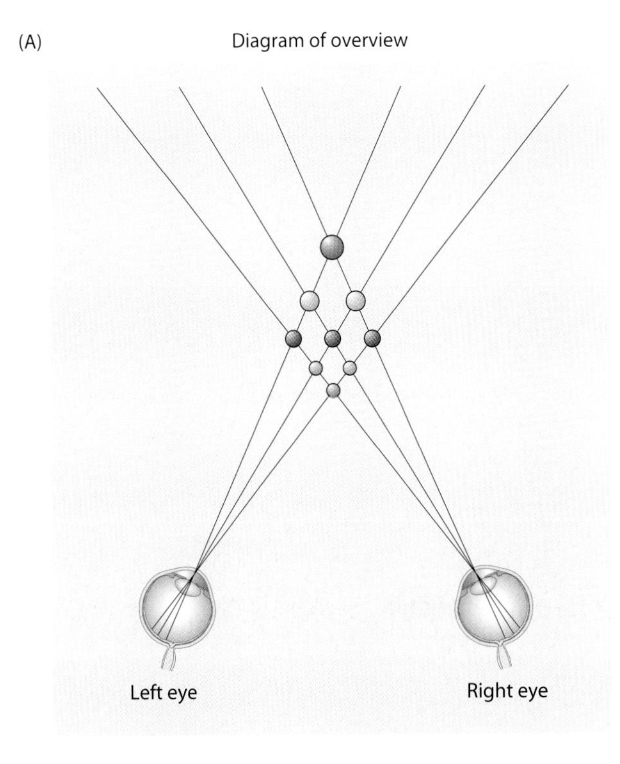

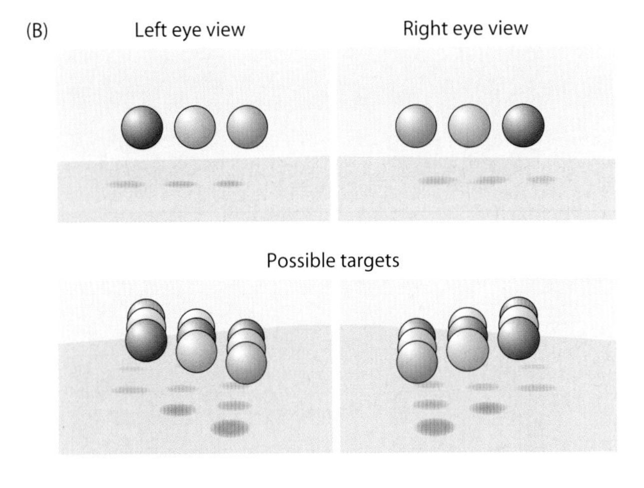

Figure 5.15 The false target problem. (A) Diagram of the "false" targets (green, yellow, and blue) that could be confused with the "real" (purple) targets. It is not obvious how, without additional information, the three purple elements could be distinguished from the other elements, a distinction that would have to precede any computation of disparity. (B) Left and right eye views of three elements in a scene, illustrating the difficulty in principle of identifying the distance of the target elements by analytical or algorithmic means. (A after Marr and Poggio, 1976.)

problem." It is not obvious how the visual system could distinguish "false" from "real" targets pertinent to generating appropriate behavior. This uncertainty is especially obvious in random dot stereograms described in the following section, where the same issue is referred to as the "correspondence problem." How does the visual system compute which dot in the left eye view corresponds to a dot in the right eye view, since the patterns of dots are randomly black or white?

In sum, there are good reasons to question the conclusion that stereoscopic percepts derive from computations of geometrical differences between the positions on the two retinas activated by light arising from the same points in space. The related observations, that (1) stereopsis can be elicited when disparity is only implied; (2) disparity doesn't necessarily give rise to corresponding sensations of stereoscopic depth; and (3) there is no analytical solution to the false target problem, all suggest that whereas stereoscopic percepts make use of retinal disparity, the subjective sense of depth based on binocular images does not represent this feature in the visual brain. This argument is of course much the same as that in earlier chapters pertinent to other visual perceptual qualities. Retinal image features such as luminance, spectral distribution, and projected geometry are likewise necessary but not sufficient causes of brightness, color, and form perceptions.

Random Dot Stereograms

Much work on stereopsis in recent decades has depended on random dot stereograms (RDSs), a fascinating type of stimulus that brings into sharper focus the question of how stereo depth is determined. RDSs were introduced more than 50 years ago by Claus Aschenbrenner (1954; see also Shipley, 1971) and adapted for experimental work in vision by Bela Julesz (1959, 1971, 1995), who showed how random dot stereograms could be readily made and manipulated by a computer. As illustrated in Figure 5.16, RDSs are essentially stereograms of an object camouflaged so completely

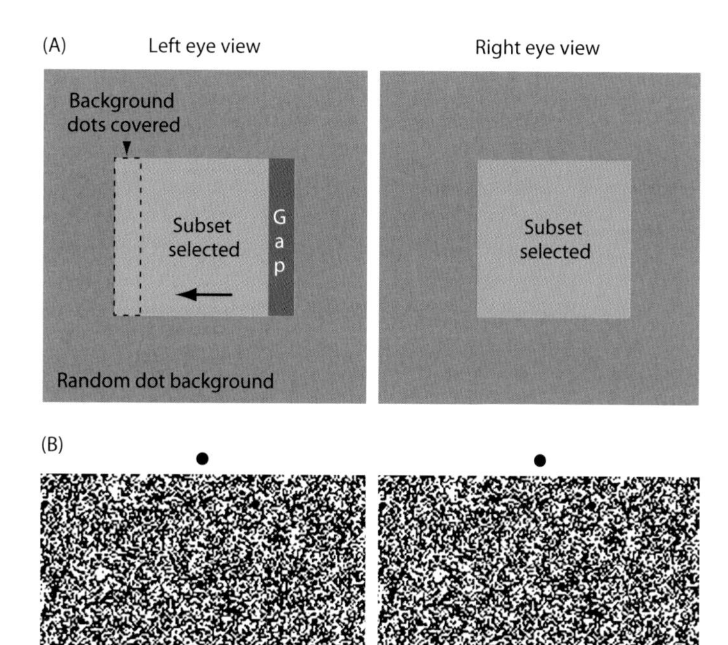

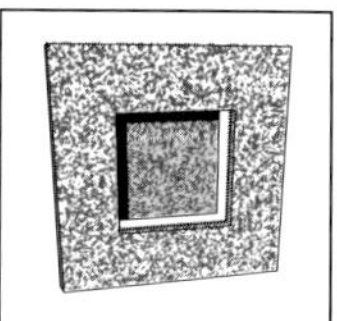

Figure 5.16 The construction of random dot stereograms. (A) A target object (a square, in this example) within a field of randomly generated black and white "dots" (pixels on a computer screen) is shifted a fraction of a degree to the left over the background dots in the left member of the stereo pair (the corresponding set of dots in the view of the right eye remains in place). The gap created to the right of the shifted set is then filled in with additional random dots; note also that another set on the left of the shifted set in the right eye view has been "covered up" in this process. (B) As a result, the shifted square appears in the back of the background array when the left and right random dot arrays are fused in divergent view. The inset below is a diagram of the perception. The stereo effect occurs because the shifted pattern of dots in the two eye views mimics what would be seen if an object in this spatial arrangement were completely camouflaged by the texture of the background.

with respect to its background that the target can be seen only when the two monocular components of the random dot pair are viewed binocularly, thus adding stereoscopic information about differences in the right and left eye views. Such stimuli eliminate all monocular depth cues (e.g., perspective, occlusion, motion parallax), and/or cognitive information (e.g., recognition of an object based on prior experience) that might obscure neural mechanisms specific to binocular depth perception.

If stereoscopic sensations of depth are generated in the same empirical manner as other visual qualities, then the binocular depth elicited by RDSs should be influenced by the frequency of occurrence of any and all differences in the left and right eye views pertinent to the possible arrangement of objects in 3-D space, including but not limited to disparity (see Figure 5.12). Considering whether some puzzling observations about the perception of random dot stereograms are explained in this framework further assesses the merits of conceiving stereopsis empirically.

One such puzzle is that random dot stereograms with the same geometry can nonetheless elicit different sensations of depth. In the conventional stereogram illustrated in Figure 5.17A, the geometrical difference between the two members of the stereo pair has been generated by shifting a selected subset of random dots to the left in the view of the left eye. As already shown in Figure 5.16, the consequence of this manipulation is the perception of a distal plane seen through an aperture. In Figure 5.17B, the same subset of dots has been shifted in exactly the same way, but on a background of dots that are uncorrelated in the left and right eye views. Although the shift of the selected dots (and therefore the disparity) in Figures 5.17A and B is the same, the "hidden object" in Figure 5.17B is now perceived to be located in front of the matrix instead of behind it when the stereograms are fused divergently. Moreover, both the square and the rest of the dots in the array appear to lie behind the plane of the printed page.

Although these different spatial perceptions are difficult to rationalize in terms of computed disparity, an empirical explanation is straightforward (Figure 5.18). When an object is seen through a proximal aperture (i.e., looking through a frame), the distal elements near the proximal object subset are typically seen by both eyes, but with a different disparity than the elements of the object; conversely, the dots near the lateral edges of the frame are seen by only one eye of the other. This typical experience is confounded in

Figure 5.17 Different binocular perceptions of depth elicited by random dot stereo pairs ▶ that have the same disparity. (A) Conventional RDS with a selected subset of dots (shown in gray in the diagrams above the stereo pair) shifted to the left in the left eye view (as in Figure 5.16; the blue surrounds indicate that the background dots are correlated in the left and right eye views). (B) The same selected subset of dots has been shifted to the left by the same amount as in (A), but now over a background of dots that are *uncorrelated* in the left and right eye views (the lack of correlation is indicated by the two different colors of the background). Insets show the perceptual consequences of each of the stereograms, seen in divergent fusion.

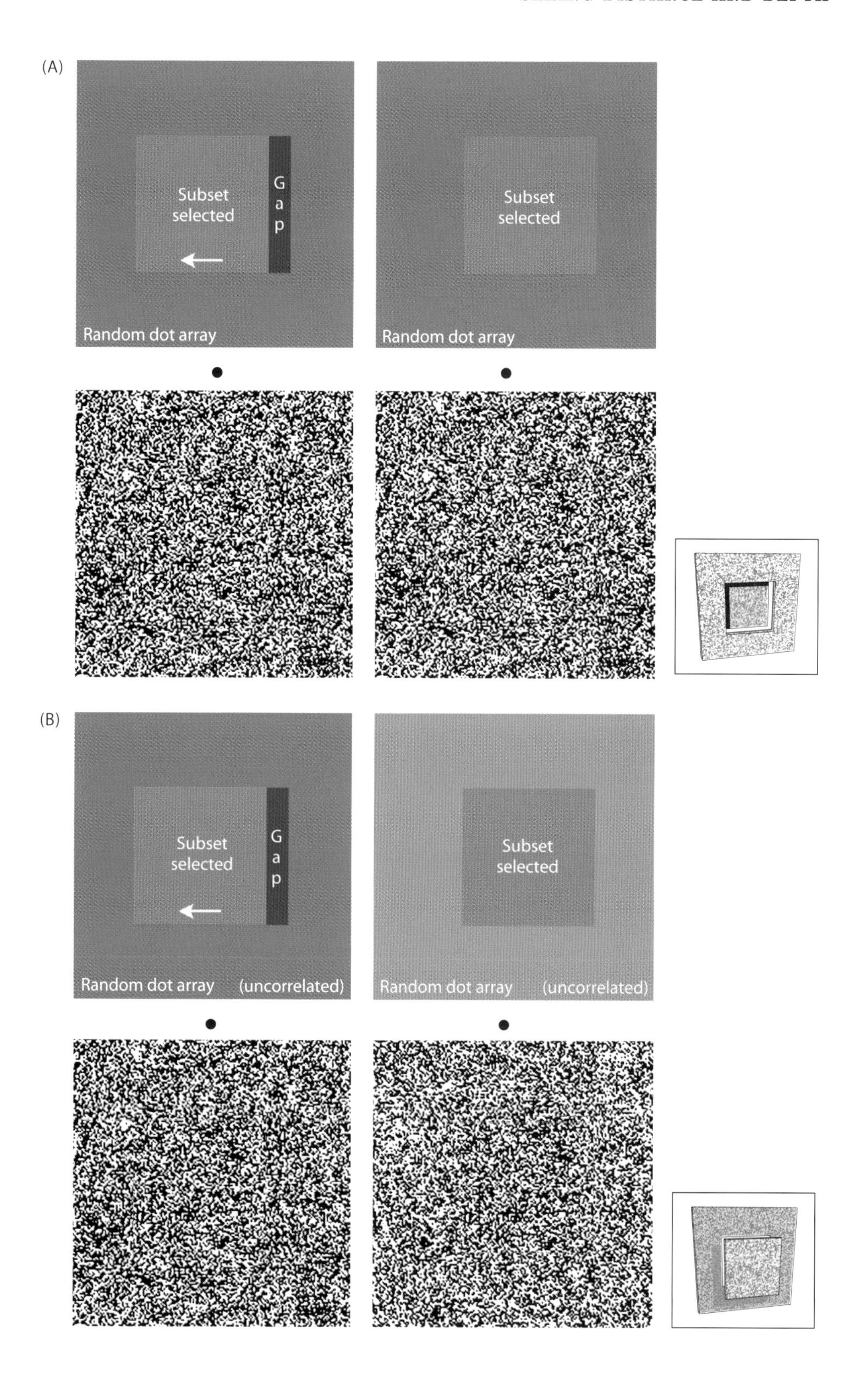
(A)
Subset selected
G a p
Random dot array
Subset selected
Random dot array
(B)
Subset selected
G a p
Random dot array (uncorrelated)
Subset selected
Random dot array (uncorrelated)

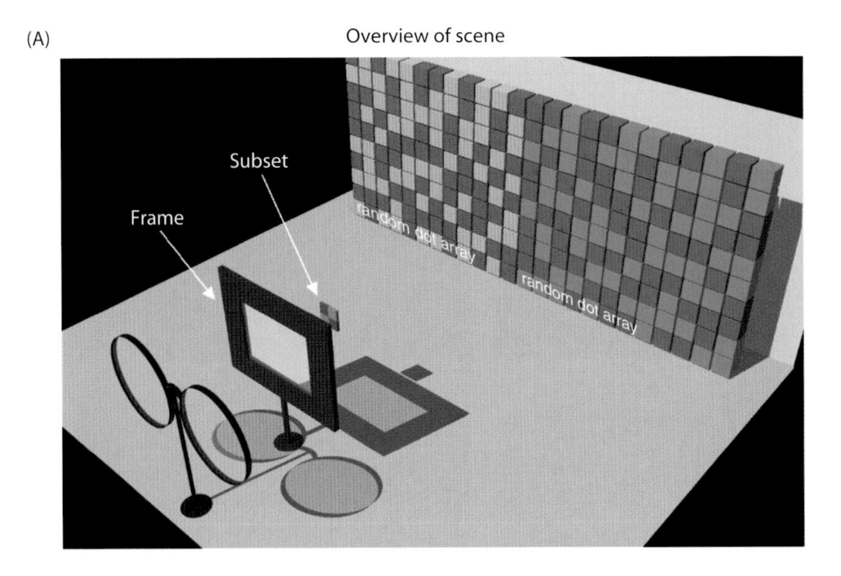

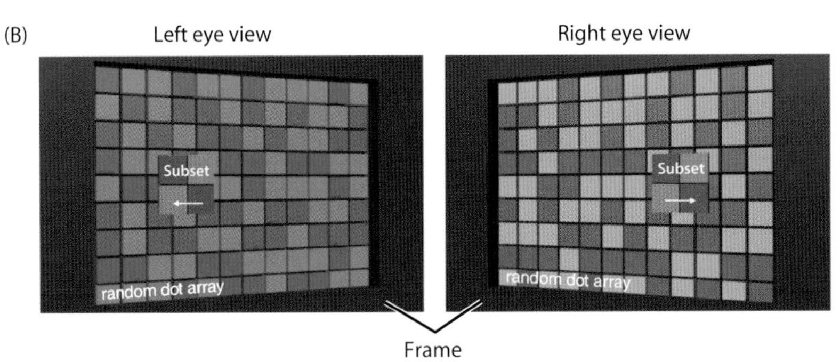

Figure 5.18 Diagram indicating why, in empirical terms, the shifted subset of dots in Figure 5.17B appears to lie in front of the background and why both background and target lie distal to the plane of the black frame. (A) Overview of the scene the observer (whose position is indicated by the spectacles) would typically have witnessed given the left and right eye views shown in Figure 5.17B. (B) Diagram of the left and right eye views of the scene in (A). As in Figure 5.17, the blue and green colors indicate the uncorrelated matrices, and the gray color indicates the selected and shifted subset. The implication is that stereo depth is induced by experience and not the disparity as such.

the stereo pair in Figure 5.17B, in which background dots are uncorrelated. Uncorrelated arrays presented without a correlated subset induce little or no sense of depth (see Julesz, 1971, pp. 157ff.). Thus, the fused stimulus in Figure 5.17B is more consistent with a square in front of the rest of the dots in the matrix, but lying behind the plane of the frame, as illustrated.

An empirical interpretation of stereoscopic depth sensations is further supported by the apparent shape of the "hidden object" in conventional RDSs when the object is seen in front of the background compared with its shape when seen through an aperture in the matrix of surrounding dots (Figure 5.19). When a selected subset of random dots is shifted to the left, as in Figure 5.19A, and the stereo pair is fused divergently, then both shifted subsets of dots and the dots covered by the shifted subset are perceived behind the plane. When, however, the same subset of dots is shifted in the right eye view, as in Figure 5.19B, only the hidden object is perceived in front of the foreground. As a result, when these RDSs are fused, the perceived shape of the hidden object changes from a horizontally oriented rectangle in Figure 5.19A to a square in Figure 5.19B.

Because it is difficult to understand this effect on the basis of the computed disparity of the two arrays, Julesz (1971, p. 261) dismissed this phenomenon as not representing "true stereopsis" because it could not be accounted for in this way. An explanation in terms of accumulated experience, with the binocular stimuli generated by objects in different spatial arrangements, is again straightforward, as indicated in Figure 5.19 (see also Shimojo and Nakayama, 1990a,b; Anderson, 1998). Although Julesz was right in concluding that this phenomenon cannot be rationalized in terms of retinal disparity, he was wrong to dismiss it as not being representative of stereopsis. On the contrary, such effects provide further evidence that binocular perceptions of depth are determined by the association of right and left eye stimuli with successful behaviors in the past, and not simply the geometrical disparity of the two retinal images. As in the earlier discussions of luminance, spectral distribution, and geometry, analysis of a feature such as disparity does not explain what we end up seeing. In stereopsis the empirical information provided by the frequency of occurrence of left and right eye images, that appear together informs behavior without requiring the visual brain to compute disparity. One retinal image provides an empirically informative context for the other, much as in the dice analogy described in Chapter 1.

Advantages of Stereopsis on an Empirical Basis

Generating binocular depth sensations empirically has several advantages. First, this strategy allows any and all differences in binocular images to be included in the generation of stereo depth perceptions, rather than depending on a single parameter such as geometrical disparity. Second, since the activation of such circuitry is reflexive, no on-line computations or algorithms need to be postulated. A third advantage is the provision of a seamless relationship between the sense of depth generated by monocular information and the augmentation of this sense by binocular information. Neurons responsive to geometrical disparity are of course involved in generating stereoscopic perception on an empirical basis, but only insofar as they contribute—along with all the other neuronal responses that convey information about differences in the binocular views at any moment—to behaviors that have been useful when interacting with particular arrangements of objects in space. If monocular and binocular depth sensations rest

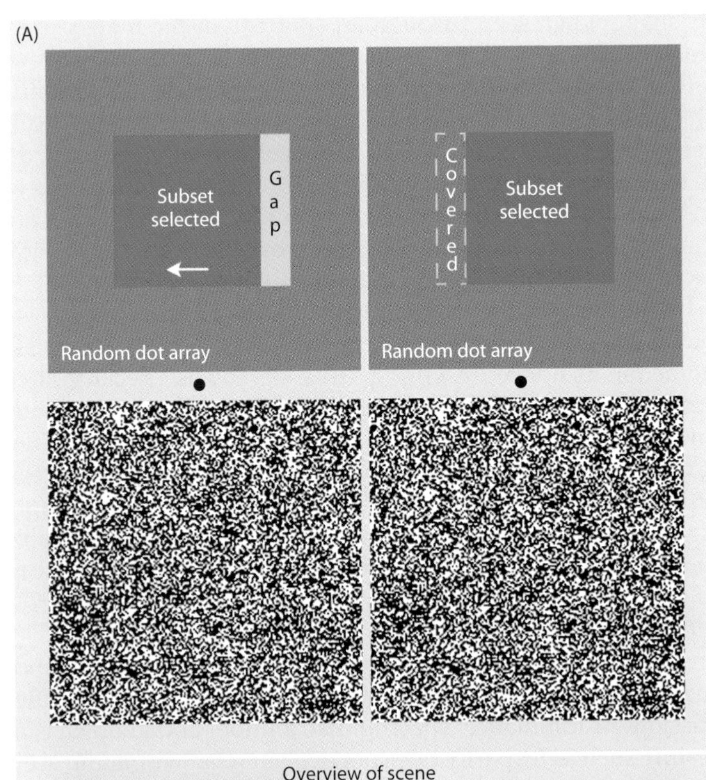

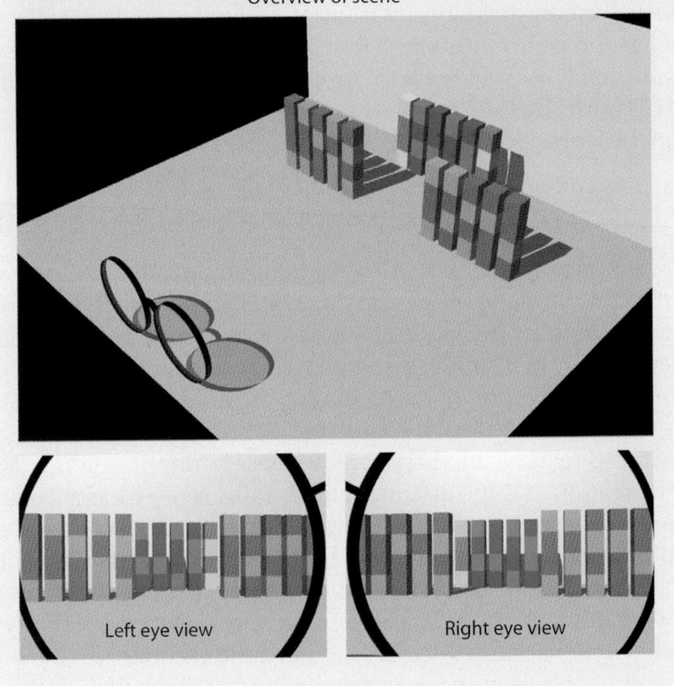

Figure 5.19 Different shape of the "hidden object" in a random dot stereogram as a function of whether the selected subset of dots is perceived in front of or behind the surrounding matrix. (A) When, as in Figures 5.16 and 5.17, the selected subset is shifted to the left in the left member of the pair and viewed in divergent fusion, the object perceived is a rectangle seen through a frame formed by the rest of the dots in the matrix. (B) When, however, the same subset is shifted to the left in the right member of the pair, subjects perceive the hidden object to be a smaller square in front of the background. The panels below the RDSs indicate the empirical basis for this difference. When the selected subset is shifted to the left in the left eye view, as in (A), the binocular scene mimics what the two eyes normally experience when a distal plane is viewed through a proximal aperture. Conversely, when the subset is shifted to the left in the right member of the stereo pair, the right eye is stimulated by a set of dots to the right of the shifted subset that does not occur in the corresponding position in the left eye view, and the opposite is true for the view of the left eye. As a result, the stimulus mimics what the two eyes see when an object lies closer to the observer than the background plane. In either case, the dots witnessed by one eye and not the other (indicated in green and blue) are seen in the background plane, thus leading to the perception of the hidden object as a rectangle in (A) but a square in (B).

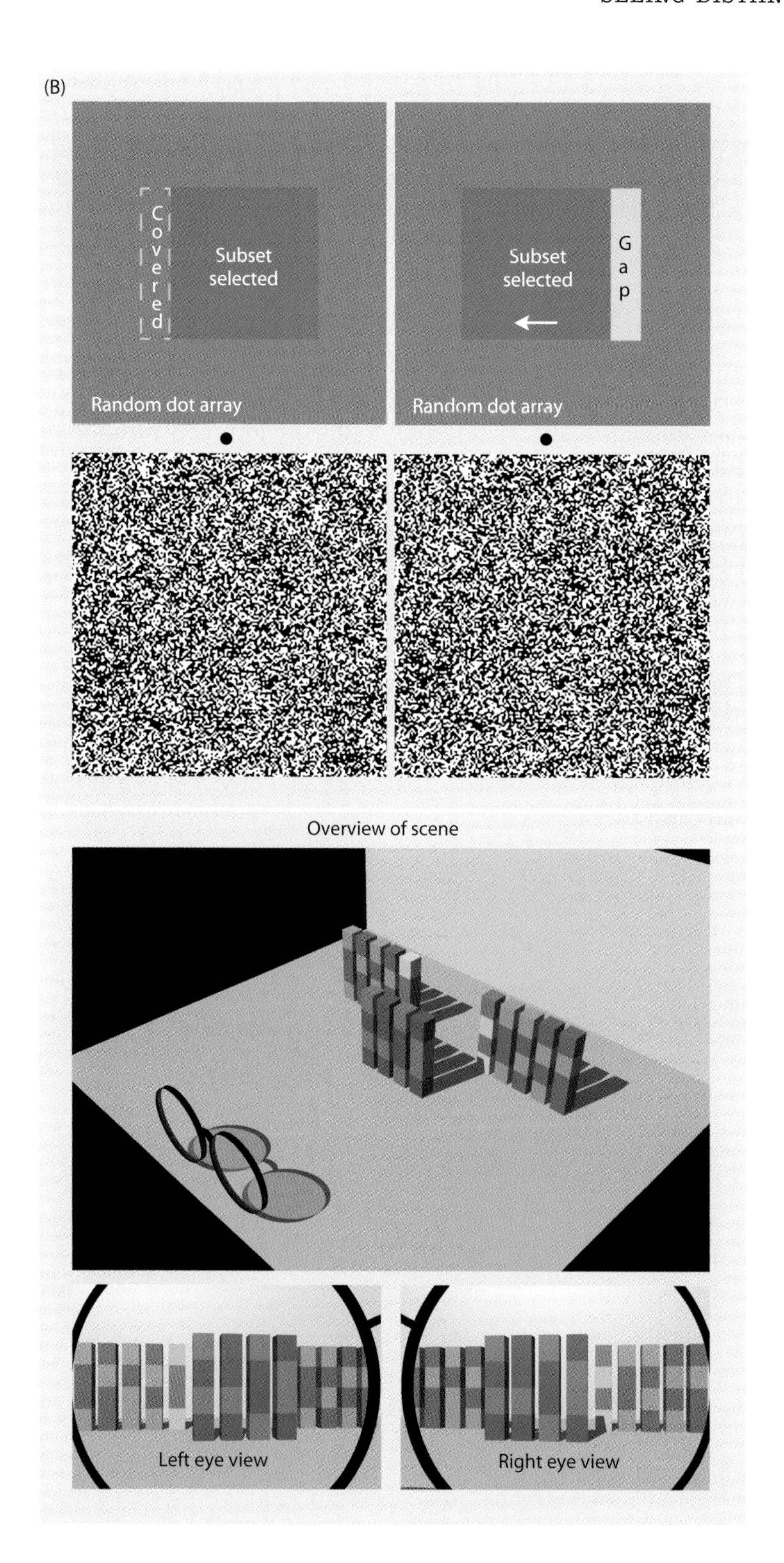
(B)
Covered
Subset selected
Random dot array
Subset selected
Gap
Random dot array
Overview of scene
Left eye view
Right eye view

on the same neural foundation, the emergence of stereopsis (or its disappearance) in a species requires no fundamental neural change. Since the circuitry that has arisen on an empirical basis is already in place, all that is required to achieve stereopsis is that the relevant neurons be increasingly influenced by binocular cues in addition to the long list of cues that influence the monocular perception of distance and depth.

Conclusion

The ability of a wholly empirical framework to explain many aspects of distance and depth perception provides further evidence that what we see must be explained empirically. In addition to the obvious contribution of learned information to the monocular perception of distance and depth, binocular phenomena including cyclopean fusion, rivalry, and stereoscopy can all be accounted for in these terms. Perceptions of distance and depth, whether monocular or binocular, are arguably determined by the frequency occurrence of stimuli in which occlusion, shading, chromatic differences, size, texture, differences in the left and right eye images, and other factors (including non-visual influences) have generated links between retinal light patterns and behaviors that contend successfully with the inverse problem. As in other perceptual domains, these links are presumably stored in visual circuitry that has accrued over evolutionary and individual time, leading to reflexive perceptions and behaviors that work in a physical world that images as such can't reveal.

6

Seeing Motion

Introduction

The perceptions considered in previous chapters all entail stationary stimuli falling on the retina of a nominally stationary observer. However, the relationships among objects, retinal images, and the observer are rarely, if ever, static. Any theory of vision must therefore deal with the perceptions of motion generated by images that change rapidly over time. As in the perception of luminance, spectral differences, geometrical form, distance, and depth, there are many puzzling observations in motion perception. Preeminent among these are the anomalous way the physical speed of an object is seen and the dramatic changes in the apparent direction of a moving object when the context is altered. The goal of this chapter is thus to consider whether perceived motion can be rationalized in empirical terms. The outcome is that the way humans see motion can also be understood in terms of visual circuitry that reflexively links image sequences to perceptions and behaviors that have contended successfully with physical speeds and directions that can't be deciphered by image analysis.

Definitions

In Newtonian physics, motion refers to the absolute speed and direction of objects in a frame of reference. The physical limits range from total immobility to the speed of light in all possible directions. In psychophysics, however, motion is defined subjectively by our ability to perceive image sequences projected on the retina as moving. The range of perceived object speeds is of course much more restricted than actual physical speeds: We don't see the hour hand of a clock moving or a bullet in flight, even though these objects are translating at physical rates that are easily measured. (Translation refers to movement in which points on an object have the same velocity with respect to a frame of reference.) The range of projected speeds that humans have evolved to see as motion is from roughly 0.1° per second to 150° to 200° per second. Below the lower end of this range, objects appear to be standing still; as speeds approach the upper end of the range, they generate a sense of blur, and then become invisible when they move faster. Likewise, the directions of movement that we see do not simply correspond to the movement of objects in Euclidian space. The reason is that neither the speed nor the direction of moving objects in 3-D space is specified by a sequence of retinal projections.

The Inherently Uncertain Meaning of Motion Stimuli

This difference between motions in the physical world and the motions we see is of course a major problem in understanding vision. Observers must respond accurately to the real-world speeds and directions of objects, but they must do so on the basis of the quite different speeds and directions projected onto the retina. The inability of projected image sequences to specify the physical motion of objects defines yet another aspect of the inverse optics problem: When the light from moving objects is projected onto a two-dimensional surface, size, distance, orientation, and speed of are conflated (Figure 6.1). As a result, the possible sources of a retinal image sequence are not indicated by the changes in 3-D space that define motion in physical terms, as has long been noted (Stumpf, 1911; Wallach, 1935; Brünswik, 1952, 1956/1997). How the visual system nonetheless produces motion percepts that lead to generally successful behavior is not known.

If the motions seen in response to stimulus sequences are explainable in the same framework that rationalizes many aspects of lightness/brightness, color, form, distance, and depth, then the perceptions elicited by image sequences should accord with—and be predicted by—accumulated human experience responding to stimuli generated by moving objects. The frequency of occurrence of any image sequence within the full range of retinal image sequences generated by objects in the world should determine the speeds and directions actually seen.

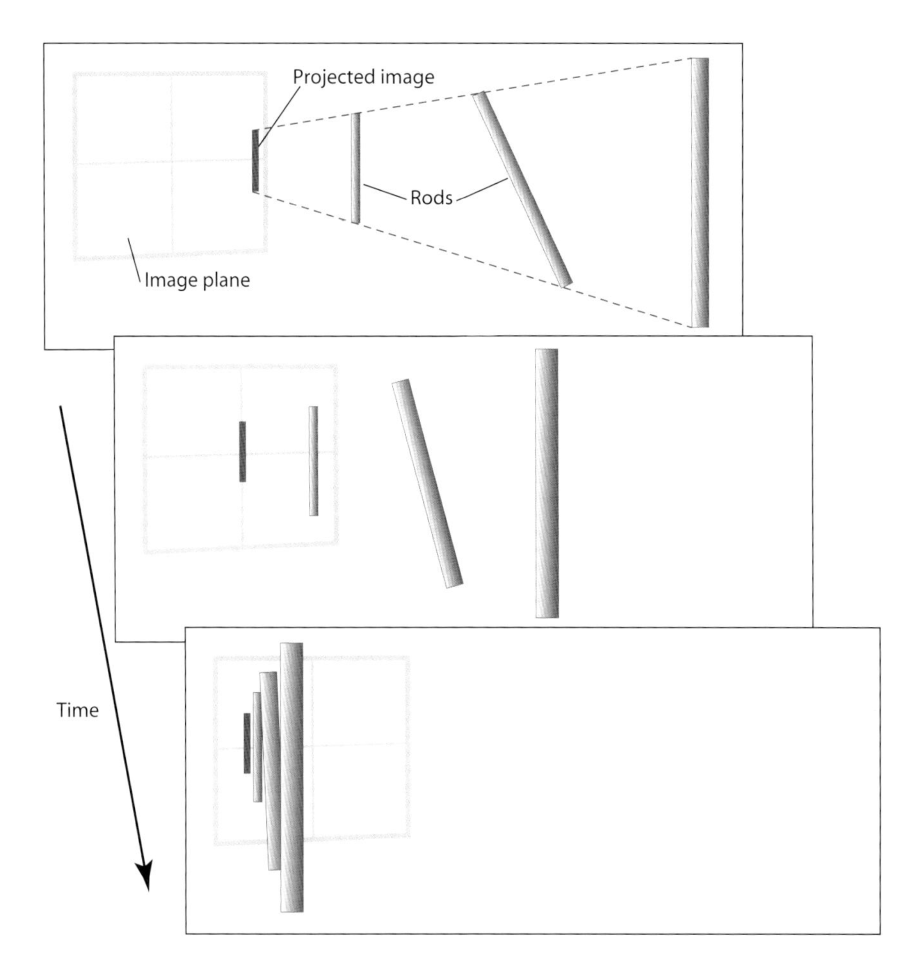

Figure 6.1 The relationship between moving objects in the world and the corresponding sequence of images projected onto a two-dimensional surface. Three objects of different sizes, at different distances and in different orientations moving at different speeds in 3-D space all project the same image sequence. Thus, image sequences cannot specify the speeds and directions of the physical motion of objects in the world.

Determining the Relationship between Image Sequences and Moving Objects

A major obstacle in testing the merits of an empirical explanation of perceived motion is determining the frequency of occurrence of the speeds and directions of moving objects in the world, which in turn determine the frequency of speeds and directions in projected images. As before, this information about the frequency of projected speeds and distances will have driven the association of image sequences with the successful behaviors. Although data relating projected and real-world geometry could readily be obtained for static scenes using laser range scanning (see Chapters 4 and 5), there is at present no technical way of collecting information about the direction, speed, and 3-D position of moving objects. In principle, getting such information is possible by range-scanning; in practice, however, such devices take many seconds to acquire each scanned image, much more time than would be needed to obtain information about the speeds and directions of moving objects in the human visual environment.

Nonetheless, human experience with the frequency of projections generated by object motion can be approximated in at least two different ways. One approach is computing the frequency distribution of all the physical displacements that could, in principle, have generated a simple moving stimulus, assuming that all physical motions are equally likely to occur. Although the assumption is patently false—the motion of natural objects is affected by gravity, the presence of a terrestrial surface, friction, and a host of other factors that bias the speeds and directions in image sequences—this approach provides at least a starting point in understanding the frequency of occurrence of various real-world motion stimuli. All the ways a given image sequence could have been generated by a moving object can be calculated, and the speeds and directions humans would be expected to perceive on empirical grounds predicted, albeit in a very limited way (Yang et al., 2001).

A much better approach is to generate moving objects in a simulated 3-D environment and tally up the frequency of occurrence of all the speeds and directions their projections produce on an image plane (Figure 6.2A). Although still grossly simplified, this surrogate for the transformation of

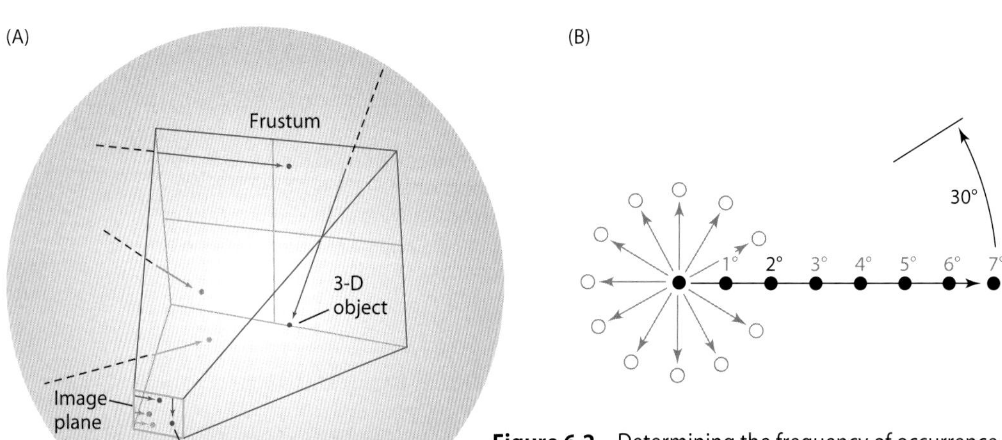

Figure 6.2 Determining the frequency of occurrence of image sequences generated by objects moving in a virtual environment. (A) A simulated space (red outline) is embedded in a larger spherical environment to approximate a physical space relevant to projections onto the retina; the projections of objects that enter this space on an image plane (blue outline) can then be tallied. (B) A template applied repeatedly to all regions of the image plane determines the frequency of occurrence of the speeds and directions of projected sequences moving over a distance of 2° in this example. By accumulating data over different distances, experience with projected speeds and directions can be tallied. (After Wojtach et al., 2008.)

movements in 3-D space into 2-D projections can be used to estimate how often humans have experienced different projected speeds and directions. By sampling the image plane over a range of spatial and temporal intervals (Figure 6.2B), these data can be used to predict the perceived speeds and directions of motion that subjects would be expected to see in complementary psychophysical studies, in this way testing the hypothesis that motion percepts are generated empirically. The following sections examine how some otherwise mysterious aspects of motion perception can be explained in terms of empirical information collected in this way.

Understanding Perceived Speed in Empirical Terms

The phenomenon pertinent to speed of motion that has most intrigued investigators over the decades is the so-called flash-lag effect (Figure 6.3) (see Eagleman and Sejnowski, 2002 for a review of this effect and its several variants). When a flash of light is presented in physical alignment with a continuously moving stimulus, the flash is perceived to lag behind the moving stimulus; moreover, the greater the speed of the stimulus, the greater the lag. Despite numerous attempts to explain this effect (see below), there is no generally accepted account. Considered in the present framework, the flash-lag effect should be another signature of an empirical strategy of visual processing as it applies to motion sequences in general and to the perception of object speed in particular.

This supposition can be tested by asking whether the amount of lag seen by subjects is accurately predicted by the empirical relationship between the frequency of occurrence of image sequences arising from typical moving sources in 3-D space. In this approach, the relevant psychophysical

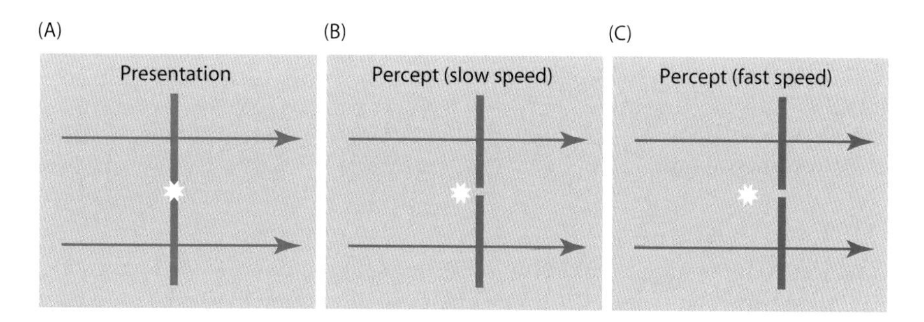

Figure 6.3 The flash-lag effect. When a flash of light (the asterisk) is presented in physical alignment with a moving object (the red bar in [A]), the flash is seen lagging behind the position of the object (B). The apparent lag increases as the speed of the moving object increases (C). The amount of lag as a function of object speed can be measured by asking subjects to align the flash with their perception of the moving stimulus at various object speeds (see Figure 6.4). (After Wojtach et al., 2008.)

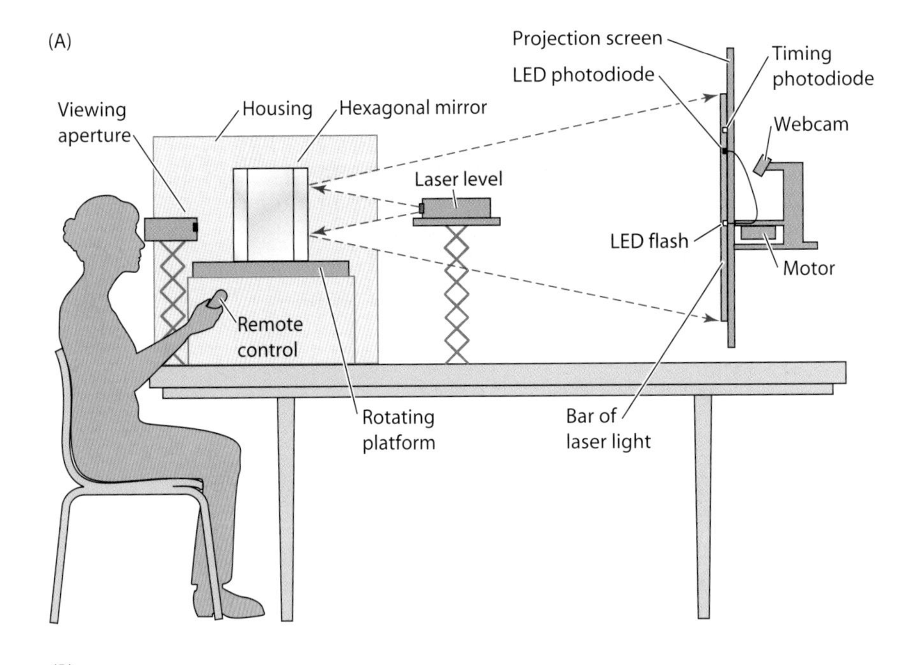

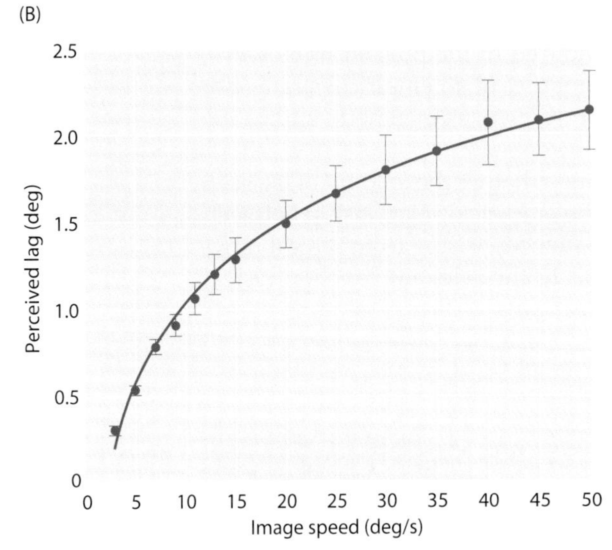

Figure 6.4 Determining the psychophysical function that describes the flash-lag effect. (A) Observers view a projection screen and reposition a flash generated by an LED until it is seen as being aligned with the center of the moving bar (see Figure 6.3). By measuring the discrepancy between the position of the perceived alignment and the physical position of the flash, the amount of lag can be determined. (B) The psychophysical function that describes the flash-lag effect over a range of image speeds. The curve is a logarithmic fit to the lag reported by observers in the testing paradigm shown in (A). Bars show standard errors. (After Wojtach et al., 2008.)

function is determined by having observers align a flash with the moving bar, repeating the test at different object speeds over the full range that elicits a measurable flash-lag effect (up to approximately 50° per second) (Figure 6.4A). The resulting perceptual function is shown in Figure 6.4B.

To examine whether the function in Figure 6.4B is consistent with an empirical explanation of perceived speed, one needs to know the frequency of occurrence of image speeds generated by objects in 3-D space moving over the range of speeds that elicit motion percepts. As indicated in Figure 6.2, this information can be gleaned by repeatedly sampling the projected speeds of generated 3-D sources moving in a simulated world (Figure 6.5A). These data can be used in turn to predict perceived speeds, since the distribution of projected image speeds in cumulative form (Figure 6.5B) defines, to a first approximation, human experience with projected stimulus speeds over the range used in psychophysical testing (see Figure 6.4). This is the same general approach that was used in studies of lightness/brightness, color, form, and distance. If the flash-lag effect signifies an empirical strategy of visual processing, then the lag reported by observers

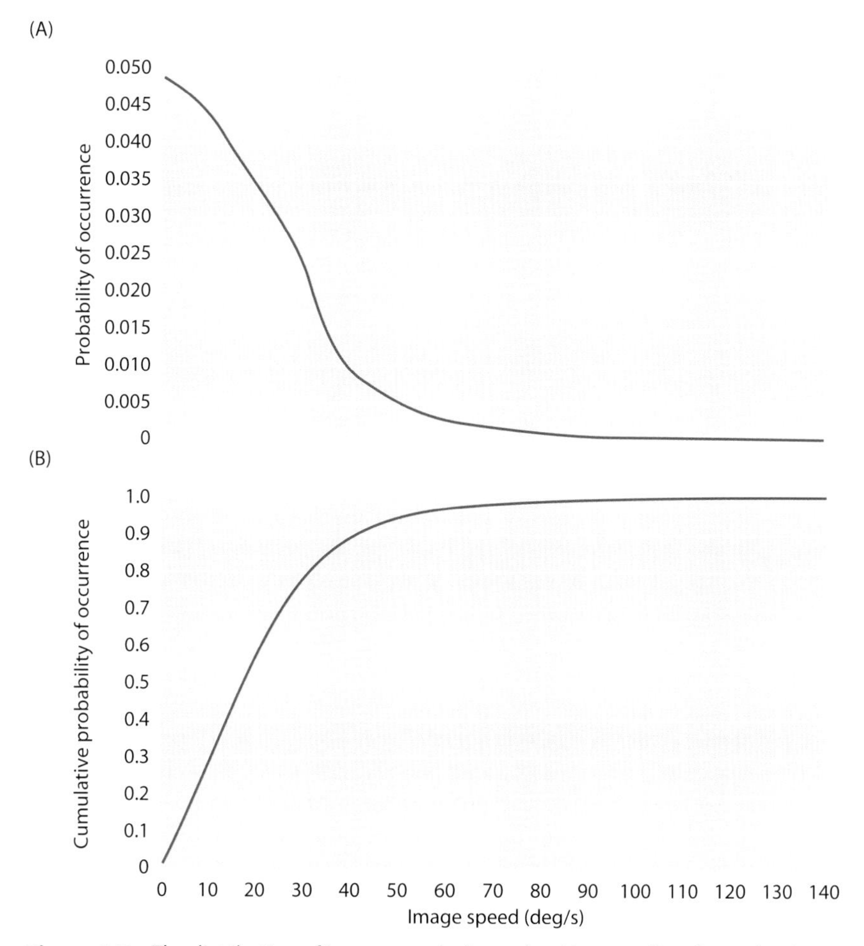

Figure 6.5 The distribution of image speeds determined by sampling the projections arising from objects moving in the virtual environment illustrated in Figure 6.2. (A) The probability distribution of the speeds on the image plane generated by the projection of the moving objects. (B) By ordering how often objects moving in 3-D space project image sequences with different speeds, the accumulated experience of humans with speeds on the retina can be approximated. If motion percepts are determined empirically, the amount of perceived lag shown in Figure 6.4B should be predicted by the cumulative ranks of the image speeds. (After Wojtach et al., 2008.)

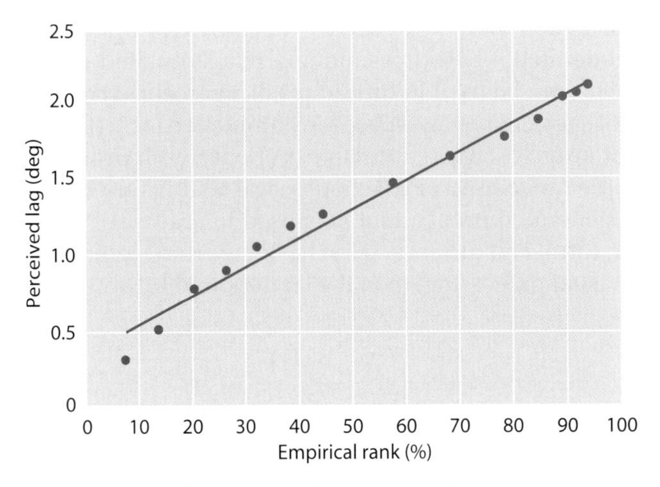

Figure 6.6 Prediction of the flash-lag effect from empirical information about the frequency of occurrence of image speeds generated by objects moving in 3-D space. Plotting the perceived lag reported by observers (red dots) over the range of stimulus speeds that elicits the flash-lag effect (~3°/s–50°/s; see Figure 6.4B) against the percentile rank of speeds in Figure 6.5B shows a good correlation (blue line) between the psychophysical data and the empirical predictions (~97% of the observed data are predicted from the empirical rankings). (After Wojtach et al., 2008.)

for different stimulus speeds in Figure 6.4B should be given by the relative ranks of different image speeds in the cumulative probability distribution.

As in the organization of perceptual experience described in earlier chapters, the higher a stimulus sequence ranks on the empirical scale in Figure 6.5B, the faster the perceived motion. The reason is that this way of seeing object speeds will have typically been successful in interactions with real-world objects, even though perceived motion does not accord with the physical speeds of the generative sources. This operational link between the frequency of occurrence of speeds on the retina and behaviors in response to physical objects whose speeds cannot be known is, in an empirical framework, what underlies the speeds seen.

With respect to the flash-lag effect, since a stationary flash has an image speed of 0° per second—and therefore an empirical rank of 0—the rank of any moving stimulus should appear ahead of a flash, as it does. In other words, the flash will lag behind the projection of a moving object on the retina because its empirical rank is lower. Since increases in image speed correspond to higher empirical ranks, the amount of lag should increase accordingly, but in the nonlinear manner indicated by the psychophysical results shown in Figure 6.4B. Figure 6.6 shows that the psychometric function is indeed predicted by the empirical ranking of image speeds in the accumulated experience of human observers. As discussed later in the chapter, the perceived lag has nothing to do with improving the alignment of perception with the actual position of objects in 3-D space or otherwise representing physical reality more "correctly."

A concern about this approach is of course the adequacy of any simulated environment in determining the frequency of occurrence of image sequences arising from physical sources. It is obvious that a variety of real-world factors can influence the speed of objects, and these are not included in the simulation (see above). However, the biases that would certainly be apparent if real-world object speeds could be measured are less important in the determination of image speeds than one would imagine. As discussed in Chapter 5 in the context of static images, the reason is that effect of perspective transformation tends to trump any other influences (see Figure 6.1), and the simulated environment captures perspective quite precisely. Changing the distribution of objects speeds in a 3-D environment has surprisingly little influence on the distribution of projected speeds in Figure 6.5 (see Wojtach et al., 2008). Since perspective transformation generates image speeds that are nearly always less than the speeds of 3-D objects, the effect of different object speeds in the production of image speeds is largely nullified, with projected speeds being strongly biased toward slower values. This bias is apparent in statistical analyses of image speeds in movies (Dong and Atick, 1995) or simply in a priori calculations (Ullman, 1979; Yuille and Ullman, 1990). Perspective transformation is thus the primary determinant of the distribution of image speeds that humans actually experience.

In summary, the inverse optics problem effectively rules out seeing speed by any spatiotemporal analysis of images as such. The correspondence of the observed and predicted results in Figure 6.6 argues that the speeds we see reflect a wholly empirical strategy of perception in which the relevant perceptual space has been shaped by the operational links between the frequency of occurrence of image speeds and successful behavior.

A More General Test of Perceived Speed

The ability to rationalize the flash-lag effect in an empirical framework is a limited test of the broad claim that speed is perceived according to accumulated experience with image sequences. It thus makes sense to examine whether the frequency of occurrence of the projected transformations of *any* moving object predicts the perceived speed elicited by simple motion stimuli. This further analysis is not as difficult an undertaking as it may seem. Since speed is the only issue and all objects comprise collections of points, the virtual environment in Figure 6.2A and the resulting data in Figure 6.5 can be used to predict the perceived speed of any point in an image sequence over the range of speeds on the retina that humans see as motion.

For example, consider the image sequence generated by a point-like object that traverses a distance of 2° on the image plane. By repeatedly sampling the image plane in all directions with a 2° template (see Figure 6.2B), the frequency of occurrence of object speeds in images sequences traversing this distance on the retina can be determined. As mentioned already, perspective transformation requires that image speeds will always be equal to or slower than the 3-D speeds of the objects that produce them. Thus, the

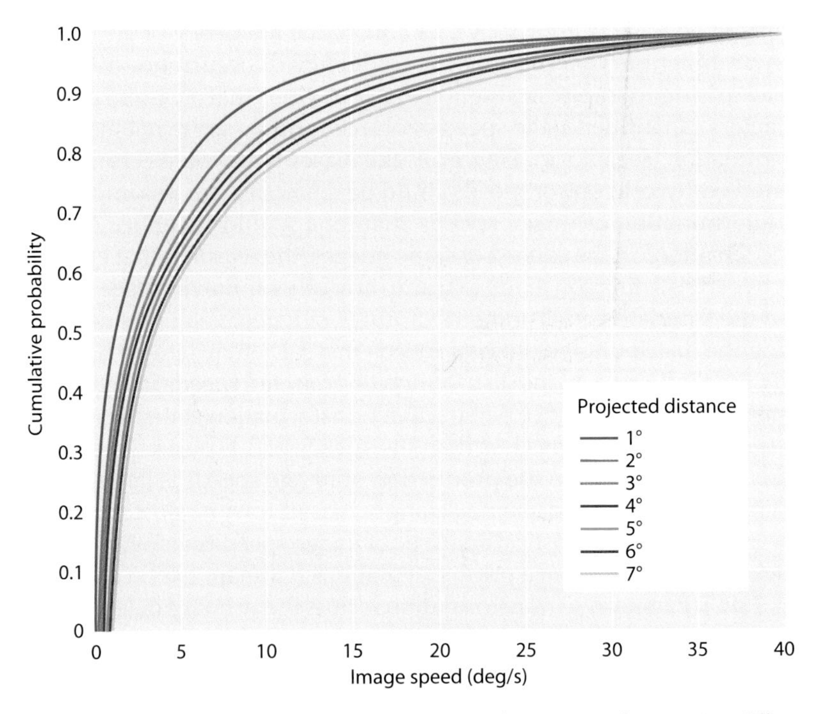

Figure 6.7 Cumulative probability distributions of image speeds traversing different distances generated by objects in a simulated 3-D environment. Each cumulative probability curve orders how often moving objects in 3-D space produce image sequences with different speeds traversing different projected distances. These functions provide an empirical basis for predicting the perceived speeds observed in psychophysical testing. (After Wojtach et al., 2009.)

projected distances traversed on an image plane are less than the distances traveled in 3-D space, and the probability distributions of projected speeds are strongly biased toward slower speeds, as indicated in Figure 6.5A.

For a more general analysis, however, one needs to know the cumulative probabilities of speeds over a full range of different distances traversed on the image plane. Data showing the frequency of occurrence of different image speeds traversing any given distance on the image plane can then be normalized and re-plotted as cumulative probability distributions to get some idea of the overall human experience with projected speeds (Figure 6.7).

The cumulative probability functions in Figure 6.7 indicate how often the possible physical sources of moving objects generated projections equal in speed to or slower than the image sequence in question moving over the corresponding distance. If motion percepts are indeed determined by accumulated experience, then the speed perceived in any circumstance should be predicted by the empirical rank of the projected speed of image sequences in accumulated experience.

The functions in Figure 6.7 imply that when two objects generate image sequences that translate over different projected distances at different speeds but have the same empirical rank, their perceived speeds should be the same. Conversely, if two moving objects generate the same image speed over different projected distances but have different empirical ranks, then their perceived speeds should be different. To examine these predictions, subjects can be asked to compare the speed of a reference stimulus with the speed of a test stimulus, the task being to indicate when the apparent speed of the test stimulus matches the reference stimulus in Figure 6.8A. The predictions made on this basis are shown in Figure 6.8B.

The accuracy of these predicted perceptions can be evaluated for reference and test stimuli traversing the retina at different speeds but having the same empirical rank. For example, a reference stimulus traversing 2°

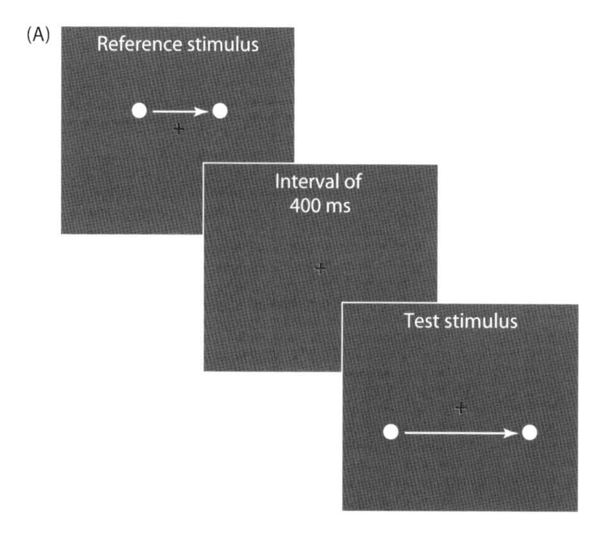

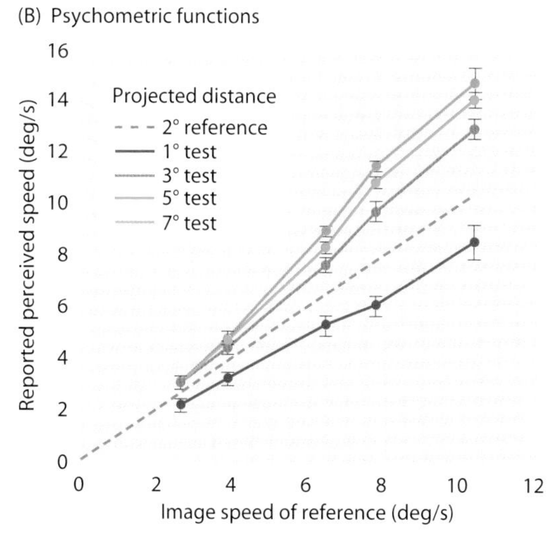

Figure 6.8 The perceived speeds of objects moving over different projected distances in psychophysical testing. (A) Subjects compared the perceived speeds of reference and test stimuli, indicating when the variable test image speed appeared to be equal to the reference speed. (B) The perceived speeds of the test stimuli relative to a 2° reference stimulus (dashed blue line), plotted as a function of the physical image speed of the reference stimulus. Bars show standard errors. (After Wojtach et al., 2010.)

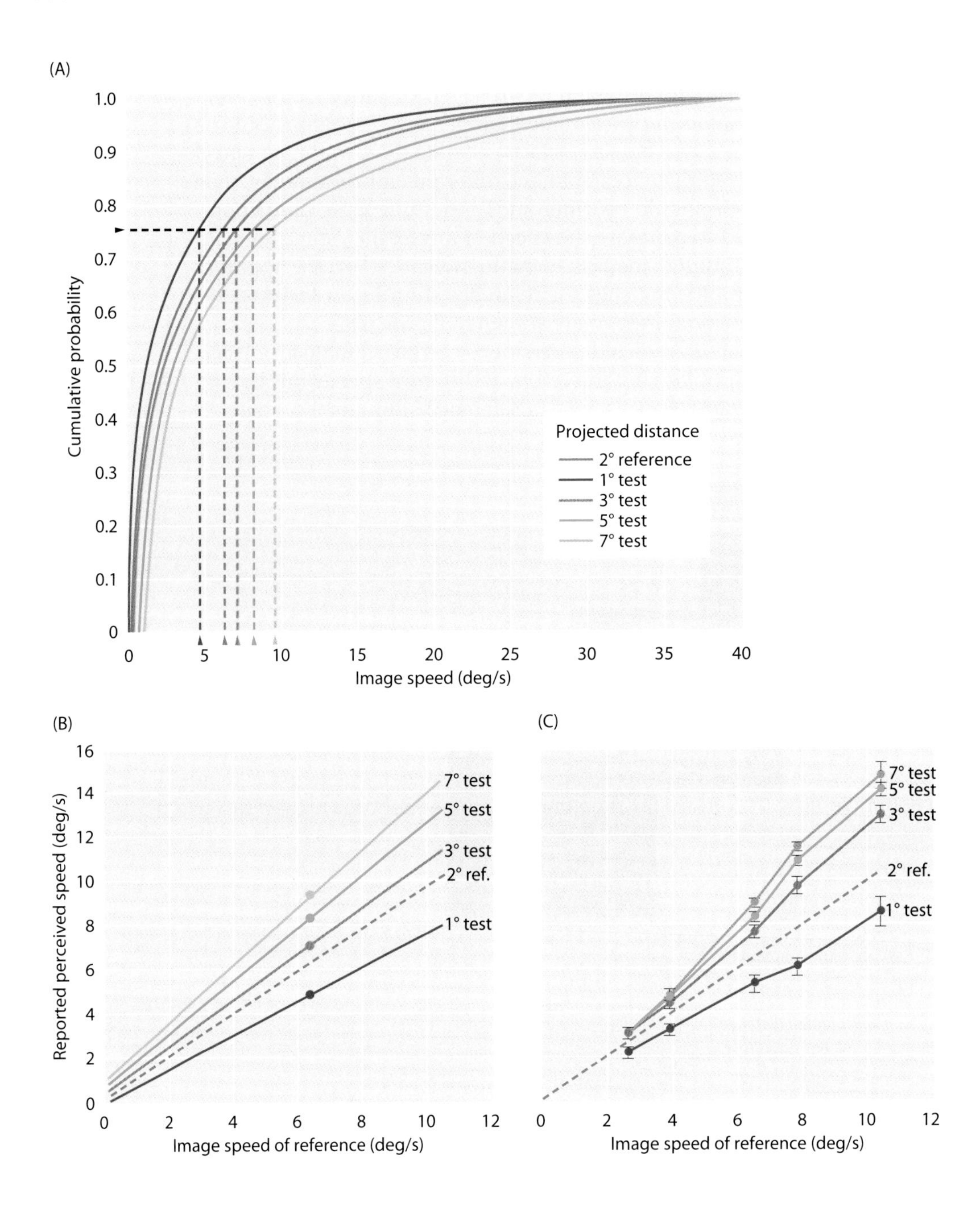

on the image plane at a speed of 6.5° per second can be compared with test stimuli traversing distances 1°, 3°, 5°, or 7° (Figure 6.9A; the curves are those in Figure 6.7). When the cumulative probabilities of the possible sources of test stimuli have the same empirical rank as the reference stimulus, the frequency of occurrence of the image sequences that traverse these different distances are, by definition, the same. If motion perception is empirically

◀ **Figure 6.9** Predicting the perceived speeds of image sequences traversing different distances on the image plane at different speeds, but having the same empirical rank. (A) As an example, a 2° reference stimulus with an image speed of 6.5°/s (dark blue arrowhead on the abscissa) has a particular percentile rank determined by its frequency of occurrence (black arrowhead on the ordinate). If motion perception is empirically based, then test stimuli with the same rank should generate the same perceived speed, despite their different actual speeds on the image plane (indicated by the other colored arrowheads along the abscissa). (B) The data in (A) are re-plotted to indicate the predicted motion percepts for the various test stimuli compared with the reference stimulus (dashed blue line). (C) The observed psychophysical functions shown again here from Figure 6.8B closely match the empirical predictions in (B). Bars show standard errors. (After Wojtach et al., 2009.)

determined, then these cumulative probability functions should predict the relative speeds seen by the observers. As shown in Figures 6.9B and C, the predictions made on this basis are in reasonable agreement with the observed results.

The cumulative probability distributions in Figure 6.7 also predict the motion percepts elicited by stimuli having the same speed on the image plane but traversing different projected distances. Consider, for instance, the different empirical ranks generated by an image speed of 6.5° per second traversing distances of 1°, 2° 3°, 5°, and 7° on the retina (Figure 6.10A). If relative empirical rank determines the perceived motion, then stimuli moving at the same speed over different projected distances should appear to be moving at different speeds. Thus, when a projected image is moving at 6.5° per second, the motion seen should appear slower when traversing 2° than when traversing 1°; conversely, the same image speed should appear faster when traversing 2° than when traversing larger projected distances (e.g., 3°, 5°, and 7°). As shown in Figure 6.10B, these predictions are also borne out.

These several observations support the conclusion that all perceptions of object speed are generated in the same empirical manner.

Understanding Perceived Direction in Empirical Terms

Motion is defined by both speed and direction, and the question remains whether the perceived direction of a moving stimulus can also be accounted for in empirical terms. Another class of motion percepts has been especially useful in exploring this issue, namely the changes in apparent direction that occur when moving objects are seen, as they often are, behind objects in the foreground that occlude a full view. The phenomena apparent in these circumstances are called "aperture effects" and have been studied for decades ("aperture" refers to a frame of some sort through which a scene is viewed). When, for example, a moving line translating horizontally from left to right at a constant speed is viewed through a circular opening that occludes its ends, the perceived direction of movement is downward and

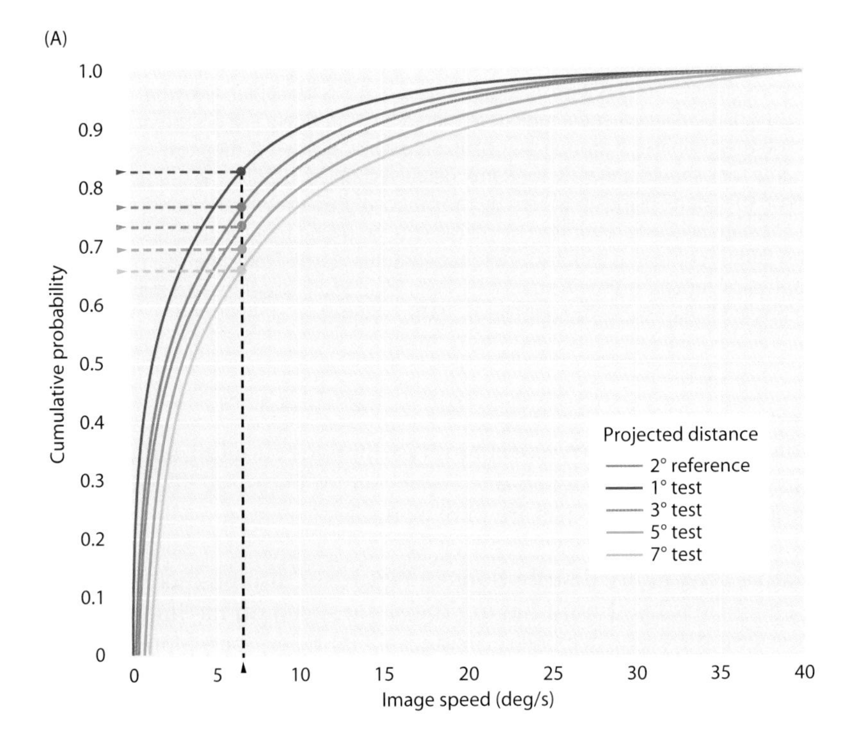

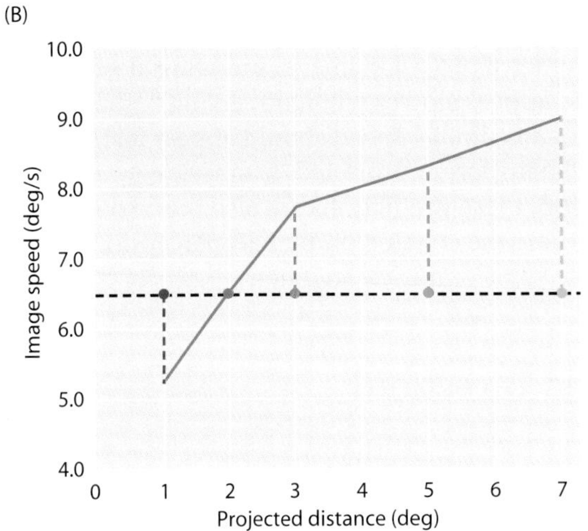

Figure 6.10 Predicting the psychophysical results for image sequences traversing different distances on the image plane at the same speed, but having different empirical ranks. (A) Test stimuli traversing the image plane at an image speed of 6.5°/s (black arrowhead on the abscissa) have ranks that range from the 65th to the 82d percentile (colored arrowheads on the ordinate). If motion percepts are generated empirically, then the same image speed should be perceived to be slower when traversing distances of 3°, 5°, or 7° in comparison with a 2° reference, but faster when traversing a test distance of 1°. (B) The blue function indicates the projected speeds at which test stimuli traversing different projected distances (1°, 3°, 5°, and 7°) appear the same to observers as a point traversing 2° at 6.5°/s (the data are re-plotted from Figure 6.8B). Test stimuli presented at 6.5°/s (dashed horizontal line) traversing distances of 3°, 5°, or 7° (dashed vertical lines) are seen as moving more slowly than a 2° reference stimulus traveling at the same speed, whereas test stimuli traversing 1° at 6.5°/s are seen as moving faster. (After Wojtach et al., 2009.)

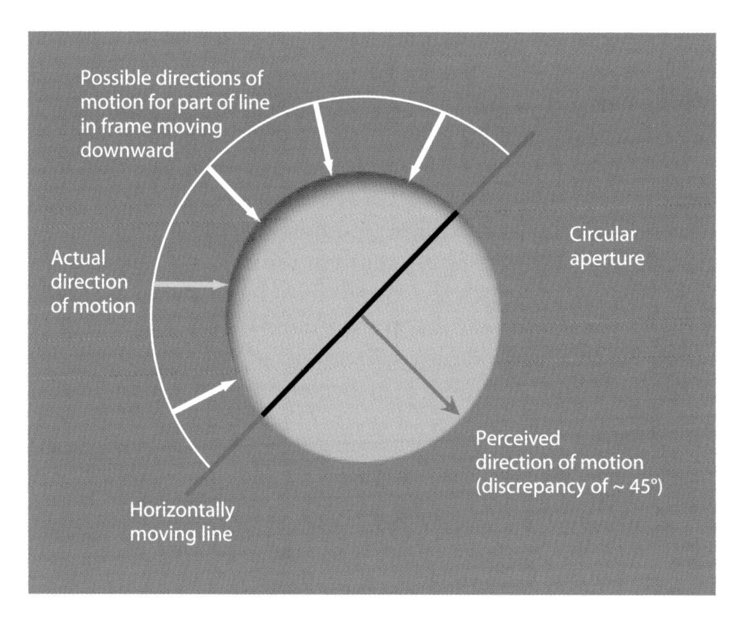

Figure 6.11 The effects of an occluding aperture on perceived direction of motion. The linear object in the aperture is moving horizontally from left to right, as indicated by the yellow arrow. However, when seen through the aperture, the line is perceived to be moving downward to the right (red arrow), a shift in apparent direction of 45°. A particularly impressive way to experience the effect of an aperture is to create a sequence in which an aperture like the one shown here can be applied or removed from the same moving line. In the absence of the aperture, the line is of course seen as moving horizontally; the moment the aperture is applied, however, the direction of movement changes (see www.purveslab.net for demonstrations of this effect in apertures of different shapes).

to the right (Figure 6.11). To make matters even more puzzling, the altered direction seen depends on the specific shape of the occluding aperture. (Although not described here, perceived speed is also affected by these shape changes.)

These changes in perceived direction as a result of occlusion were first studied in detail by Hans Wallach (1935/1996) more than 70 years ago. Their dramatic nature has attracted much attention, not least because of the relevance of these phenomena to understanding the neural basis of motion processing (see, for example, Adelson and Movshon, 1982; Hildreth, 1984; Nakayama and Silverman, 1988; Shimojo et al., 1989). How, then, can these effects be rationalized in an empirical framework, and how does an empirical explanation compare practically and theoretically with others that have been offered over the years?

A useful starting place in thinking about aperture effects in empirical terms is to consider the frequency of occurrence of the directions of unoccluded line projections generated by objects moving in 3-D space. This exercise makes it easier to understand how these "baseline" distributions are affected by occlusion. As illustrated in Figure 6.12A, the unoccluded distributions can be derived using a virtual environment that reproduces the perspective transformation of moving lines experienced by humans (see Figure 6.2), to a very rough approximation, and with all the provisos alluded to earlier in relation to speed.

The average directions of unoccluded line movements on the image plane for the full range of projected orientations are shown in Figure 6.12C. As expected, lines of a given length and orientation have about the same probability of moving in any direction on the image plane. This roughly uniform probability distribution of image directions in full view describes the approximately uniform frequency of occurrence of linear image

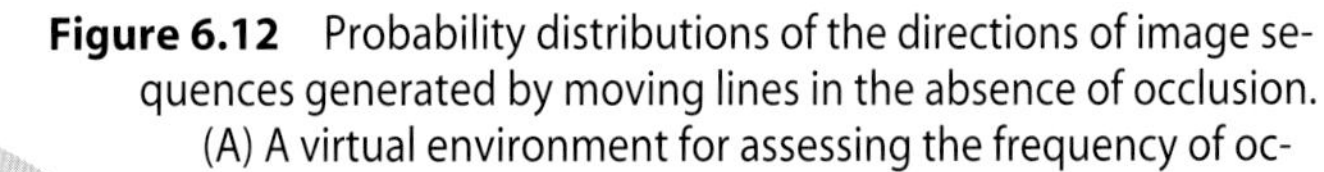

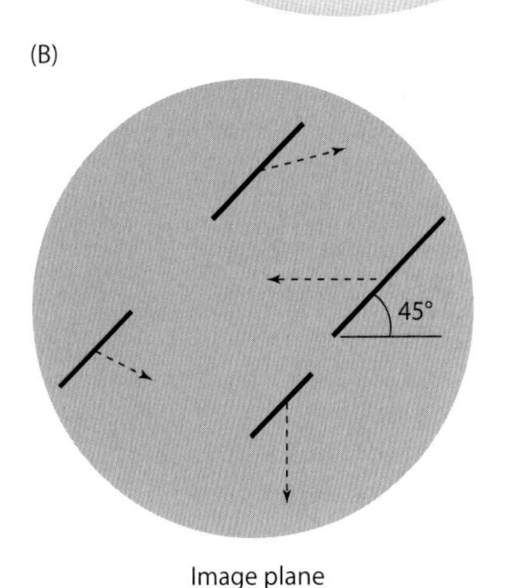

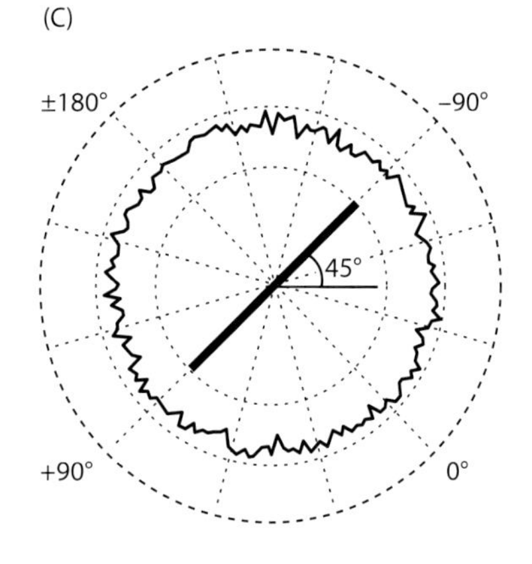

Figure 6.12 Probability distributions of the directions of image sequences generated by moving lines in the absence of occlusion. (A) A virtual environment for assessing the frequency of occurrence of the projected directions of randomly oriented lines moving in all possible directions in the simulated 3-D space. (B) Enlargement of the circular image plane in (A). Projected lines with any particular orientation (45° in this example) generated by 3-D lines moving in any given direction on the image plane can be sampled to determine the frequency of their occurrence. (C) The probability distribution for the projected sequences of unoccluded lines oriented at 45°. Angle notations around the perimeter of this polar coordinate graph represent the projected directions of movement, 0° indicating the direction perpendicular to the orientation of the line. The distance from the center to any point in the coordinate system represents the probability of image sequences moving in that direction. The frequency of occurrence of projected directions is approximately uniform. (After Sung et al., 2009.)

sequences moving in different directions arising from linear objects that humans would have witnessed over evolutionary and individual time.

When a line moves behind an aperture, however, the probability distribution of possible image directions necessarily includes only a subset of 2-D directions illustrated in Figure 6.12C. This altered distribution of image directions depends on the shape of the aperture and the orientation of the projected line being considered (see below). The nature of these differences in the experience of projected directions humans have always had as a result of occluding apertures thus offers a way to examine whether the perceived directions elicited by different aperture shapes and line orientations can be accounted for in empirical terms. If the perceived direction of motion is determined by accumulated past experience, then all aperture effects should be predictable on this basis.

The simplest phenomenon to explore in this way is the effect of a circular aperture (see Figure 6.11). In psychophysical testing (Figure 6.13), observers invariably report that the direction of the moving line is orthogonal to its orientation (Figure 6.14). How and why, then, does a circular aperture affect the uniform frequency of occurrence of the projected directions of unoccluded moving lines (see Figure 6.12C)? When a circular template is systematically applied to the image plane to address this question, the probability distributions of the 2-D directions of differently oriented lines capable of generating image sequences within the aperture are changed in two ways. First, lines moving in half the possible directions on the image plane are irrelevant if the sequence of images of the line is moving from left to right; i.e., projected lines moving from right to left cannot contribute

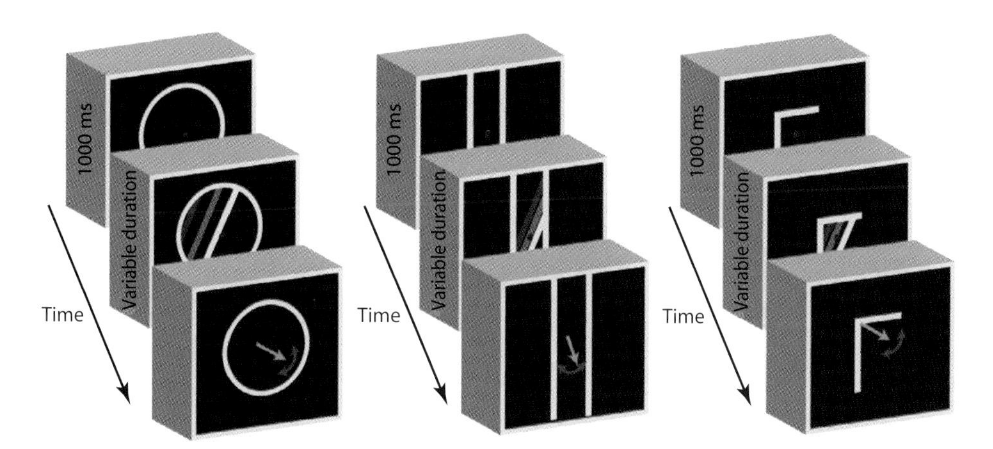

Figure 6.13 Testing the direction seen by observers when a line moves across an aperture. Subjects view a circular aperture (left), a vertical slit aperture (middle) or a triangular aperture on a computer screen; after they have fixated the red dot, a line moves across the aperture in an orientation determined by the experimenter. Subjects then rotate the green arrow to indicate the direction of motion they perceived. The results of such testing are shown in Figures 6.14–6.16. (After Sung et al., 2009.)

to the relevant distribution of directions, as is apparent in Figures 6.11 and 6.12. Second, and more importantly, the frequency of occurrence of lines moving from left to right that can satisfy the aperture will—for reasons explained below—be strongly biased in favor of the direction orthogonal to the line, as indicated by the ovoid probability distributions in Figure 6.14. The shapes of these distributions give the predicted direction and, since only one direction can be seen, the direction humans have experienced most frequently (the mode of each distribution, indicated by the red arrows in Figure 6.14B) should accord with the direction perceived by observers. The predicted directions closely match the directions seen by observers, as indicated by the green arrows in the figure.

This result, however, is less impressive than it might seem, since the perceived direction of motion perpendicular to the orientation of a moving line in a circular aperture can be accounted for in several other ways (Wallach, 1935/1996; Adelson and Movshon, 1982). The empirical argument would be more telling if it predicted aperture effects that are more difficult to explain. A case in point is viewing moving lines through a vertical slit (see Figures 6.13 and 6.15A), which generates the perception of an

Figure 6.14 Comparison psychophysical results and empirical predictions for moving lines seen through a circular aperture. (A) The aperture. As in Figure 6.12, line orientation (θ) is measured from the horizontal axis; the direction of movement is measured as positive or negative deviations from the perpendicular direction (indicated as 0°) with respect to the line orientation. (B) The distributions of the projected directions constrained by the aperture. The green arrows indicate the directions reported in psychophysical testing (see Figure 6.13); the red arrows indicate the directions predicted by empirical analysis. The directions of the red and green arrows are nearly identical. (After Sung et al., 2009.)

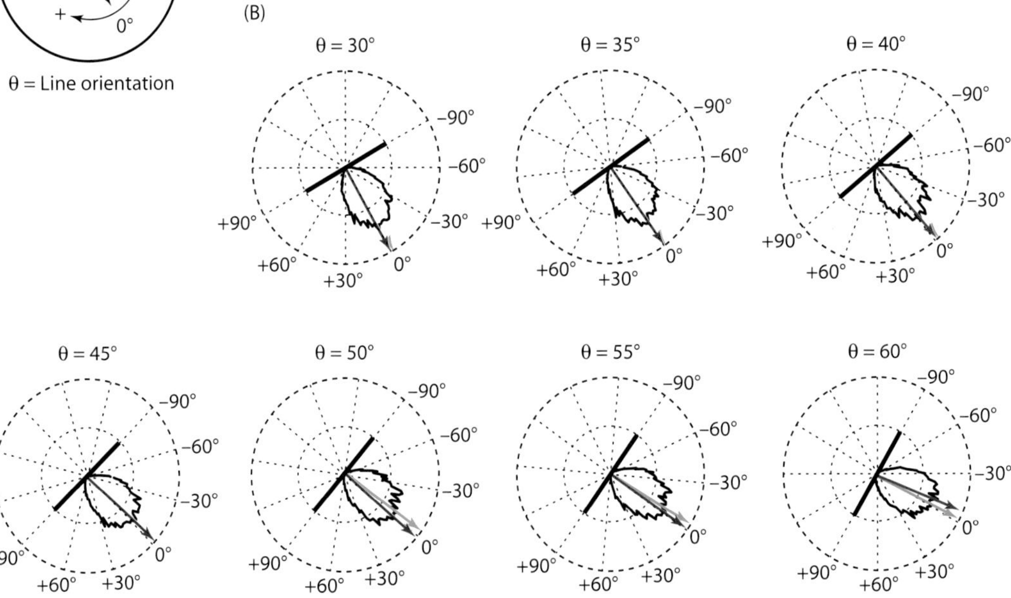

approximately vertical movement of a line moving from left to right (see www.purveslab.net for a demonstration). In contrast to the effect of viewing through a circular aperture, the perceived direction varies with the orientation of the projected line (Figure 6.15B; see the green arrows). The perceived directions are biased counterclockwise, and the bias is greater for larger angles. All this presents a far more challenging set of observations to explain in empirical or any other terms.

The frequency of occurrence of the movement directions of differently oriented lines projected through a vertical slit can be obtained, as before, using a template applied to all possible locations in the image plane, this time in the form of a vertical slit. The empirical predictions made on the basis of these data can then be compared to the perceived directions determined psychophysically, as shown in Figure 6.15B. The variation in the perceived direction as a function of the orientation of the stimulus line again accords with the frequency of occurrence of moving line directions in the context of a vertical slit (compare the directions of the red and green arrows). Prediction of these more subtle effects offers stronger support for the conclusion that perceived direction is determined empirically than does prediction of direction in a circular aperture.

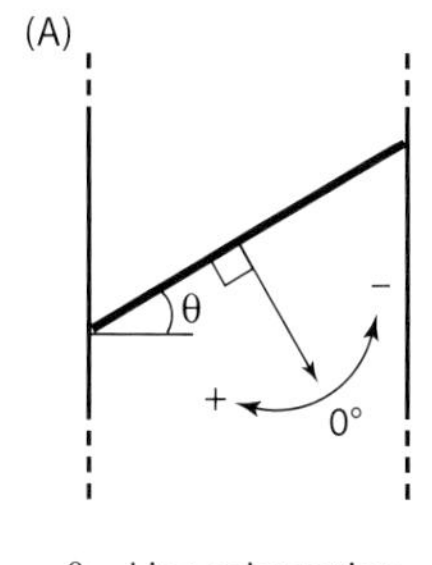

Figure 6.15 Comparison of empirical predictions and psychophysical results for moving lines seen though a vertical slit. (A) The aperture. (B) Distributions of the projected directions of moving lines in different orientations constrained by a vertical slit. As before, the green arrows are results of psychophysical testing and the red arrows are the empirical predictions; the gray arrows indicate vertical. The predictions are again in agreement with psychophysical results. (After Sung et al., 2009.)

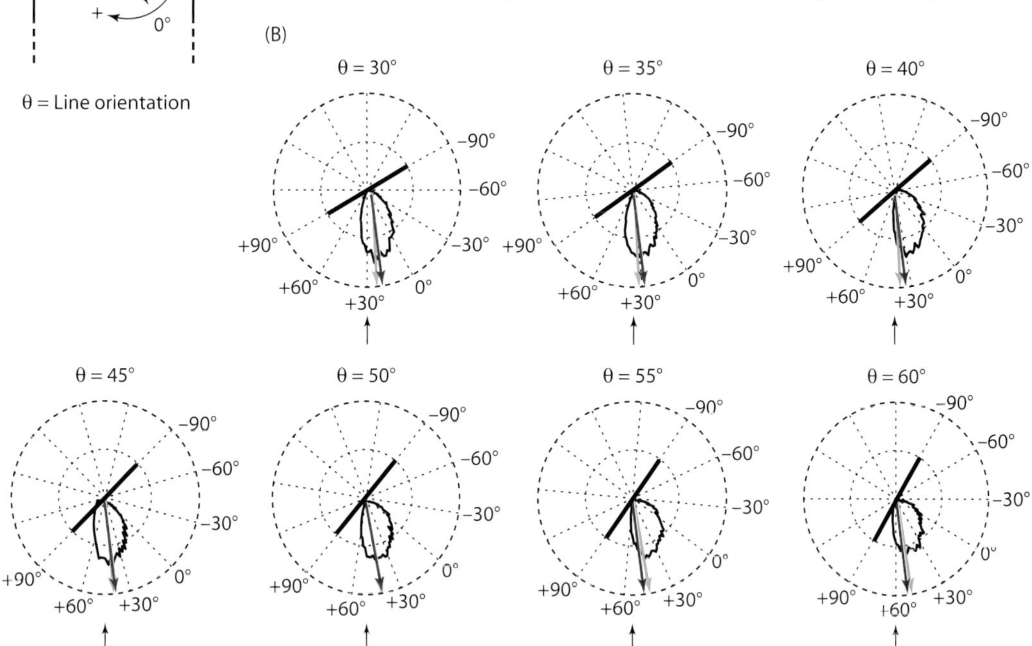

A triangular aperture (see Figure 6.13) presents even greater challenges for any explanation of perceived direction. Wallach (1935/1996) reported that when the orientation of a moving line in a triangular aperture was either more or less than 45°, the perceived direction of movement deviated in an opposite manner. To examine how a triangular aperture affects the frequency of occurrence of lines projected in different orientations, a triangular template can be used to sample images (Figure 6.16). The results of psychophysical testing (green arrows) quantitatively confirm Wallach's qualitative observations that directions seen in a triangular aperture vary systematically as a function of the orientation of the moving line (Sung et al., 2009). For example, when the orientation of a moving line is 30°, the observed direction is biased by approximately 24° counterclockwise with respect to the direction normal to the line. On the other hand, when the orientation of the line is 60°, the perceived direction is biased by approximately 24° clockwise. The empirical predictions in Figure 6.16B based on the frequency of occurrence of the projected directions (red arrows) correctly capture these perceptions of motion direction.

Thus, the frequency of occurrence of image sequences produced by 3-D sources in a virtual environment accounts for a range of aperture effects,

Figure 6.16 Comparison of empirical predictions and psychophysical results for moving lines seen in a triangular aperture. (A) The aperture. (B) The frequency of occurrence of projected directions of line sequences constrained by a triangular aperture. As before, the green arrows indicate the results of psychophysical testing and the red arrows indicate the empirical predictions. The perceived directions differ from the direction normal to the line orientation for orientations that are not 45°. These biases are again predicted by the empirical data. (After Sung et al., 2009.)

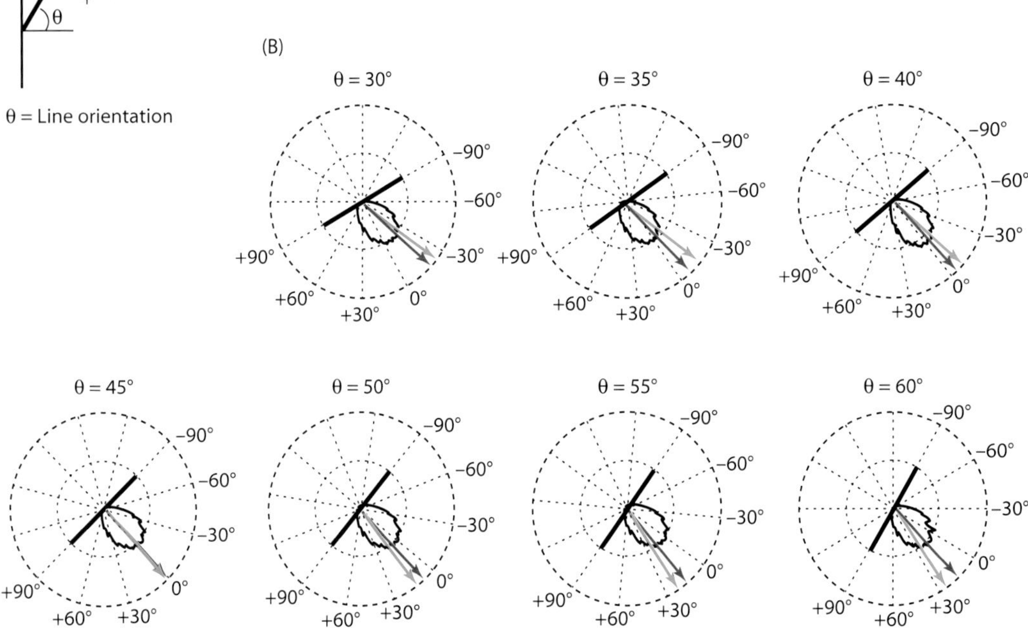

supporting the conclusion that perceived directions of motion, like perceptions of speed, are determined empirically.

Why Projections through Apertures Are Biased

Why, then, does the frequency of occurrence of image sequences observed through apertures change in particular ways, leading to the biased experience that appears to determine the directions seen?

To understand the probability distributions in Figures 6.14 through 6.16, take the biased directions of image sequences projected through a circular aperture. As shown in Figure 6.17A, there is a particular direction of motion (black arrow) that entails the minimum projected length (red line) that can fully fill the aperture during its traverse. A projected line traveling in any other direction must be longer if it is to fill the aperture (blue line); shorter lines in either case will fall inside the aperture boundary at some point in its course, producing a different stimulus and a different perceived

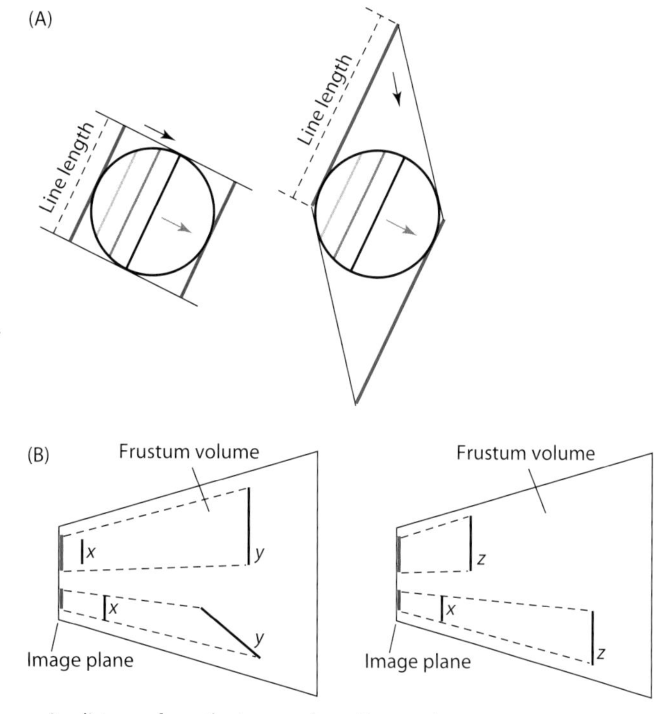

Figure 6.17 Reasons for the biased distributions of directions of lines projected through apertures. (A) Because longer lines include shorter ones, the occurrence of projections generated by the red line moving in the direction indicated by the black arrow on the left will always be greater than the occurrence of projections that satisfy the aperture generated by the blue line moving in any other direction (black arrow on the right). As a result, the most frequently experienced projected direction when viewing linear objects that satisfy a circular aperture will always be the direction orthogonal to the orientation of the projected line (gray arrows). (B) Other biases generated by perspective. Two physically different lines, *x* and *y*, can generate projected lines of the same (red) or different (red vs. blue) lengths on the image plane. Because the longer line, *y*, can, in a different 3-D orientation, generate the same projected length (red) as the shorter line, *x*, projected lines of different lengths have different probabilities of occurrence. The same linear source, *z*, can generate different projected lengths (red vs. blue) depending on its distance from the image plane. Due to the nature of human visual space (broadly similar to the diagram in Figure 6.2), a greater number of relatively distant real world sources can generate projection onto the retina, leading to a further bias in favor of shorter line projections. Both of these geometrical considerations lead to more short than long line projections. (After Sung et al., 2009.)

direction. Because a line of any length includes all shorter lines, far more projected lines moving orthogonally to their orientation will satisfy the aperture compared with the number of lines moving in other directions. As a result, the distribution of directions that can occupy a circular aperture is strongly biased toward the direction normal to the orientation of any line, as indicated by the probability distributions in Figure 6.14B. These simple facts about geometry are the basis of the major bias observed for any aperture and thus for how humans will have experienced retinal projections of objects moving behind occluding frames.

The fact that longer lines include shorter ones, however, is not the only geometrical factor contributing to the biased distributions. An additional influence is that objects, when viewed in perspective, project lines that are smaller than (or at best equal to) their physical dimensions (Figure 6.17B). As a result, the ovoid probability distributions for projections in a circular aperture are narrower at their base than at their apex but otherwise symmetrical (Figure 6.14B). For a circular aperture, this further influence of perspective does not affect the mode of the distribution, which remains orthogonal to the orientation of the line. In a vertical slit or triangular aperture, however, the biases arising from the relationships illustrated in Figure 6.17B cause the frequency of occurrence of the projected directions to change as a function of the projected orientation of the line sequence, such that the distributions are asymmetrical (see Figures 6.15B and 6.16B). The reason is that the projected line length and distance traveled needed to satisfy the apertures in these cases are not the same, as they are for circular aperture. The biased generation of shorter than longer line projections arising from perspective thus alters the distribution as the orientation of the line in the aperture changes, and by the same token the mode. This alteration of the mode of the distribution explains the specific directions seen when moving lines are presented in the context of vertical slits or triangular apertures.

The differences in the probability distributions generated by the geometry of different apertures describe, again to a first approximation, the way the human visual system will always have experienced the projection of objects in the context of the sorts of occlusions that routinely occur in natural viewing. These frequencies of occurrence determine, in turn, the directions of motion that people see in different circumstances.

Perceptions Elicited by Lines That Are Translating and Rotating

The projected directions of moving lines or other objects in the real world are, more often than not, rotating as well as translating. Recall that translation, which describes the stimuli considered so far, refers to movement in which a point or points on the object have the same velocity with respect to a frame of reference. Rotation refers to the movement of points around an axis, which entails differential movement of elements that comprise an object.

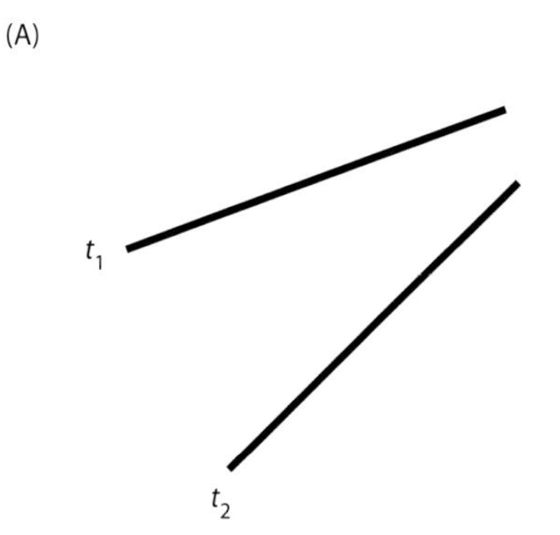

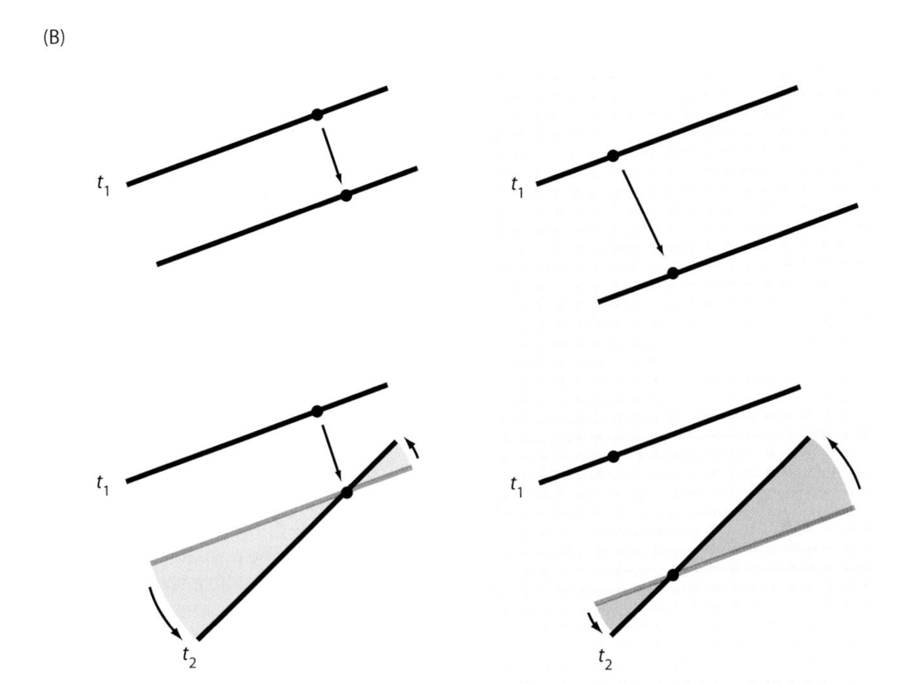

Figure 6.18 The inverse problem as it applies to the projected motion of an object that is translating and rotating. (A) Sequential positions of the same moving line projected onto an image plane at two different times (t_1 and t_2). (B) Two examples of the large number of translational and rotational movements that could underlie the sequential positions of the projected line illustrated in (A). Black dots indicate axes of rotation. (After Yang et al., 2002.)

This aspect of real-world motion adds another challenge to the problem of explaining perceived motion in empirical terms. Physical motions that entail rotation as well as translation produce more complex image sequences, and the physical motion of the generative objects in 3-D space is equally uncertain. As shown in Figure 6.18, such stimuli can arise from many possible translation and rotation pairs. The question is thus whether the perceptions of motion elicited by such stimuli can also be accommodated in the framework used to rationalize the flash-lag and aperture effects.

The computations entailed in determining the projections of objects that are translating and rotating in a simulated environment are more difficult than for translation alone. In consequence, rationalizing the perceptual

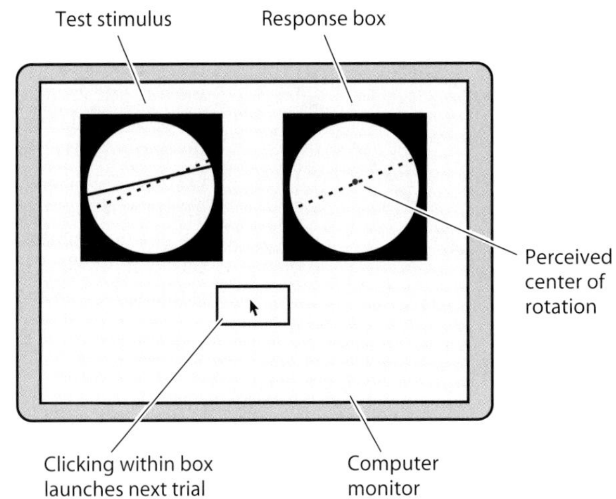

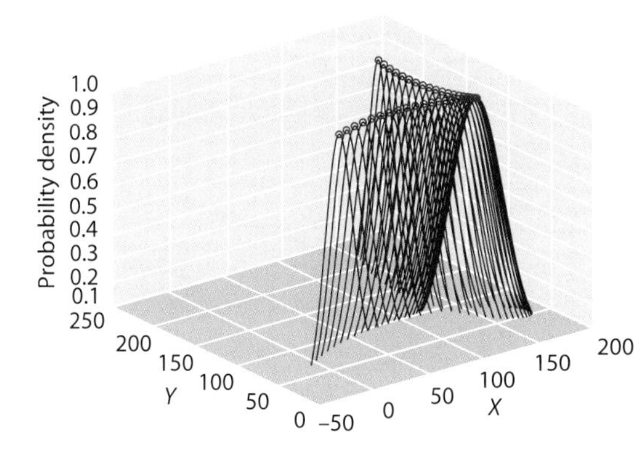

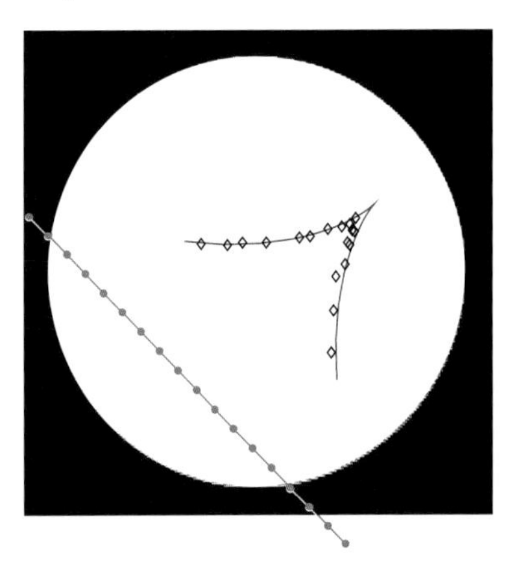

Figure 6.19 Comparison of the predicted and perceived movement of a line that is translating and rotating. (A) Example of a test stimulus (left) in which a line moves from the upper left downward and to the right across a circular aperture while rotating around a fixed center on the line. Subjects are asked to report the perceived center of rotation at the moment the moving line became coincident with the dotted line by marking the apparent center with a dot in the response box (right). (B) The probability distributions of the possible sources of the stimulus in (A), plotted over time; the maxima of the distributions (red dots) describe a cycloidal trajectory. (C) The sequential positions of the centers of rotation reported by subjects likewise fall along a cycloidal trajectory. The diamonds indicate the average position of the perceived centers of rotation at different times in the presentation; the solid red line shows the responses predicted empirically. The blue line indicates the actual trajectory of the center of rotation of the stimulus line, and the blue dots indicate the times at which subjects made their judgments. (After Yang et al., 2002.)

phenomena elicited by such stimuli has so far been limited to an empirical argument based on geometrical principles as they apply to the projections of linear objects translating and rotating in an aperture (Yang et al., 2002). Nonetheless, since the frequency of occurrence of image sequences that entail both translation and rotation is quite different from those that occur during translation alone, different motion perceptions are predicted.

The translating and rotating stimuli illustrated in Figure 6.19A generate a perception of rotation in which the trajectory of the perceived rotational centers follow a cycloidal curve. A cycloid is a type of curve defined by the trajectory of a point on the radius of a circle that is rolling along a straight line; imagine, for example, the curve traced by a mark on the perimeter of a wheel as it crosses a flat surface. Empirically, this perception again makes sense. Based on the frequency of occurrence of the projected image sequences generated by the possible sources of such stimuli (Figure 6.19B), observers should see rotation along the cycloidal path defined by the apparent rotational centers. As indicated in Figure 6.19C, this is what subjects report.

A further aspect of the perception of the stimulus presented in Figure 6.19A is that the shape of the aperture used does not affect the apparent direction of motion. Figure 6.20 compares the position of the perceived centers of rotation when translating and rotating stimuli are presented in circular and rectangular apertures, respectively. Little or no difference in

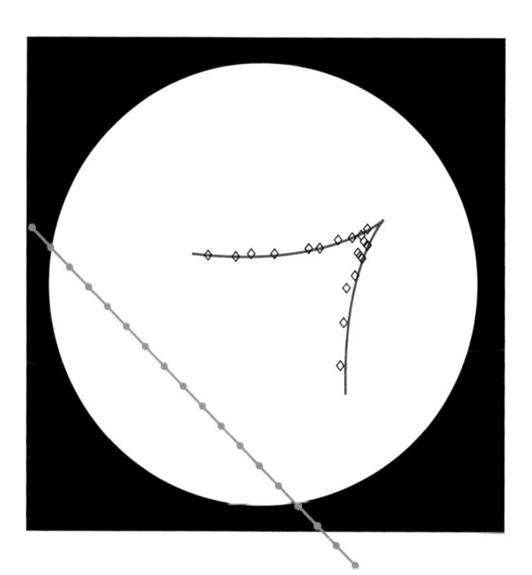

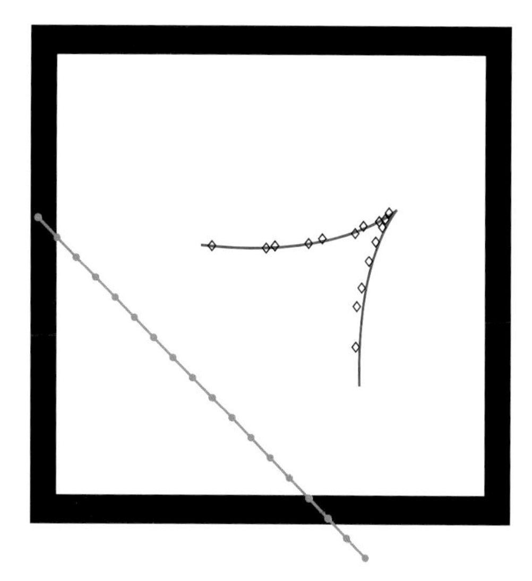

Figure 6.20 The absence of an effect of aperture shape on the perception of motion in response to the stimulus in Figure 6.19A. In both a circular aperture (left) and a rectangular aperture (right), the perceived centers of rotation trace out the same cycloidal path. As in Figure 6.19B, the diamonds indicate the average position of the perceived centers of rotation; the solid red lines are the results predicted by theory. The blue lines show the actual trajectory of the axes of rotation, and the dots the times at which the psychophysical judgments were made. (After Yang et al., 2002.)

the motions perceived in the two types of apertures is reported. This result is quite different from the "classical" aperture effects described in the preceding section, in which the shape of the occluding frame profoundly influences the direction of motion seen. The observation is also consistent with the frequency of occurrence of image sequences in human experience. Any particular sequence of speed, direction, acceleration, and/or longitudinal deformation projected onto the retina from the aperture boundaries will have occurred rarely; that is, the portions of the moving line at the boundaries are far more likely to have been associated with successful behaviors directed toward linear objects moving across the boundaries of the frame than along them. Thus, on empirical grounds, the translational component of a linear stimulus should not be perceived, and the shape of the aperture should have little influence on the motion seen.

Psychological Models of Perceived Motion

Given these explanations of a range of motion percepts, it is important to compare the approach here with other explanations that have been offered for some of these phenomena, and for motion perception generally.

With respect to the apparent speed of moving objects, many accounts have focused on the flash-lag effect. Some theories that have been proposed are: (1) that the visual system compensates for neuronal latencies to "predict" the position of a moving object (Nijhawan, 1994; Khurana and Nijhawan, 1995; Nijhawan, 1997; Khurana et al., 2000); (2) that vision employs information derived from the immediate future to "postdict" the position of moving objects (Purushothaman et al., 1998; Whitney and Murakami, 1998; Whitney et al., 2000a,b); (3) that the flash-lag effect is the consequence of "anticipation" in early retinal processing (Berry et al., 1999); and (4) that the effect occurs because stimulus processing entails shorter latencies for moving stimuli than for static flashes (Eagleman and Sejnowski, 2000a,b).

These proposals have at least two deficiencies. First, for technical and historical reasons, they assume that psychophysical function to be explained is linear. Thus, none of these theories addresses the actual nonlinear function shown in Figure 6.4B. Second and more important, these interpretations assume that the perceived lag derives directly from the features of image sequences on the retina and their subsequent processing by the visual system. This interpretation ignores the fact that any direct analysis of images, moving or otherwise, is inevitably meaningless for visually guided behavior because of the inverse problem (see Figure 6.1).

With respect to rationalizing the apparent direction of motion image sequences, most explanations have focused on aperture effects. The most popular approach has been to suppose that the visual system calculates the local velocity vector field in an image sequence (Adelson and Movshon, 1982; Horn and Schunck, 1981; Hildreth, 1984). Some problems with the idea that aperture effects arise as a result of visual computations that minimize variation in local vectors are also apparent. As generally acknowledged (Hildreth, 1984), solutions that match the perceptions of human observers require ad hoc assumptions to constrain the "solution space," including the assumption that moving objects produce a constant velocity field (Limb and Murphy, 1975; Fennema and Thompson, 1979; Marr and Ullman, 1981), that object motion is "rigid" (Ullman, 1979; Bennett et al., 1989; Yuille and Ullman, 1990), and/or that object motion is "smooth" (Hildreth, 1984; Yuille and Ullman, 1990). Whereas these assumptions are all reasonable, they fail to capture the biases that are actually elicited by projections through various apertures (see above). For example, Adelson and Movshon's (1982) explanation based on vector constraint wrongly predicts that the perceived direction in a triangular aperture should be normal to the line orientation (because all possible local direction vectors on a line are normal to the line orientation). Similarly, Hildreth's (1984) calculation of a vector field under the assumption of least variation (i.e., the "smoothness" of motion assumption) does not account for the increasing biases as the angle of orientation increases in a vertical slit (see Figure 6.15B; the Hildreth model predicts that the perceived direction should be strictly vertical). Another problem with all such explanations is the absence of any biological reason why the visual system should generate such percepts.

Understanding these phenomena in empirical terms does away with the need to calculate the local features of image sequences such as velocity vectors (Fennema and Thompson, 1979; Adelson and Movshon, 1982; Hildreth, 1984) or spatiotemporal energy (Adelson and Bergen, 1985). A wholly empirical understanding of motion perception also abrogates the need for unconscious inferences or Bayesian computations (Knill and Richards, 1996; Stocker and Simoncelli, 2006).

Physiological Models of Perceived Motion

Physiological models of motion processing generally are couched in terms of a hierarchy in which lower-order receptive field properties of motion-sensitive neurons in V1 generate higher-order neuronal properties in cortical areas MT and MST in the non-human primate brain and what is referred to as MT+ in humans (Hubel and Wiesel, 1962, 1974, 1977; De Valois et al., 1982a,b; Movshon et al., 1986; Livingstone and Hubel, 1987, 1988; Nichols and Newsome, 2002). The culmination of this process is presumably the motion perceived. This "bottom-up" approach has been amended by two-stage (Braddick, 1974) or three-stage (Lu and Sperling, 1995) processing schemes, as well as by the addition of "component cells" and "pattern cells" that could explain further details of visual motion perception (Movshon et al., 1986; Rust et al., 2006), and remains a popular conception of how motion percepts are generated.

Explaining perceived motion on this basis is problematic, however, because the physical properties of moving stimuli routinely fail to correlate with perception. Although motion-sensitive neurons are of course necessary causes of motion percepts, their receptive field properties do not, in themselves, explain the speeds and directions we actually see.

In empirical terms, perceived motion is determined simply by the distribution of the speeds and directions in the 2-D projections experienced in the past and instantiated in the evolved and developed connectivity of the visual system. This historically derived circuitry would then respond reflexively to any image sequence by eliciting perceptions of motion and behaviors that have been successful in coping in a world in which 3-D object movements are hidden from the senses.

Conclusion

The problem rationalizing motion perception, as in other domains of vision, is the inevitably uncertain meaning of images or, in this instance, image sequences. Psychologists have sought to explain perceptual anomalies of apparent speed (e.g., the flash-lag effect) and direction (e.g., aperture effects) in terms of ad hoc models. Physiologists have quite naturally sought to explain motion perception in terms of the receptive field properties of motion-sensitive neurons. But neither approach has been particularly successful in rationalizing the phenomenology of the motions we see. A more radical conception is that motion percepts are determined entirely by the frequency of occurrence of the image sequences that humans have always experienced. The fact that key aspects of perceived speed and direction can be accounted for in this way supports the conclusion that the visual system is indeed using a wholly empirical strategy to generate the apparent speeds and directions of moving objects. As in lightness/brightness, color, form, and distance perception, the relationship between projection frequency and behaviors directed towards unknowable sources has presumably been instantiated in visual system circuitry by natural selection during evolution and modification of this inherited circuitry by activity-dependent neural plasticity during life.

7

Making Sense of the Visual System in Empirical Terms

Introduction

The central problem in understanding biological vision is that light stimuli provide no direct access to the physical world. Given the inherently uncertain provenance of retinal images and the multiplicity of factors that determine behavioral success, it is difficult to fathom how vision can usefully guide what we do or think. Although research on receptive fields has dominated modern vision research and made enormous contributions, overcoming the uncertain meaning of images cannot be achieved by encoding image features. If feature analysis as a mode of operation is ruled out in principle by the impossibility of apprehending the physical world by operations on images as such, then it follows that the information that vision depends on must be of a different sort. The observations described in the preceding chapters indicate that visual system uses a wholly empirical strategy; the relevant information, and therefore what we see, is the history of human and individual experience. Accepting this idea requires a radical change in how one thinks about the structure, function, and ultimate purposes of vision.

Equally difficult is setting aside the subjective sense that a mental represention of images provides a facsimile of the world, and that visual qualities must be analytically derived from reality. Nonetheless, if the evidence supports these difficult conclusions, there is little choice but to make the best of the functional, anatomical, psychological, and philosophical challenges that a wholly empirical concept of vision entails. The purpose of this last chapter is to sum up the argument and suggest how to move forward in this framework.

The Receptive Field Properties of Visual Neurons

The focus of vision research over much of the last 50 years has been on the response characteristics of individual neurons—their receptive field properties—at various stations of the primary and higher order visual pathways. These properties are defined by (1) the area of visual space a stimulus must occupy to elicit a response from the neuron in question and (2) the specific characteristics of the stimuli required to change the neuron's baseline firing rate (Figure 7.1). Since the initiation of such studies in the

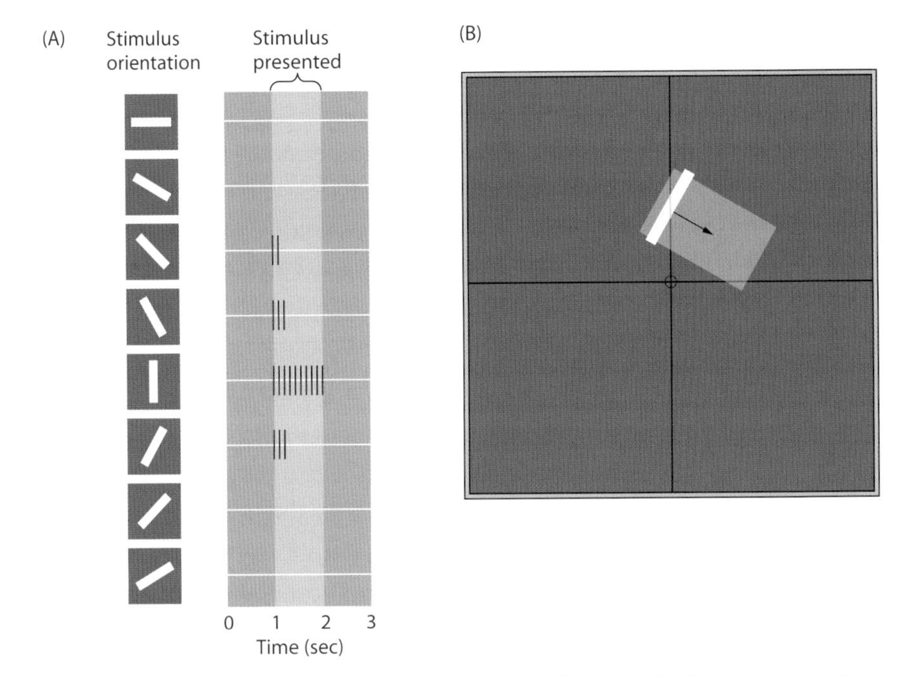

Figure 7.1 Example of the receptive field properties of a neuron in the primary visual cortex of an experimental animal. (A) The series of panels shows the neuronal responses (action potentials elicited per unit time) as a function of the location and orientation of the stimulus on the retina. (B) The stimulus is a bar of light moved across a screen in different locations in different directions. The area at a particular location in visual space (indicated in gray) and the qualities of the stimulus (e.g., the orientation and direction of the bar) required to maximally excite (or inhibit) the baseline rate of firing define the neuron's "classical" receptive field and its properties.

late 1950s, a variety of neuronal receptive field properties in the visual systems of experimental animals have been described. Earlier chapters used examples of these properties to make specific points.

Work on the receptive field properties of visual cortical neurons was pioneered by David Hubel and Torsten Wiesel (see Hubel and Wiesel, 2005, for a complete account of this remarkable work). Based on the results of electrophysiological recordings in the visual thalamus, the primary visual cortex, and higher order (extrastriate) visual cortical areas, Hubel and Wiesel surmised that the characteristics of lower order neurons generate the more complex responses of higher order neurons by virtue of the convergence of the outputs of the lower order cells onto higher order ones. Implicit in this arrangement is the idea that visual perceptions correspond to the activity of neurons in the relatively higher stations of the system. For example, orientation-selective neurons in primary visual cortex receive inputs from lower order neurons in the thalamus that are not orientation selective but are topographically arrayed in a manner that generates this receptive field property (Hubel and Wiesel, 1962, 1974, 1977). It seems plausible that, in the aggregate, orientation-selective neurons in V1 create a central representation of the contours in a stimulus, leading ultimately to the recognition of object forms in higher order areas of the visual system. Similarly, the output of low level neurons responsive to luminance would give rise to sensations of lightness or brightness when combined with other features in appropriate cortical regions, neurons responsive to particular spectra would generate the perceptions of color, and so on for other visual perceptual properties. Evidence that supports this general interpretation is the discovery of visual cortical neurons whose activity is associated with perceptions of lightness/brightness (Paradiso and Hahn, 1996; Rossi et al., 1996; Rossi and Paradiso, 1999; MacEvoy and Paradiso, 2001), color (Zeki, 1983a,b; Bartels and Zeki, 2000), object size (Dobbins et al., 1998), and stereoscopic depth (DeAngelis et al., 1998; Cumming and Parker, 2000). Remarkably, some neurons in the temporal cortex respond selectively to faces (Desimone et al., 1984), and some to images of specific people (Quiroga et al., 2005).

The link between the activity of visual cortical neurons and perception in awake, behaving animals with chronically implanted recording electrodes also accords with the same general scheme of visual processing. This approach allows trained animals to report what they see in response to visual stimuli by a subsequent behavior, typically by making a saccade to a target. For instance, work carried out by William Newsome and colleagues has shown that the activity of neurons in extrastriate area MT (see Chapter 6) is closely related to perceived motion (Figure 7.2). In these studies, rhesus monkeys are shown a display of dots moving in different directions. If a sufficient fraction of the dots move in the same direction, experimental animals (or human subjects) perceive an overall direction of motion in the display. Monkeys can be trained to move their eyes to a target that signifies the direction of dot movement, in this way reporting the direction of movement they see and the time at which they see it. Recordings in MT show that the activity of single neurons is often correlated with the direction of

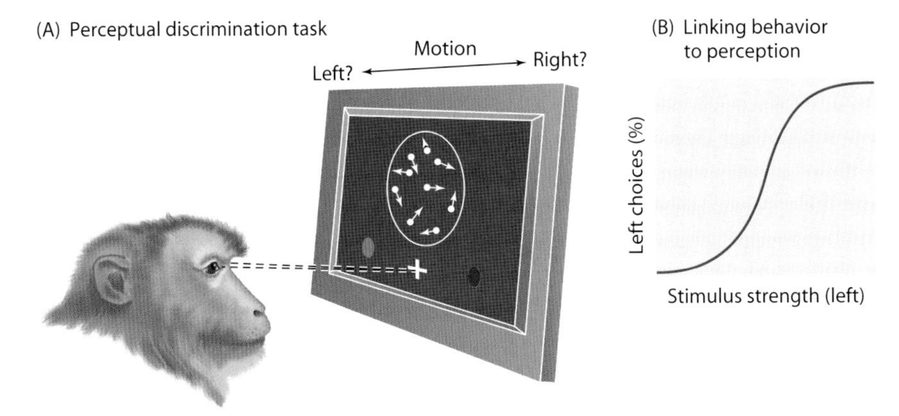

Figure 7.2 Relating the activity of motion-sensitive neurons in area MT to motion perception. (A) A monkey is trained to report what it sees by the saccades it makes to either the red or green target, depending on whether the animal perceives rightward or leftward motion in response to a pattern of moving dots. (B) By changing the amount of coherent motion in the moving dot pattern, a psychophysical function is obtained that plots perceptual accuracy against the amount of motion coherence among the dots. Electrical stimulation of small populations of MT neurons (not shown) shifts this curve in a systematic way, showing that the stimulated neurons influence motion perception. (After Sugrue et al., 2005.)

dot motion. Indeed, the activity of neurons in the population is sometimes a better predictor of the direction of dot motion than the behavior of the monkey (i.e., its eye movements). Neurons that showed selective activity for a particular direction of motion could then be stimulated electrically to ask whether MT neurons play a causal role in these perceptual discriminations (Sugrue et al., 2005). In about half the cases, this "microstimulation" increased the probability that the monkeys would move their eyes in the expected direction.

The fact that the activity of neurons in the relevant area of the extrastriate visual cortices correlates with the perceived qualities of visual stimuli supports the concept of a hierarchical sequence of increasingly complex neuronal receptive field properties that underlie what we see. A corollary is that neural processing "decodes" retinal image features to yield perceptual representations of the world that are good enough to guide behavior. These ideas seem so straightforward that it is hard to imagine any other explanation of visual perception.

Problems Linking Receptive Field Properties to Perception

Despite the implicaion that an increasingly detailed understanding of the properties of visual neurons and their interactions will explain perception, the observations described in earlier chapters argue otherwise. Whereas the activity of visual system neurons obviously underpins visual perception, it should be apparent that the neuronal receptive field properties do not explain what observers see, at least in the terms in which they are presently conceived. For example, with respect to the studies summarized in Figure 7.2, the speed and direction of a dot projected on the retina can correspond to many different speeds and directions of points moving in 3-D space, as explained in Chapter 6. Thus, while the results in awake behaving monkeys observing moving dots are both plausible and instructive, any suggestion that the stimulus speeds and directions perceived represent the speeds and directions of the generative objects is neither.

If neuronal receptive fields cannot represent the features of the physical world or even retinal images, then what do they signify? The answer at this point is that no one really knows. As more work has been done over the last 25 years, it has become increasingly apparent that the initial concept of neuronal receptive fields and their purposes in visual processing is not compelling. The growing concern is based on a variety of neurophysiological and anatomical evidence (see, for instance, Lennie, 1998; Engel and Furmanski, 2001; Lewen et al., 2001; Reinagel, 2001). One important observation is that neurons in awake animals show a much wider range of responsiveness to stimuli than do neurons in anesthetized animals. In these more "normal" circumstances, neuronal activity is influenced by stimulus characteristics well outside the "classical" receptive fields defined by Hubel and Wiesel and others (Figure 7.3). Moreover, both neuroanatomical and electrophysiological studies have made clear that the vast majority of inputs to neurons in both the lateral geniculate nucleus and the various stations in the visual cortex come not from the retinal input but from interactions with other neurons at the same and higher levels of the system. Even the output of the retina is strongly affected by interactions among neurons in the different retinal layers. Many examples of these wider influences on the response properties of visual neurons, particularly visual cortical neurons, have

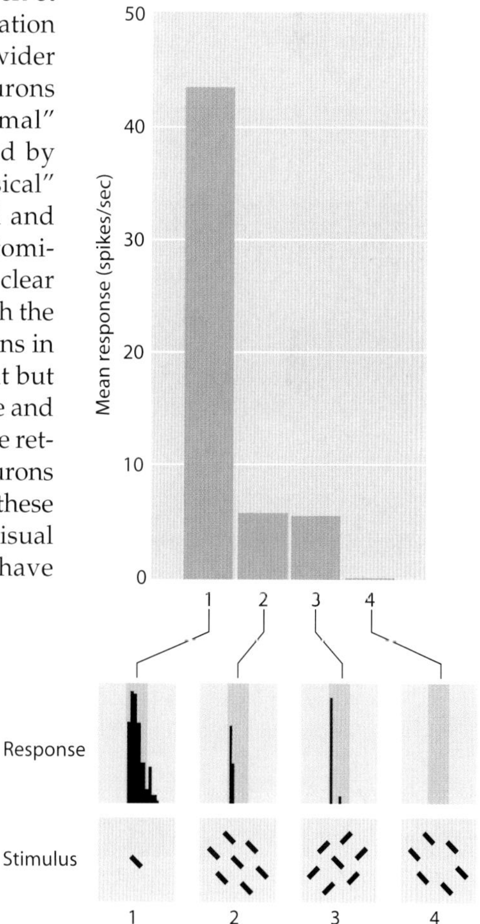

Figure 7.3 Changes in receptive field properties as a result of contextual information. In this example, the response of the neuron to a moving bar in a particular orientation defines the cell's "classical" receptive field (1), as in Figure 7.1. The response is suppressed, however, by presentation of moving bars in any orientation outside this area (2 and 3); (4) shows the absence of a significant response when the stimulus features outside the "classical" receptive field are presented alone. (After Knierim and Van Essen, 1992a.)

been documented (see, for instance, Maffei and Fiorentini, 1976; Nelson and Frost, 1978; Gilbert and Wiesel, 1990; Knierim and Van Essen, 1992a; Li and Li, 1994; Kapadia et al., 1995; Zipser et al., 1996; Lennie, 1998; Dragoi and Sur, 2000; Fitzpatrick, 2000; Worgotter and Eysel, 2000; Douglas and Martin, 2007). A similar dependence of receptive field properties on context has been demonstrated in the rodent olfactory (Kay and Laurent, 1999) and somatic sensory systems (O'Keefe and Dostrovsky, 1971; O'Keefe, 1979; Nicolelis et al., 1997; Ghazanfar and Nicolelis, 1999), and the malleability of sensory receptive field properties is now assumed to apply generally.

The complex organization of neuronal groups at virtually all levels of the visual system has also been made clearer by advances in tract tracing methods and ways of imaging of cortical activity. As described in the Appendix, the primary visual cortex of primates and many other mammals is organized in iterated columns, in which the neurons tend to have similar response properties to visual stimulation. An especially effective way of seeing the columnar organization of neuronal responsiveness to the orientation of stimuli is optical imaging of the "intrinsic signal" (Figure 7.4). This signal is generated by the local changes in the ratio of oxyhemoglobin to deoxyhemoglobin; imaging neuronal activity is possible because more active cortical regions absorb long wavelength light better than less active areas when the cortical surface is illuminated with red light (see Blasdel and Salama, 1986; Bonhoeffer and Grinvald, 1993, 1996; Obermeyer and Blasdel, 1993; Weliky et al., 1996). By using a sensitive video camera and averaging over many trials (the changes in the oxygenated hemoglobin ratios are only 1 or 2

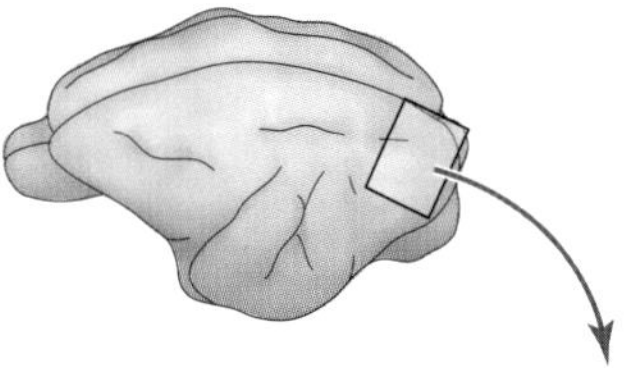

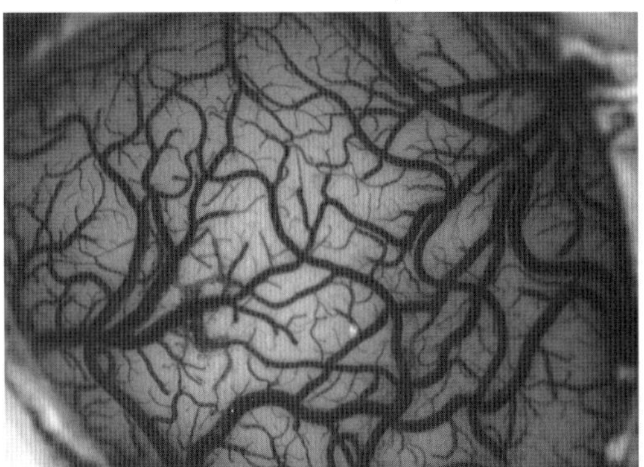

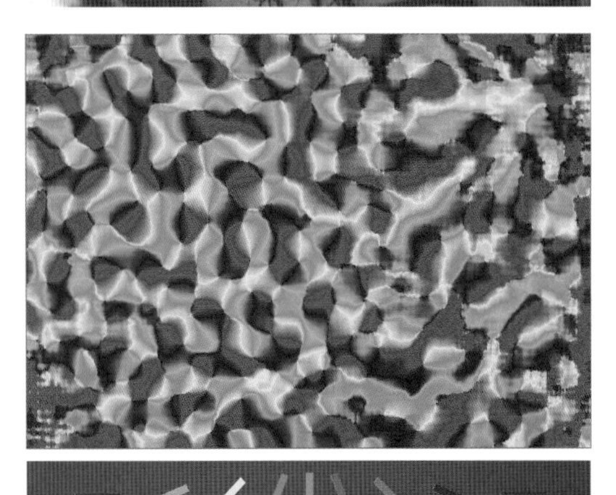

Figure 7.4 Optical imaging as a means of visualizing patterns of cortical activity in response to differently oriented stimuli. A video camera records light absorption by the primary visual cortex as an experimental animal views stimuli on a video monitor. The images of the cortex obtained in this way are digitized and stored to subsequently construct and compare the patterns ("maps") associated with the different stimuli. The example shown here is the "orientation preference map" in the primary visual cortex of a tree shrew. Each color represents the orientation of the stimulus that was most effective in activating the cortical neurons at that site. (Courtesy of Len White and David Fitzpatrick.)

parts per thousand), it is possible to map the patterns of cortical activity elicited by various visual stimuli. The activity elicited by edges or lines in different orientations is mapped in a continuous fashion such that adjacent positions on the cortical surface tend to have only slightly shifted orientation preferences.

The importance of the pattern of orientation columns in the argument here is that the same pattern of cortical activity can be elicited by differently oriented stimuli (Figure 7.5) (Basole et al., 2003). This result is unexpected since the activity elicited in primary visual cortex by oriented lines should, on the basis of classical receptive field studies, correspond to a given orientation in visual space. It appears, however, that orientation-selective neurons actually respond to a variety of oriented stimulus patterns and presumably contribute to a variety of orientation percepts. A similar effect of context on the response properties of single cells within V1 to luminance contrast has also been documented (MacEvoy and Paradiso, 2001).

Finally, very different organizations of neurons and their connections can generate vision across species (see Nilsson and Land, 2001 for a wealth of remarkable examples). Moreover, the consequences for behavior can be similar in widely different animals; color constancy, for example, is apparent in both primates and bees (Lotto and Chittka, 2005; Clarke and Lotto, 2009). Consistent with this evidence, there are many ways in which a behavior can be generated by interactions among the same set of neurons, even in very simple neural circuits (Prinz et al., 2004).

Rationalizing these observations is not easy. To salvage the traditional interpretation of receptive fields in relation to perception, one could imagine that detecting, encoding, and representing stimulus features in terms of receptive field properties is basically the way vision operates, but in a

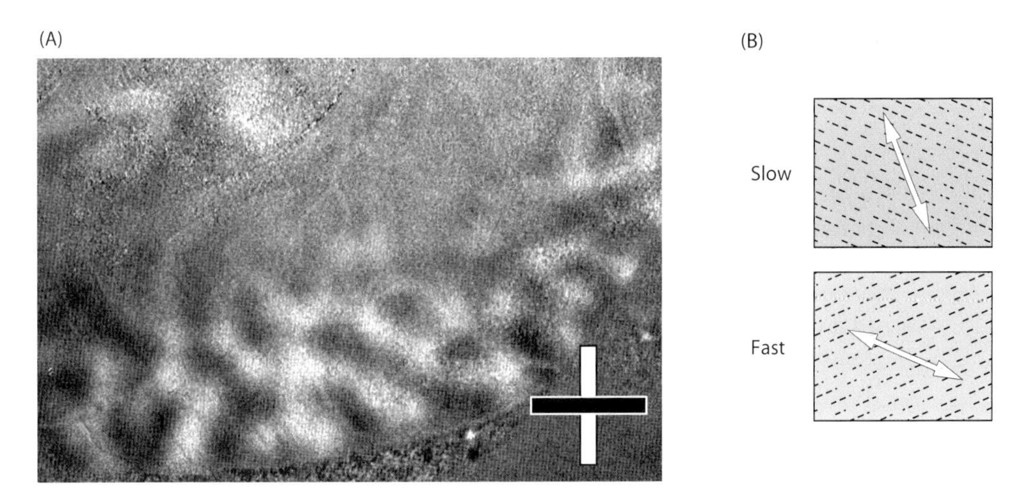

Figure 7.5 The same pattern of optical activity in the ferret visual cortex (A) can be elicited by either of the stimuli in (B). (See text for further explanation.) (After Basole et al., 2003.)

more complex way than initially thought. For instance, suppressive surrounds of the sort shown in Figure 7.3 might indicate the detection of feature categories rather than features as such (Das and Gilbert, 1999). Another approach would be to shift the task of representing of image features and/or real-world sources onto larger and/or more diverse ensembles of neurons under the supposition that a "population code" and/or a "temporal code" are embedded in the response properties of neuronal groups (see, for example, Barlow, 1995; Olshausen and Field, 1996; Pouget et al., 2000; Vinje and Gallant, 2000; Engel et al., 2001). Others have rationalized some receptive field properties in terms of the statistical qualities of images, the idea being that it is computationally efficient to encode the statistical regularities of images (Olshausen and Field, 1996).

The Empirical Alternative

Whether a better understanding of the relationship between neural activity and perception might still arise from additional studies of neuronal receptive fields remains to be seen. Based on the arguments here, however, a different framework for understanding vision will be needed. The evidence discussed in previous chapters indicates that because the significance of images per se can't be known, visual perceptions are determined by the empirical links that have been gradually forged between images and successful behavior. As a result, what we see is not some simulacrum of the physical world, but a subjective universe in which perceptions are predicted by the history of experience.

To appreciate a strategy that may seem bizarre and/or implausible, it may be helpful to reiterate some aspects of the loaded dice analogy from Chapter 1. The analogy described how a player could succeed when direct analysis of a physical state is precluded. The point was that an empirical evaluation of the dice and the implications for behavior could nonetheless be made by tallying the frequency of occurrence of the numbers rolled over the course of many throws. More specific information about how to behave could be gleaned by taking one die as a "target" and the other as "context"; the more frequently a target number on one die is associated with a context number on the other, the greater the chance that a given behavior in response to the combination will succeed. The frequency of occurrence of different stimulus patterns arising in nature is similar to the historical outcome of repeatedly rolling loaded dice. Sampling a database that relates images with their physical correlates can thus stand in for the relative success of the trial and error behaviors that would have progressively modified visual system connectivity over the course of evolution and life. In this way useful percepts and behaviors can be generated despite the inverse problem. A corollary, however, is that visual perceptions will never correspond to the properties of the physical world. What we see is the guide that enabled successful behavior in the past given a world that is hidden from the senses, although measurable with the instruments of physics and chemistry.

Accordingly, the perceptual qualities of lightness, brightness, color, form, distance, depth, and motion are defined not by the image characteristics, physical sources, or even the relationships between them, but by information derived operationally from the history of trial and error interactions with the world. In addition to contending with the inverse problem, this strategy seamlessly incorporates influences arising from of the full range of brain systems pertinent to behavioral success (other sensory modalities, attentional systems, memory systems, emotion systems, etc.) without having to resort to the awkward idea that top-down (cognitive) influences act on bottom-up (sensory) input. Given the problems confronting the emergence of biological vision, this strange way of describing what we see and how we see it seems inevitable.

Mechanism Underlying Visual Perception

How, then, does the visual system actually work? The answer may be surprisingly simple. If empirical associations between light stimuli and useful behaviors determine what we see, then the mechanism for generating visual percepts is the result of the input (a retinal image) eliciting an output (a percept and associated behavior) that has already been determined by species and individual experience. In other words, vision is reflexive.

The concept of a reflex, which is not precisely defined, generally alludes to behaviors that entail a minimum of "higher order" processing. The textbook paradigm is one of the spinal reflexes studied by Charles Sherrington in the early twentieth century, or an autonomic reflex such as salivation in response to a food stimulus, studied by Ivan Pavlov in his work on conditioning at about the same time. For Sherrington, a reflex meant an involuntary response that depends on relatively simple and well defined neural pathways (Figure 7.6). Sherrington was well aware, however, that the idea

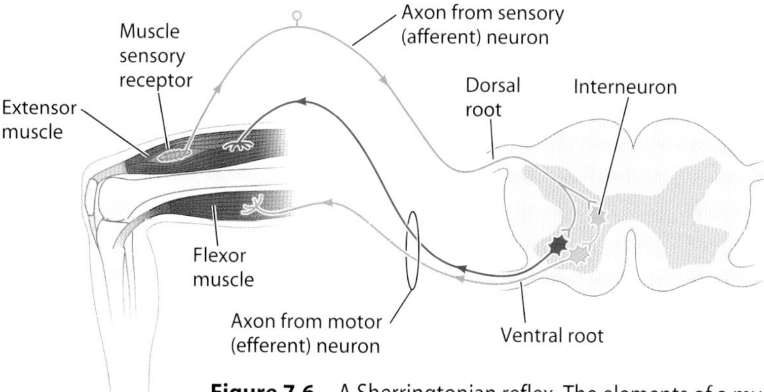

Figure 7.6 A Sherringtonian reflex. The elements of a myotatic reflex are shown in this diagram. Omitted, however, are ascending pathways that carry information from the spinal cord to the brainstem and cerebrum and the descending pathways that modulate the execution of this and other spinal reflexes. (From Purves et al., 2008b.)

of an isolated reflex is a "convenient… fiction," since "all parts of the nervous system are connected together and no part of it is ever capable of reaction without affecting and being affected by other parts…" (Sherrington, 1906b, p. 7).

Despite Sherrington's astute caveat, a reflex is still thought of as a neural response that is free of "top down" influences. But this interpretation does not fit the facts. For example, the visual system organizes eye movements by the interaction of neural activity in the brain regions indicated in Figure 7.7. Saccades, the conjunctive eye movements that direct the point of gaze to objects of interest, are normally made about 3 times per second. The vast majority of saccades thus occur in response to aspects of retinal stimuli of which we unaware, and they involve sensory–motor transformations in the brainstem (the superior colliculus) that operate much the same as spinal reflexes. Saccadic eye movements can also be generated by a conscious desire to look at something. There is no reason to suppose that when an observer wants to look at an object and does so anything very different is happening, even though additional (and less well understood) circuitry is involved. As every neurologist knows, interruption of ascending and descending spinal cord pathways shows that spinal reflexes are also determined by higher order neural circuitry.

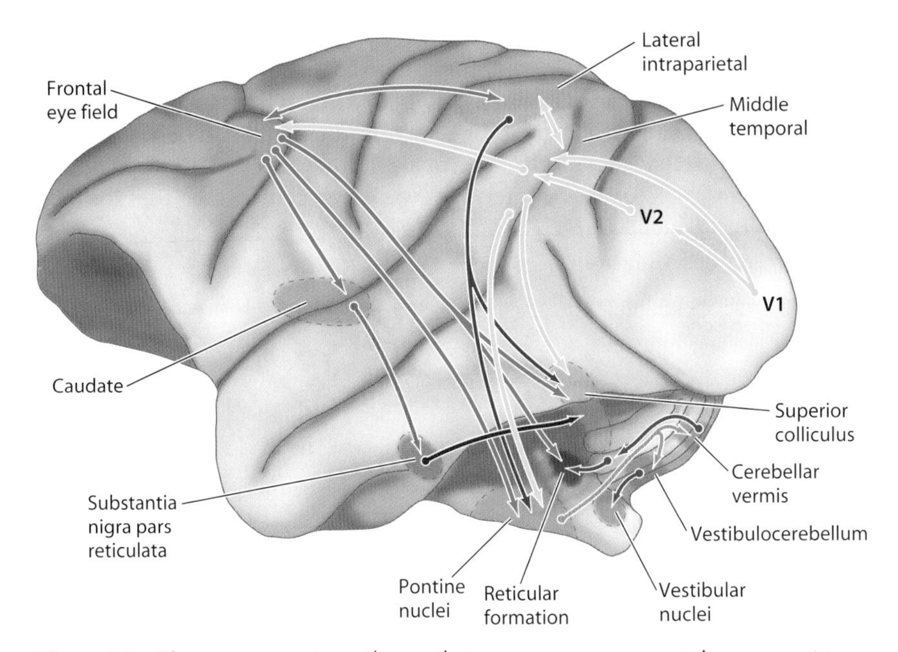

Figure 7.7 The sensory–motor pathways that govern eye movements in response to visual stimuli, illustrated in the brain of a rhesus monkey where this circuitry has been well studied. Despite the routine and usually reflexive nature of eye movements, many regions of the brain are involved in producing this important visual behavior that can also be voluntary. (From Purves et al., 2008a.)

Thus, Sherrington's definition of reflexes—automatic responses fully determined by extant neural circuitry whose function can nonetheless be influenced by many brain regions—seems is an apt description of the mechanism that generates visual percepts. In this conception, vision entails no neural computations or special mechanisms. The work has already been done by the connectivity established by previous evolutionary and individual experience.

How the Links between Images and Behavior Are Established

The driving force that instantiates the reflexive links between light stimuli and behaviors during the evolution of a visual species is natural selection: Random changes in the structure and function of the visual systems of ancestral forms have persisted—or not—in descendants according to how well they serve the reproductive success of the animal whose brain harbors that variant. Any configurations of neural circuitry that mediate more successful responses to visual stimuli increase among the members of the species, whereas less useful circuit configurations do not. The significance of this conventional statement about the phylogeny of any biological system for present purposes is simply to emphasize the existence a universally accepted mechanism for instantiating and updating neural associations between light stimuli and behavioral success.

Although the mechanisms of inheritance and the epigenetic rules of early development determine the organization of the visual system prior to an animal's interaction with the outside world, much work in neurobiology over the last century has shown that neural circuitry is further shaped in postnatal life according to individual experience using mechanisms of neural plasticity (Purves and Lichtman, 1985; Hubel, 1988; Daw, 1995; Hubel and Wiesel, 2005). Taking advantage of experience accumulated during postnatal development and adult life allows individuals to benefit from circumstances in innumerable ways that contend with the challenges in the world more successfully than would be possible using inherited circuitry alone.

These changes of neural connectivity arising in phylogeny and ontogeny are, in the end, two different mechanisms (natural selection vs. neural plasticity) and time frames (eons vs. days, weeks, or years) for building the empirical consequences of experience into the wiring of the visual system.

Understanding the Organization of the Visual System in These Terms

Most thinking about vision in recent decades has focused on processing that is essentially logical, depending on algorithmic computations. In most applications, computer algorithms lay out a series of defined steps in series or parallel to address problems that have already been solved in principle. The success of this general approach—referred to generically as "artificial intelligence"—hardly needs emphasis.

There is, however, another way that computers can solve complex problems, as first pointed out in an influential paper by Warren McCulloch and Walter Pitts in 1943. Rather than depending on a series of logical steps that dictate each operation, McCullough and Pitts suggested that problems might also be solved by a computing device comprising units ("neurons," in their biologically inspired terminology) whose interconnections would progressively change according to feedback about the success (or failure) in dealing with the problem at hand (Figure 7.8). The key feature of such systems—which, as a consequence of the biological basis for this idea came to be called "artificial neural networks" (or just "neural nets")—is that the computer can solve a problem without a priori knowledge of the answer, the steps needed to reach it, or even a clear conception about how the problem might be addressed. In effect, neural nets generate solutions by trial and error, i.e., by gradually producing more and more useful responses according to operational success. As a result, the architecture of the "trained network" (or "evolved network" in the case of genetic

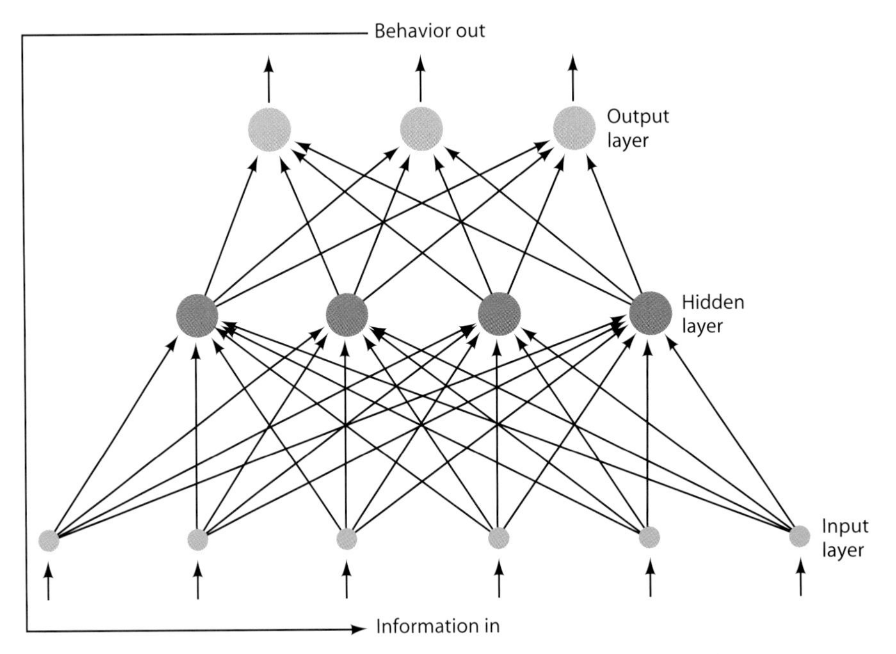

Figure 7.8 The structure of a simple artificial neural network. The network comprises an input layer, an output layer, and a "hidden layer." The common denominator of neural network architectures is a richly interconnected system of nodes or "neurons." The system is programmed such that the strengths of the connections between the nodes are changed progressively according to the results of the trial-and-error responses produced at the output. As a result, the system deals ever more and more successfully with the task or problem it has been given. (From Purves et al., 2008b.)

optimization algorithms; see below) is a manifestation of the history of its experience. Because the operation of such systems is not easily reduced to a series of logical steps, an aphorism is that neural networks provide the second-best solution to any given problem. Although this alternative strategy for carrying out computation has experienced a series of ups and downs since McCulloch and Pitts's paper appeared in the early 1940s, the approach is an effective way of dealing with complex problems, especially ones that entail many contingencies.

An extension of this approach that ties it even more closely to biology is the evolution of neural networks in virtual environments. This method uses genetic algorithms that mimic gene replication, mutation and recombination (Figure 7.9). Under these circumstances, the properties and abilities of a population of virtual organisms (neural networks) gradually change as a result of the selection and reproduction of "offspring" networks that perform better. Thus, random changes in the connectivity in "parents" persist or not in their descendants as a result of behavioral outcomes, continually updating network circuitry according to empirical success. Configurations of neural circuitry that better exemplify behavioral success thus wax in the evolving population of networks, whereas circuit configurations that are less useful wane. Although such "artificial life" paradigms are vastly simplified versions of the biological systems they represent, when artificial neural networks are trained or evolve to contend with images, they begin to respond to visual input in much the same way we do (see Chapters 2 and 3).

This way of solving problems closely matches the wholly empirical interpretation of perception and behavior. If the link between visual stimuli

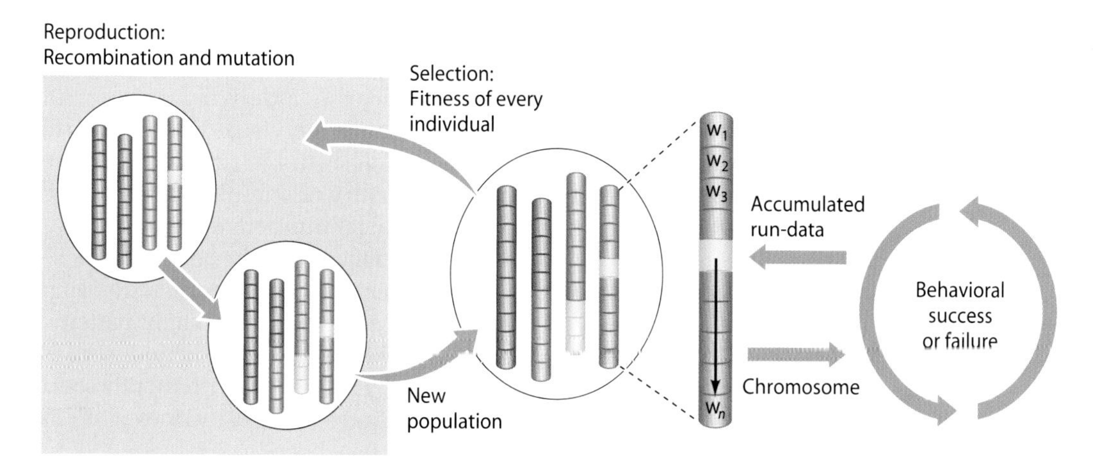

Figure 7.9 Evolution of neural networks using genetic algorithms. The "weights" of the connections in a neural network can be modeled as genes that mutate and rearrange during meiosis in ways that mimic biological reproduction. "Genetic" changes lead to corresponding changes in the behavior of the networks, which can then be selected for according to behavioral success ("fitness") in a virtual environment. (From Purves et al., 2008b.)

and behavior requires an instantiation in neural circuitry of the relative success of trial-and-error responses, then artificial neural networks and their operation provide a way of exploring the physiological and anatomical underpinnings of vision (although certainly not an easy way).

Conceiving visual processing in a wholly empirical framework can also explain why neuronal response properties have been difficult understand in logical terms. Because the circuitry is established historically rather than by analysis of image features, no logical processing would be expected, just as none is expected in artificial neural networks and, arguably, the human visual system.

In short, the characteristics of neural networks, whether artificial or biological, are well suited to deal with the inevitably uncertain meaning of visual stimuli. If vision does operate on a wholly empirical basis, then mimicking the history of visual evolution and development in highly simplified environments may be the best way to understand why biological visual systems are organized the way they are.

Conclusion

The fundamental problem for biological visual systems is that retinal images cannot reveal the physical world, much less its significance for behavior. Light reaching the retina inevitably conflates the contributions of illumination, reflectance, and transmittance (and many subsidiary factors); the spatial variables of size, distance, and orientation are likewise intertwined and cannot be disentangled in images as such. As a result, the provenance of any visual stimulus—and therefore its significance for visually guided behavior—is unknowable by means of image information alone. These facts present an obvious quandary: Successful visually guided behavior depends on responding to physical objects and conditions, but this information about the world is not conveyed by light stimuli.

If analysis of retinal stimuli cannot specify the underlying reality with which observers must contend, how then does the visual brain generate percepts and behaviors that typically work? The studies reviewed in these chapters indicate that the behavioral ambiguity of stimuli created by the inverse problem is resolved by having retinal stimuli activate circuitry that has been determined by accumulated experience with the behavioral consequences of environmental interactions over phylogenetic and individual time. The result is visual system wiring that associates retinal light patterns with successful perceptions and behaviors. Thus, what we actually see is the behavioral significance of "sense data." The ability to predict the odd way we perceive the world is the evidence that vision really does work in this counterintuitive way.

Understanding what observers see and the underlying circuitry in wholly empirical terms is different in both concept and consequence from attempts to rationalize visual perception and its neural underpinnings in terms of detecting, filtering, analyzing, and representing the physical properties of objects, or even the properties of retinal images. In an empirical interpretation of vision, there is no series of computations that ultimately represents images, or statistical inferences about stimulus sources. Although seeing in this way may seem maladaptive, it deals wonderfully well with the otherwise intractable problems confronting the evolution of biological vision.

Finally, to reiterate the what was said in the Preface, this concept of vision could be falsified if: (1) It could be shown there is a plausible alternative to this way of contending with the inverse problem; or (2) it could be shown that the range of visual perceptions dealt with in earlier chapters can be predicted in some other way. If a wholly empirical theory of vision cannot be rejected on one of these grounds, it would be best to forge ahead and explore vision in this framework.

A Primer on Visual System Structure and Function

The Non-Neural Components of the Eye

All sensory systems have a pre-neural apparatus of some sort that collects, filters, and amplifies the relevant energy in the environment so that it interacts more effectively with the receptor cells, where the neural phase of sensory processing begins. In the case of vision, this initial process entails the cornea, lens, and ocular media that focus and filter light before it reaches the neural retina (Figure A.1).

Since photons that emanate from a point diverge, they must be focused on the retina or any other detector if they are to accurately represent the arrangement of the points in space from which they derived. A sharp image is thus created only when these diverging "rays" are brought back together by the refractive properties of the cornea and lens so that the pattern of activation on the retina corresponds to the pattern of points on the surfaces of the objects from which the light was reflected or emitted (a reasonably sharp image can be made in the absence of refraction by passing light through a pinhole, but only at the expense of image intensity, since the majority of diverging rays are necessarily lost in the process). Because the cornea and lens are denser than air, light entering the eye is refracted (i.e., the slower speed of photons in these denser media changes the direction of all rays that are not normal to the air–eye interface). As a result of both the density and the surface curvature of these elements, the rays diverging from each point in a scene converge such that, in an eyeball

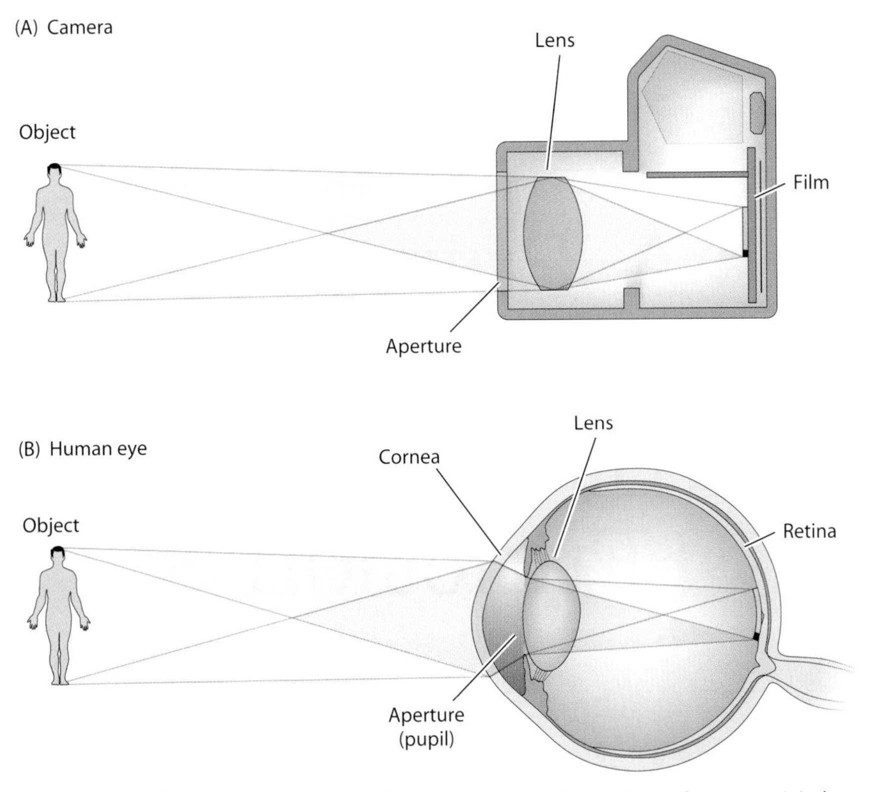

Figure A.1 The optical properties of the eye. Like the optical system of a camera (A), the light rays diverging from each point on the surfaces of the objects in a scene are brought back to corresponding points on the retina by the refracting power of the cornea and lens (B), thus creating a focused 2-D projection of the 3-D scene being witnessed by the observer.

of normal length, they meet again at the retinal layer that comprises the photoreceptor cells (discussing this process in terms of rays is a geometrical convention made possible by the fact that, for purposes of vision, photons travel in straight lines).

Although most of the focusing is achieved by the refractive power of the cornea, the lens contributes to this process in a critical way. Because the lens can change its shape under neural control, the relatively small but adjustable refracting power that it provides allows the observer to focus light rays arising from objects at different distances (i.e., to cope with the highly divergent rays that reach the pupillary aperture from relatively nearby objects, as well as the more nearly parallel rays that reach the pupil from distant objects). Even when focused, however, the eye has generated an obvious transformation: The arrangement of points in three-dimensional space has been "flattened" onto the two-dimensional plane of the projection. This transformation is one of the major factors (but only one) underlying the inevitable ambiguity of visual stimuli and the inverse problem that plays such a critical role in the theory of vision proposed in the main text.

This description of the eye is, of course, particular to humans. There is a wonderful diversity of eye structure among vertebrates and an even wider range of visual specializations among invertebrates (see, for example, Polyak, 1957; Bowmaker, 1998; Rodieck, 1998; Nilsson and Land, 2001). As might be expected, different animals respond to wavelength ranges that are especially pertinent to their ecological niche, and these differ from the range that humans have evolved to see.

The Retina

The retina is a laminated neural tissue derived from an out-pocketing of the embryonic brain. The purpose of the retina is to transduce light energy into neural signals (receptor potentials, synaptic potentials, and action potentials; see Purves et al., 2008a, for an introduction to these basic aspects of neurobiological signaling).

The events leading to visual perception are initiated when photons are absorbed by pigment molecules in the photoreceptor cells (a full account of this process can also be found in Purves et al., 2008a). Photoreceptors in humans and other mammals are structurally and functionally divisible into two broad categories: rods and cones (Figures A.2 and A.3). The rods, which contain the photopigment rhodopsin, initiate activity in a functional component of the visual system (the scotopic, or rod, system) that is especially sensitive to light and operates to greatest effect at night (Figure A.4). The scotopic system, however, is incapable of high resolution (i.e., detecting the separateness of two closely spaced points) and cannot by itself discriminate the spectral composition of light stimuli; indeed, no single type of receptor can be a basis for responding appropriately to the spectral differences that elicit sensations of color, which requires a comparison of the activity of at least two different receptor types (see below). The exquisite sensitivity of the rod system (threshold sensitivity is the reciprocal of the minimum stimulus energy required to elicit a perceptual response) arises from the relative efficiency with which photon capture generates a voltage change across the membrane of these cells as well as from the fact that many rods converge to contact the next cell type in the retinal pathway, the bipolar cells (see Figure A.2B). This convergence also explains why the rod system has little acuity: Any stimulus within the territory of the rods converging onto a bipolar cell will, as far as the visual system is concerned, have an equivalent location in space. As a result these factors, the rod system is very effective at minimal light levels but ineffective in daylight because rod responses "saturate" quickly as light levels increase.

The cone (or photopic) system, in contrast, operates best at greater light intensities (see Figure A.4) and initiates the high acuity perceptions that discriminate visual detail. The three human cone types are distinguished by

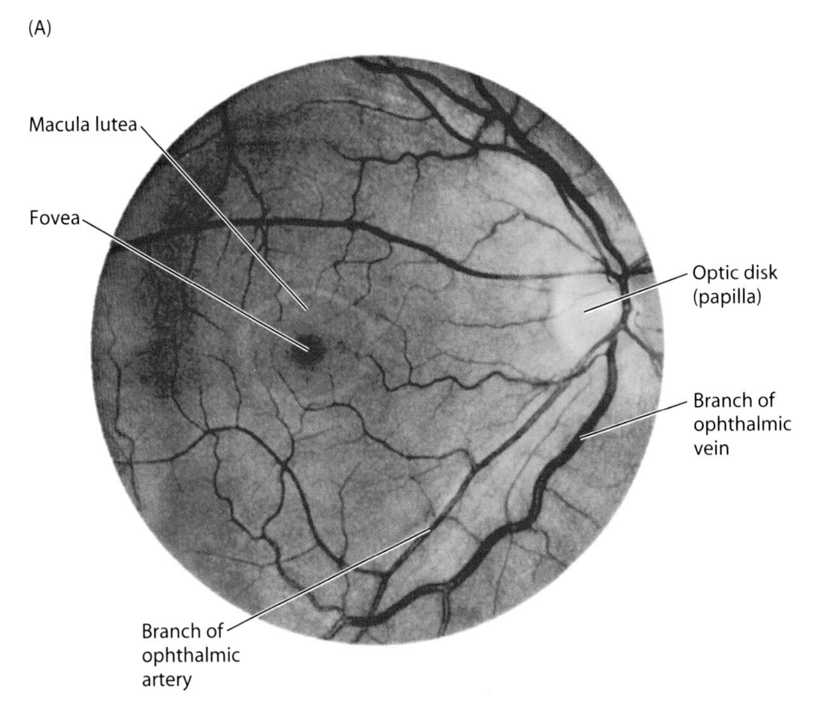
(A)
Macula lutea
Fovea
Optic disk
(papilla)
Branch of
ophthalmic
vein
Branch of
ophthalmic
artery

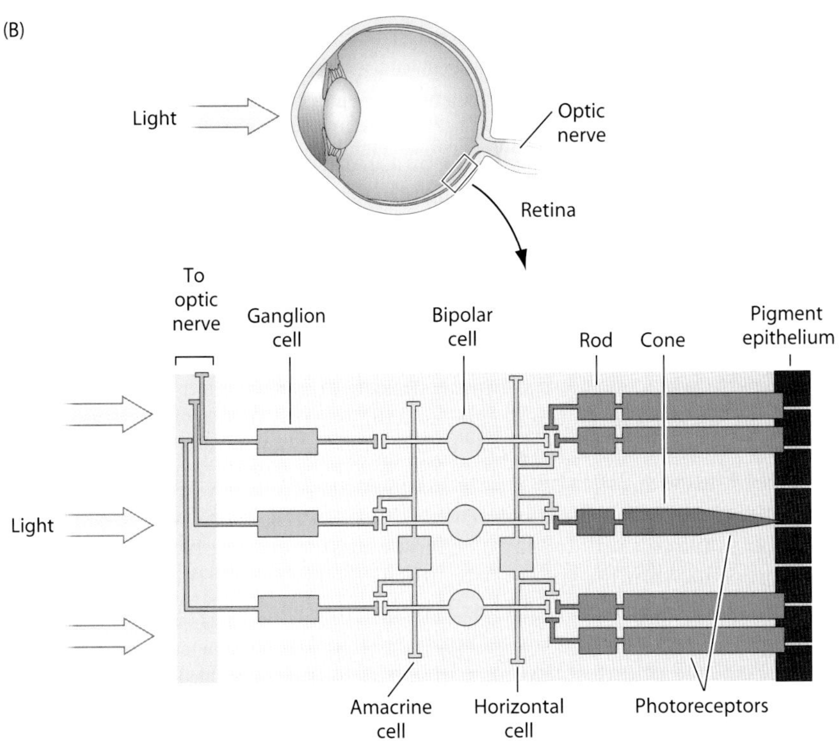
(B)
Light
Optic
nerve
Retina
To
optic
nerve
Ganglion
cell
Bipolar
cell
Rod
Cone
Pigment
epithelium
Light
Amacrine
cell
Horizontal
cell
Photoreceptors

◀ **Figure A.2** The human retina. (A) Photograph of a human retina (right eye) as seen through an ophthalmoscope, indicating the major features referred to in the text. (B) Diagram of the elements of the retina shown schematically in a blowup of the region indicated. The neuronal activity elicited by a visual stimulus flows mainly from the receptors to the output neurons (the retinal ganglion cells) whose axons make up the optic nerve. However, the activity elicited by a stimulus also spreads laterally via amacrine and horizontal cells, a fact that figures prominently in some attempts to explain the perceptual effects described in Chapters 2 and 3. (A is from Purves et al., 2008a.)

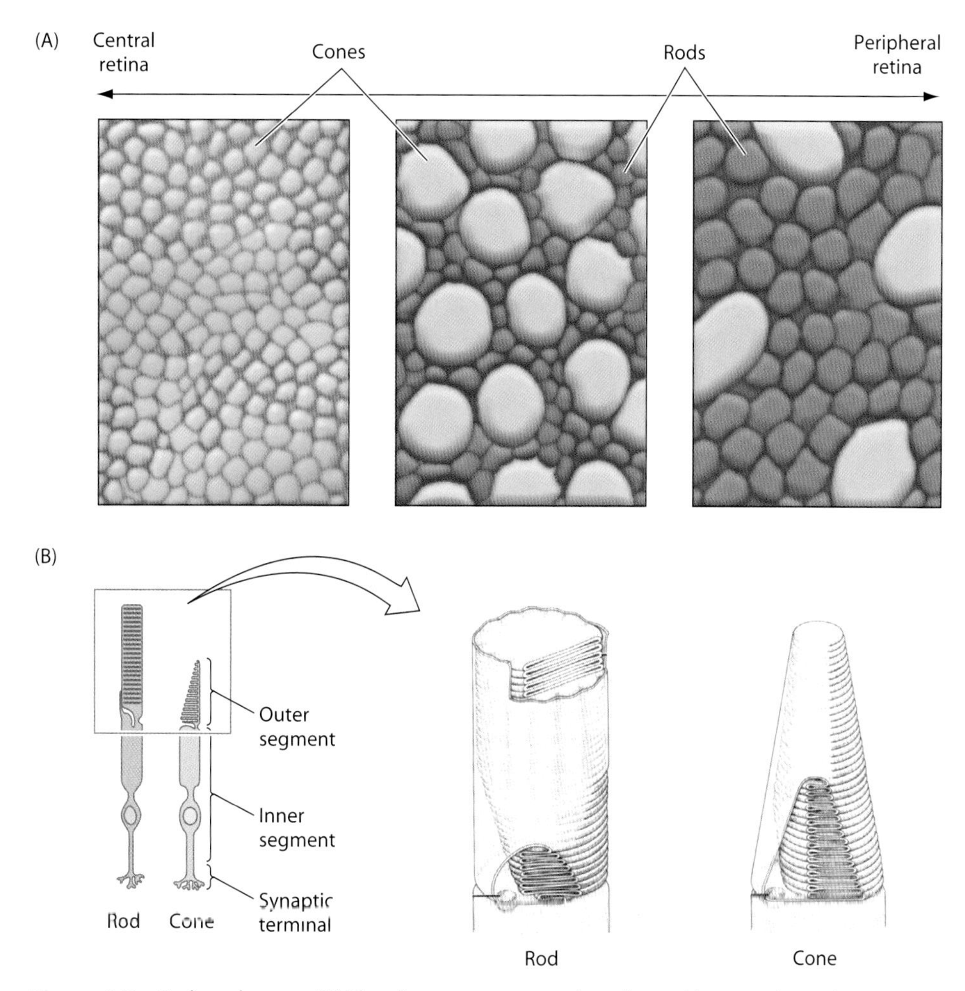

Figure A.3 Rods and cones. (A) The photoreceptor mosaic, as it would appear in a primate (e.g., rhesus monkey or human) retina. The rods are shown in blue and the cones in green. The fovea in the central retina comprises mostly cones, and its central region (called the foveola) comprises only cones. The prevalence of rods increases at progressively more eccentric retinal locations (see also Figure A.5A). (B) Diagram showing the different structural features of the two photoreceptor types; blowup shows the arrangement of the disks in the outer segments, which harbor the photopigment proteins in their membranes.

the different photopigments they contain, and thus their different absorption properties (which, as described below, is the basis for color vision). The relative insensitivity of the cone system to low light levels arises not from any deficiency in the ability of cones to capture photons (they do so as efficiently as rods) but because each captured photon generates less current change across the receptor cell membrane, and because only one or a few cones converge on the relevant bipolar cells (see Figure A.2B). The more or less one-to-one arrangement of cones, bipolar cells, and, ultimately, the ganglion cells in the central region of the retina (the fovea) provides much greater spatial resolution than the highly convergent rod system. The greater resolution arises for the same reason that increasing the density of pixels in a computer graphic improves the detail of the image: The high packing of cones in the fovea and their minimal convergence greatly increases the "density" of information available per unit area in this part of the retina (see Figures A.3 and A.5A).

The responsiveness of rods and cones enable the visual system to detect and discriminate between light intensities that differ by many orders of magnitude: The dimmest light that humans can detect is billions of times less intense than the strongest light levels we can experience without incurring physical damage to the retina (Figure A.4). This enormous range of sensitivity implies that the visual system must adapt its function to the prevailing light intensity. The firing rate of neurons, which encodes stimulus intensity, varies by only 2 or 3 orders of magnitude, whereas the sensitivity of visual system encompasses more than 10 orders of magnitude.

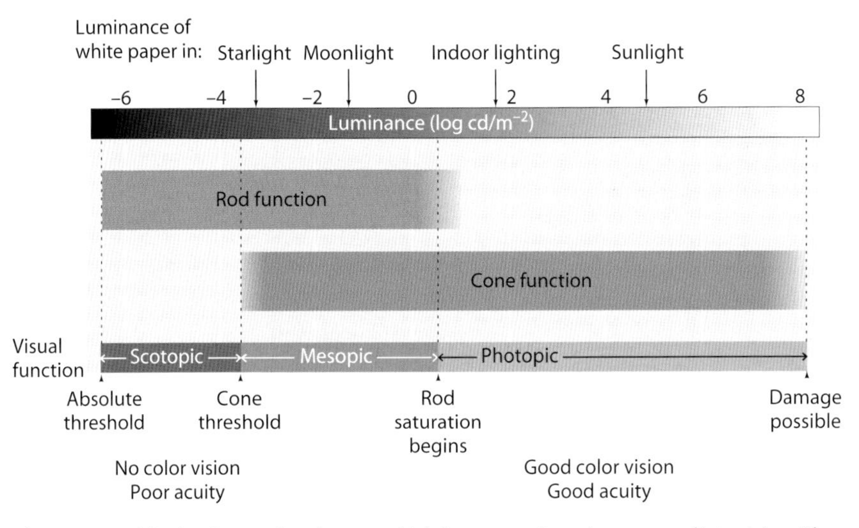

Figure A.4 The luminance levels over which human rods and cones mediate vision. The scotopic (rod) system is specialized to detect light at low levels of illumination, whereas the photopic (cone) system is specialized to process detail at what humans regard as normal levels of light. Note the broad range of luminances—the mesopic range—at which both the rod and cone systems are functional.

The adaptation is made obvious by the experience of entering a dark environment in the daytime. When moving from outdoors to a dark interior (a movie theater, for example), the differences in light intensities inside can't be distinguished at first. Over some minutes, however, objects become increasingly visible as the visual system adapts to the ambient light level. Such "dark adaptation," which depends on restoring the biochemical integrity of the rod photochemical system (rhodopsin is bleached by light), can take as long as 20 to 30 minutes if the transition is from photopic to scotopic conditions. When the visual system adapts to increased light intensity (e.g., coming out of the theater into daylight), the much more rapid phenomenon in the reverse direction is called "light adaptation"; the time course of light adaptation is faster because cones adapt relatively quickly to changes in illumination. An important qualification is that adaptation involves not only retinal mechanisms, but also poorly understood changes in the central stations of the visual system, as well as ancillary responses such as the adjustment of pupil size to a diameter appropriate to the light level.

The functional specialization of rods and cones is equally apparent in the distribution of receptor types across the retina (Figure A.5A, B). Although people don't often notice, the acuity and color sensitivity of human vision falls off rapidly as a function of eccentricity (i.e., the distance in degrees of visual angle away from the line of sight). Consequently, vision beyond the central few degrees of the visual field is surprisingly poor. Without an intact central retina (a deficiency all too common as a result of the macular degeneration that afflicts many aged individuals), peripheral vision operates at levels that qualify as legal blindness, which is defined as uncorrected acuity of 20/200 or worse on the standard Snellen eye chart test (20/200 means being able to read correctly at a distance of 20 feet only the line on the chart that a normal individual would be able to read at a distance of 200 feet). When staring at the center of a checkerboard pattern of red and blue squares that are, say, a centimeter on each side located a meter away, it is difficult or impossible for a person with normal vision to resolve individual squares or to tell the color of any particular square at even moderately eccentric locations in the visual field (to demonstrate this falloff with eccentricity, check how well you see the words at the beginning or end of one of these lines of text when you fixate on a word in the middle of the line). As a result of this inadequacy, frequently moving the direction of gaze to different locations in visual space is essential for seeing objects in detail, which is of course what we normally do during the inspection of a scene (Figure A.5C).

The reason for the inability to see well when not looking directly at an object has primarily to do with the distribution of rods and cones in the retina. As diagrammed in Figure A.5A and B, cones greatly predominate in the specialized central region of the retina called the fovea, which corresponds to the line of sight and the couple of degrees around it. The prevalence of cones, however, falls off sharply in all directions as a function of distance from this locus. As a result, high acuity vision is limited to the fovea and its immediate surround. Conversely, sensitivity to a dim

(A)

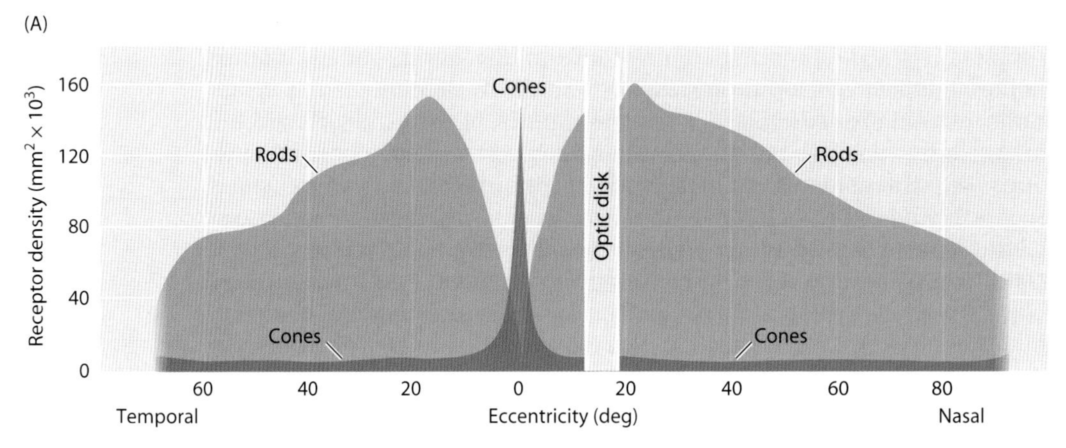
Receptor density (mm^2 × 10^3)
160
120
80
40
0
Cones
Rods
Rods
Optic disk
Cones
Cones
60
40
20
0
20
40
60
80
Temporal
Eccentricity (deg)
Nasal

(B)

(C)

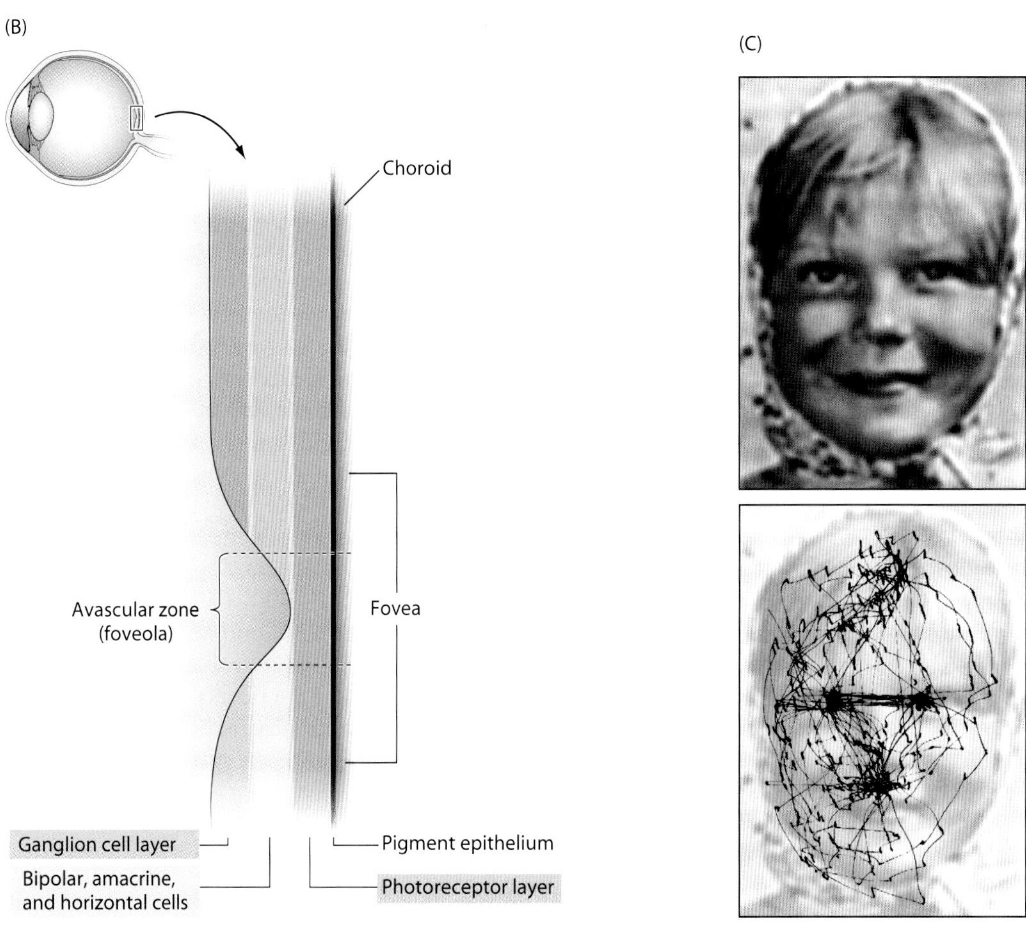
Choroid
Avascular zone
(foveola)
Fovea
Ganglion cell layer
Bipolar, amacrine,
and horizontal cells
Pigment epithelium
Photoreceptor layer

◀ **Figure A.5** Why humans see detail and color poorly when the object of interest is viewed eccentrically. (A) Density of rods and cones as a function of retinal eccentricity. The density of cones falls off rapidly moving away from the fovea; notice the complementary fall in rod density moving toward the center of the fovea (which is at 0°). (B) Diagram of the foveal region of the retina (the fovea is about 1.5 mm in diameter and takes in about 2° of visual space; see Figure A.3A). Foveal cones are tightly packed and rods virtually absent in a small region in the middle of the fovea, called the foveola (which is about 0.3 mm in diameter). The foveola is further specialized for high acuity by a thinning of the overlying retinal layers and the absence of blood capillaries. (C) To overcome the poor acuity outside the fovea, humans normally shift their direction of gaze several times a second by executing ballistic eye movements called saccades. In this example, taken from the work of the Russian physiologist Alfred Yarbus (1959), the eye movements of a subject freely inspecting this photograph of a face were recorded for about 3 minutes. The dots indicate the points of fixation; the lines show the trajectories of the saccadic eye movements to new points of interest, which, not surprisingly, are the salient facial features and other regions of light/dark contrast in the image.

stimulus is actually slightly higher off the line of sight because of the paucity of rods in the fovea and their preponderance a few degrees away. Indeed, the very center of the fovea (the foveola) is blind to dim flashes of light because rods are completely lacking in this small area (the foveola is also blind to a spot of "blue" light even at photopic levels because of a paucity of cones sensitive to the relevant wavelengths). People are generally unaware of the poor quality of eccentric visual performance simply because whenever it is important to see something clearly, the object of interest is brought onto the fovea by eye, head, or body movements (see Figure A.5C).

A great deal of additional visual processing takes place in the retina, carried out by the cells intervening between rods and cones and retinal ganglion cells, which are the output neurons of the eye. In addition to the bipolar cells, horizontal cells and amacrine cells are critically involved in retinal processing (see Figure A.2B). Although much is now known about the connectivity of these neurons and their physiology (see, for example, Sterling, 1997; Rodieck, 1998), the purposes they serve are, for the most part, not well understood. In general, however, they mediate horizontal integration of information in the plane of the retina, which is especially important in considering how one part of a scene could affect the perception of another part. The retinal neurons primarily responsible for this lateral information flow are the horizontal cells, which act to modulate neurotransmitter release from photoreceptors as a result of activity in neighboring regions.

Central Visual Pathways

Despite the functional differences between rods and cones, the two systems converge onto the retinal ganglion cells, whose axons make up the

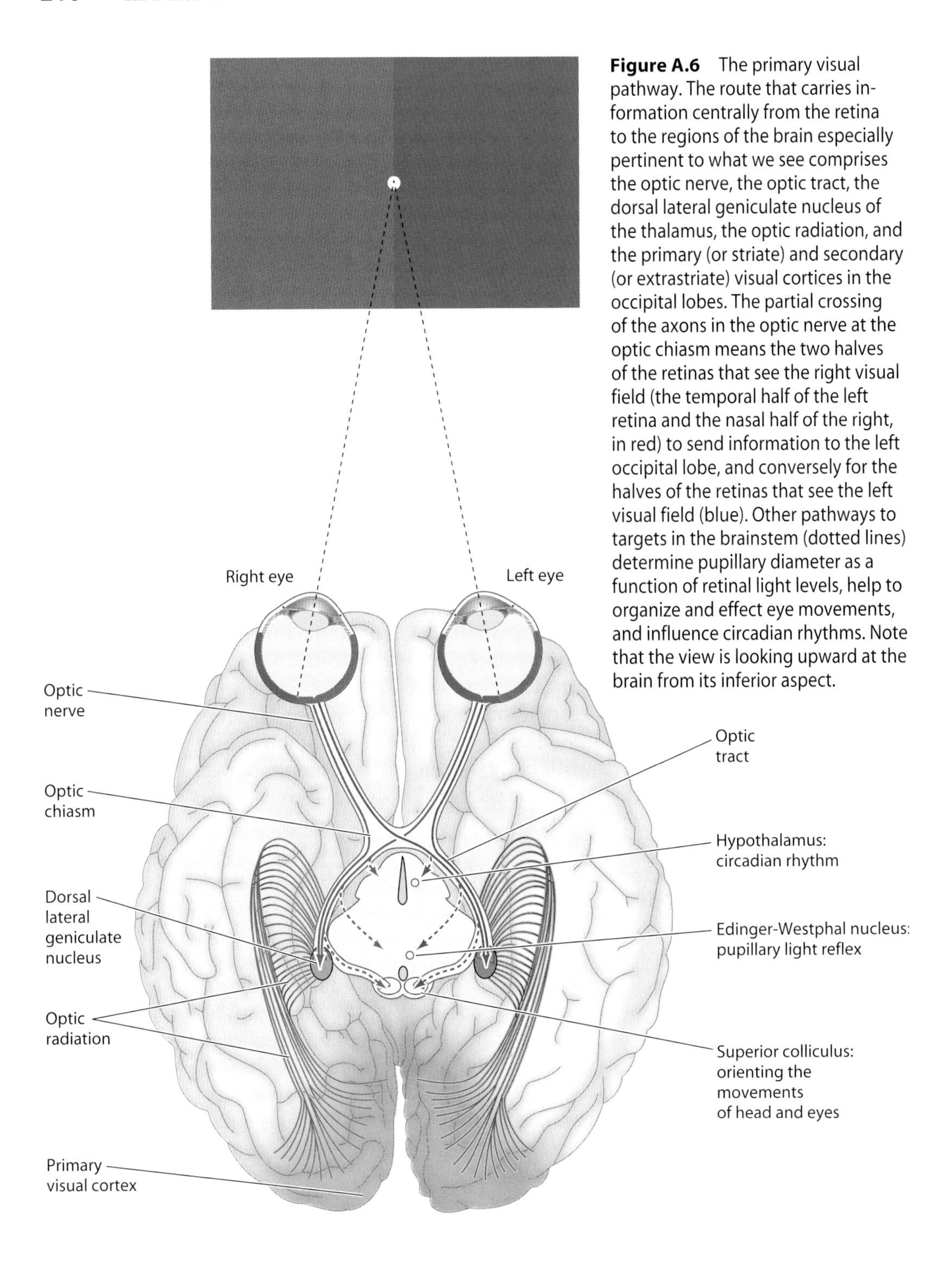

Figure A.6 The primary visual pathway. The route that carries information centrally from the retina to the regions of the brain especially pertinent to what we see comprises the optic nerve, the optic tract, the dorsal lateral geniculate nucleus of the thalamus, the optic radiation, and the primary (or striate) and secondary (or extrastriate) visual cortices in the occipital lobes. The partial crossing of the axons in the optic nerve at the optic chiasm means the two halves of the retinas that see the right visual field (the temporal half of the left retina and the nasal half of the right, in red) to send information to the left occipital lobe, and conversely for the halves of the retinas that see the left visual field (blue). Other pathways to targets in the brainstem (dotted lines) determine pupillary diameter as a function of retinal light levels, help to organize and effect eye movements, and influence circadian rhythms. Note that the view is looking upward at the brain from its inferior aspect.

optic nerve (Figure A.6). This early convergence indicates that the idea of a "rod system" and a "cone system" should be taken in an operational rather than an anatomical sense. Whereas the confluence of information from these fundamentally different receptor types onto the same target cells is no doubt a matter of economy (recall that the rods work primarily in dim light and the cones primarily at normal light levels), this sharing of the same pathway also indicates something of general importance in the organization of the visual system. Although vision certainly entails parallel processing of qualitatively different information (see below), functional parcellation does not occur in the neat way implied by the use of terms such as "channels" (a word widely used in the visual literature).

The major target of the retinal ganglion cell axons in the optic nerve is the thalamus, a walnut-sized nuclear complex in the diencephalon of humans and other mammals (the diencephalon is a major subcortical division of the brain that comprises the thalamus and hypothalamus). The thalamus, among many other functions, relays sensory information from the retina to the cerebral cortex along the pathway that carries the visual information pertinent to perception (see Figure A.6). With respect to the visual system, the relevant thalamic relay is the dorsal lateral geniculate nucleus, a laminated and therefore highly distinctive structure about which a great deal is now known (Figure A.7A). The layers of the geniculate comprise two magnocellular layers (so named because of the relatively large neurons they comprise) and four parvocellular layers containing smaller neurons. The "parvo" and "magno" layers are innervated by distinct classes of retinal ganglion cells (Figure A.7B): The magno layers are contacted by the axons of larger retinal ganglion cells called M cells, and the parvo layers by smaller ganglion cells called P cells (in fact, there are many other subclasses of ganglion cells with differential purposes that remain poorly understood).

The neuronal associations between specific classes of retinal ganglion cells and lateral geniculate neurons reflect, as might be expected, significantly different functions. Generally speaking, the smaller P neurons in the retina and the related parvocellular neurons in the lateral geniculate are concerned with spatial detail that underlie form and color perceptions. The larger M retinal ganglion cells—and the magnocellular neurons they innervate in the thalamus—process information about changes in the stimulus that underlie attention and the perception of motion. The confluence of information from rods and cones onto the same ganglion cells notwithstanding, these structural and functional differences between the magno and parvo "pathways" indicate that processing different kinds of visual information in anatomically parallel systems is also an important (but certainly not a universal) theme in the organization of the visual system.

The arrangement of lateral geniculate nucleus is actually much more complicated than implied by Figures A.6 and A.7, since neurons in both the magno and parvo layers are extensively innervated by axons descending from the cortex and other brain regions, as well as from retinal ganglion

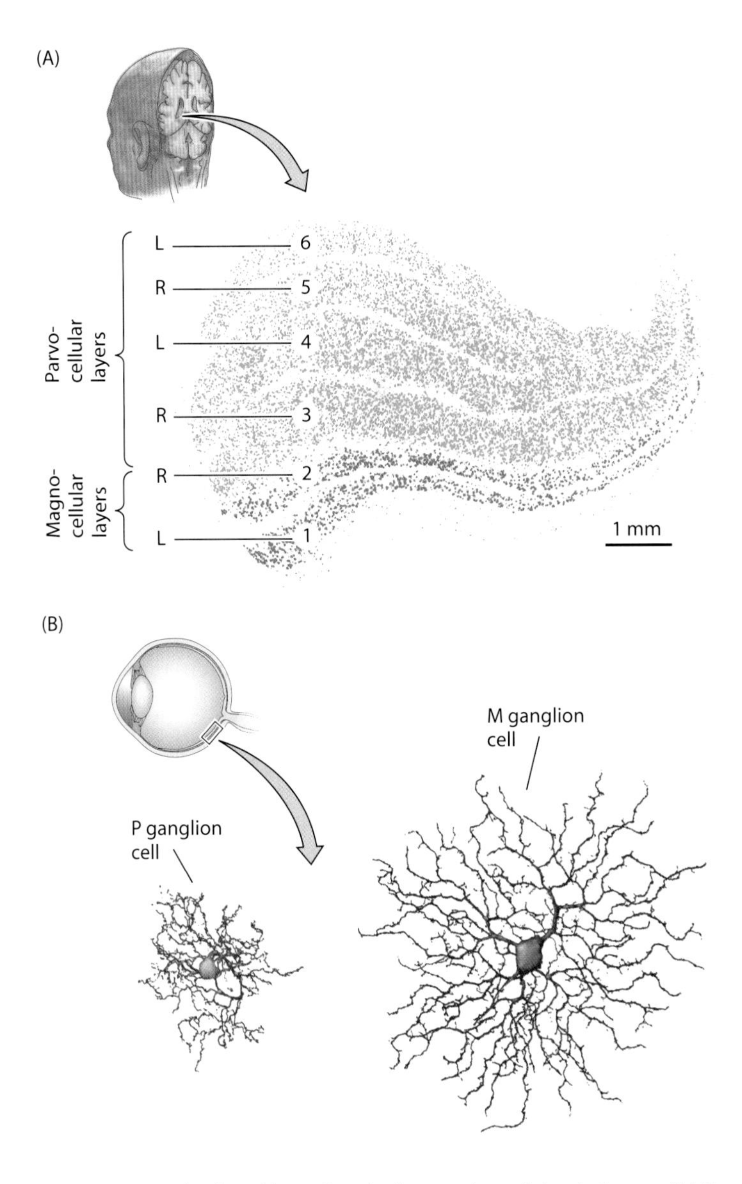

Figure A.7 The dorsal lateral geniculate nucleus of the thalamus. (A) Cross-section of the human thalamus showing the six layers of the lateral geniculate nucleus. Each layer receives input from only one eye or the other (indicated by R and L) and is further categorized by the size of the neurons it contains. The two layers shown in blue contain larger neurons and are therefore called the magnocellular layers; the four layers shown in green are called the parvocellular layers because of their smaller cells. (B) Tracings of representative M and P retinal ganglion cells, as seen after staining by the Golgi method; these neurons innervate, respectively, the magno and parvo cells in the thalamus. (A is from Andrews et al., 1997; B is after Watanabe and Rodieck, 1989.)

cells. Indeed, these thalamic neurons are estimated to receive about 10 times more innervation from the cortex and locally than they do from the retina. Although the function of this descending and/or local input is not known, the geniculate nucleus is clearly a station for processing visual information, and not the simple relay implied by the diagram in Figure A.6.

Finally, retinal ganglion cell axons project to a variety of other sites in addition to the thalamus, including the brainstem nuclei that regulate the diameter of the pupil as a function of the intensity of the light falling on the retina at any moment (thus mediating the so-called light reflex), the superior colliculus (which mediates the organization of eye movements that keep objects of interest in the line of sight), and the hypothalamus (which, among numerous other functions, organizes circadian physiological rhythms based on the daily cycles of light/dark) (see Figure A.6). Although these additional systems are extraordinarily important, they play little if any part in visual perception as such. It is thus the information carried by the retino-geniculate pathway, together with the subsequent projection from the geniculate to the visual cortex in the occipital lobes of the cerebrum that determines what we see. This is the reason the overall route from retina to cortex is called the primary visual pathway.

Visual System Topography

Visual physiologists have long recognized that abstraction of information from the retinal image is fundamental to the further processing that intervenes in the retina and the subsequent central stations of the visual system (the thalamus and striate and extrastriate cortices, where perception presumably occurs) (see Hubel, 1988 for a superb review of the classical work in this field). If the point of visual processing is to represent the retinal image centrally, why would the visual system so thoroughly deconstruct the stimulus in a way that makes putting it back together so difficult to imagine?

The idea that the visual system serves the goal of image representation has been difficult to get beyond. In addition to the obvious fact that an image is formed on the retina and that percepts do correspond in many ways to the retinal projection, a major factor in the persistence of image representation as a guiding principle is the salience of topography in the visual system (topography in this context refers to the maintenance of retinal adjacency in the higher stations of the visual system). Although both anatomy and physiology indicate that the retinal image is in most ways deconstructed, in one way it is assiduously maintained: Neurons at successive central stations of the primary visual pathway maintain their neighbor relationships. For example, when electrophysiological recordings are made from neurons in the lateral geniculate nucleus, adjacent sites in a given layer typically respond to stimulation of adjacent retinal sites (see Figure A.7A). Moreover, when a recording electrode is passed from one geniculate layer to another, the position on the retina (or in

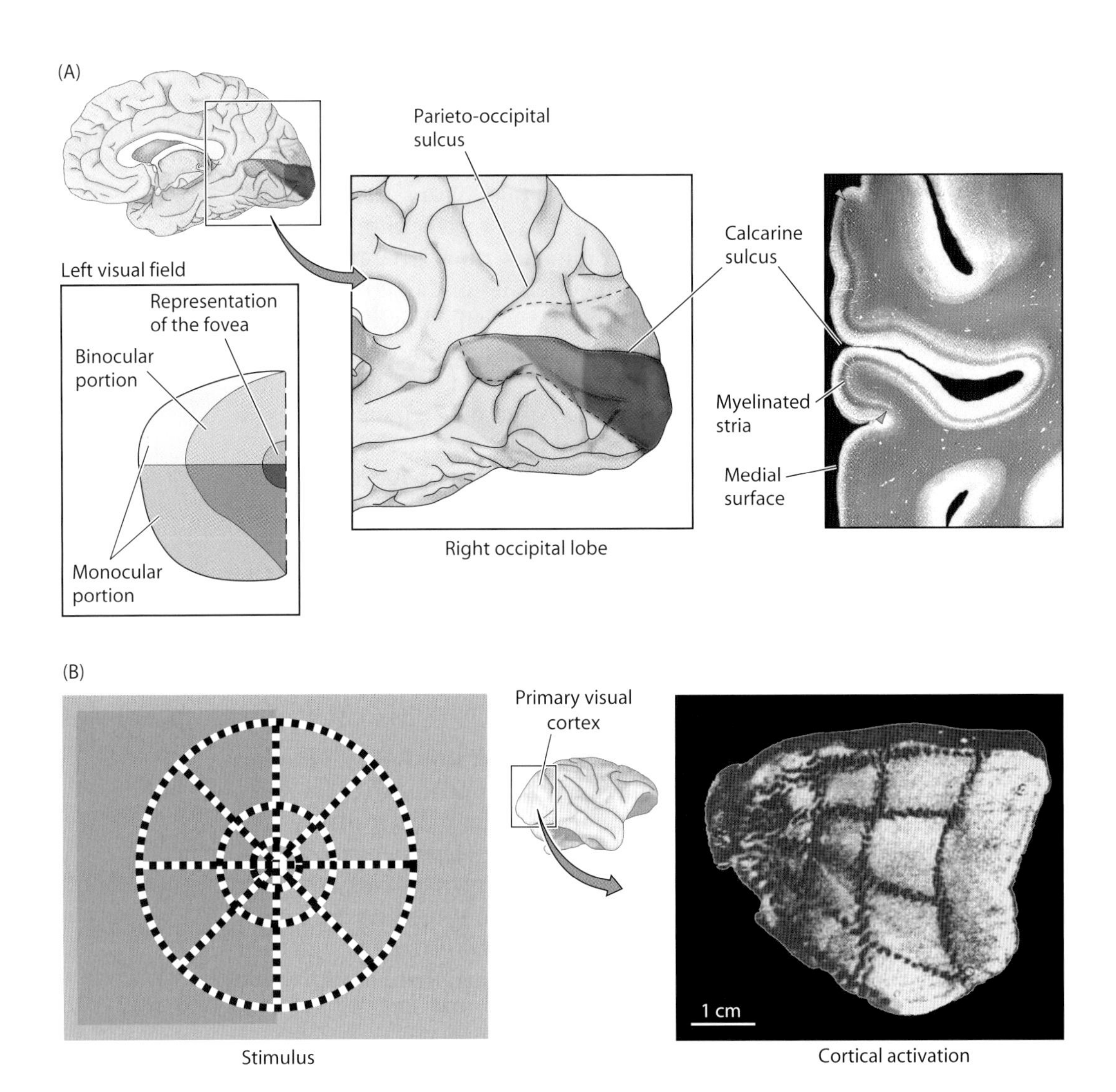

visual space) in one layer is in register with the position determined by recording from the neurons in the subjacent layer. The maintenance of this topographical (or retinotopic) pattern in the lateral geniculate and beyond (see below) has reinforced the idea that visual processing must in some sense be image processing.

The Primary Visual Cortex

The topographical organization of the retina and dorsal lateral geniculate nucleus is equally apparent in the primary visual cortex, the region in the

◀ **Figure A.8** The primary (striate) visual cortex. (A) Colored areas show the location of the human primary visual cortex and its retinotopical organization. The left visual field has been color-coded to show its overall projection via the thalamus to the contralateral (right) occipital lobe, seen here on the medial surface of the cerebral hemisphere (recall that the axons from the appropriate halves of the retinas cross at the optic chiasm such that the left visual field is eventually represented in the right visual cortex and vice versa; see Figure A.6). The blowup of the occipital region also shows that much more cortical space is devoted to processing the central area of the left visual field than to peripheral vision. The stained microscopical section shows the appearance of the primary visual cortex seen as if looking forward from the occipital pole of the hemisphere. The primary visual cortex is readily distinguished by a stripe (running between the arrowheads) in cortical layer IV; hence the alternative name, "striate cortex." The stripe, or "stria," is due to the increased number of myelinated nerve axons in this region. (B) Cortical pattern elicited by prolonged presentation of a particular stimulus pattern (shown on the left) to a macaque monkey. The technique used here entails the uptake of radioactively labeled 2-deoxyglucose; because this sugar cannot be metabolized, it accumulates in active neurons, thus revealing in an autoradiogram of the flattened cortex the regions that were particularly active during the period of stimulation (shown on the right, looking down on the flattened cortex). (A is courtesy of Tim Andrews and Dale Purves; B is from Tootell et al., 1988.)

occipital lobes that forms the major target of the axons projecting from the thalamus to higher visual centers (Figure A.8A; see also Figure A.6). Electrical recordings from this region of the cerebral hemispheres show that, like the lateral geniculate nucleus, adjacent sites are activated by stimulation of neighboring points on the retina, and therefore neighboring points in visual space.

Techniques that reveal the brain regions activated by a stimulus demonstrate the preservation of retinal topography in a particularly compelling way by allowing a direct comparison of the pattern of neuronal activation in the primary visual cortex with the pattern of retinal stimulation (Figure A.8B) (see Tootell et al., 1982, 1988; Sereno et al., 1995). The only obvious distortion of the retinal image arises from the fact that each unit area of the image on the retina is represented by a variable amount of space in the plane of the visual cortex; as might be expected, a square degree of visual space in central vision (and thus related to the fovea; see above) is represented by proportionally more cortical area (and therefore more visual circuitry) than the same unit area in the visual periphery. This disproportion, referred to as cortical magnification, makes good sense: The visual detail seen following stimulation of the fovea presumably requires more neuronal machinery in the cortex than does the less detailed information generated by stimulation of eccentric retinal regions. In any event, cortical activity patterns such as those shown in the example in Figure A.8B leave no doubt about the intimate relationship between the pattern of retinal activity and its topographical representation in the cerebral cortex.

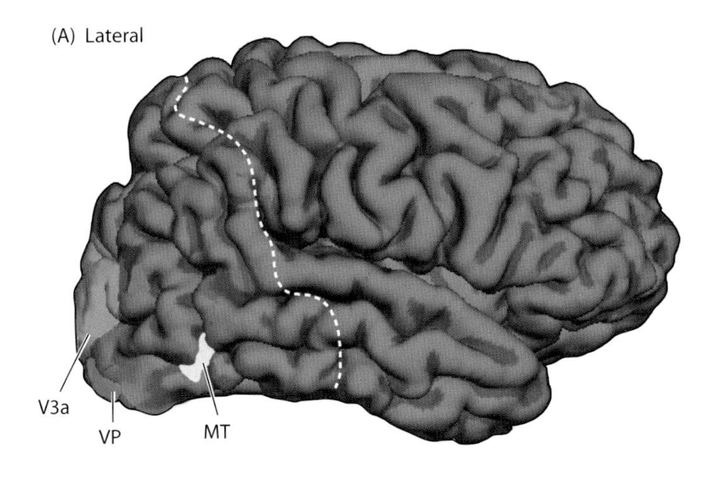

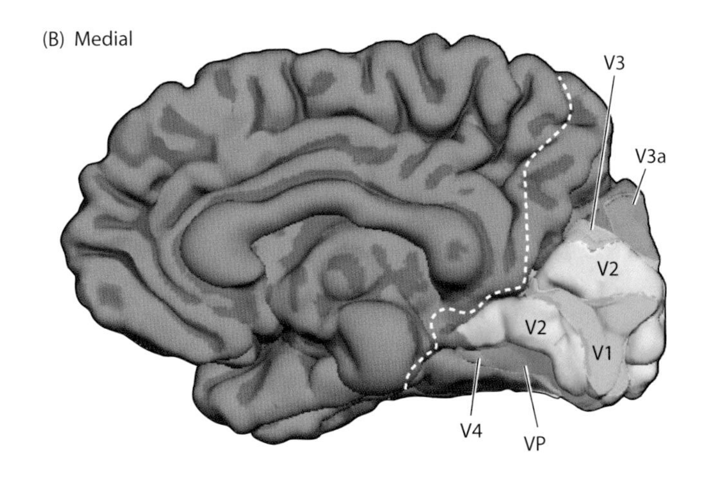

Higher Order Visual Centers: Topography and Hierarchy

Not until information from the primary visual cortex projects to "higher order" visual processing areas (called extrastriate visual cortical areas, or visual association cortex) does this sort of topography begin to break down. These areas are located in the occipital, parietal, and temporal lobes of the human (and nonhuman primate) brain, adjacent or near to the primary visual cortex in the occipital lobe (Figure A.9). Their location and function, however, remain controversial (see, for example, Bartels and Zeki, 2000).

One cause of the degradation of the topography so carefully preserved up through the primary visual cortex is the larger size of the neuronal receptive fields in the extrastriate cortex (Figure A.10). The receptive

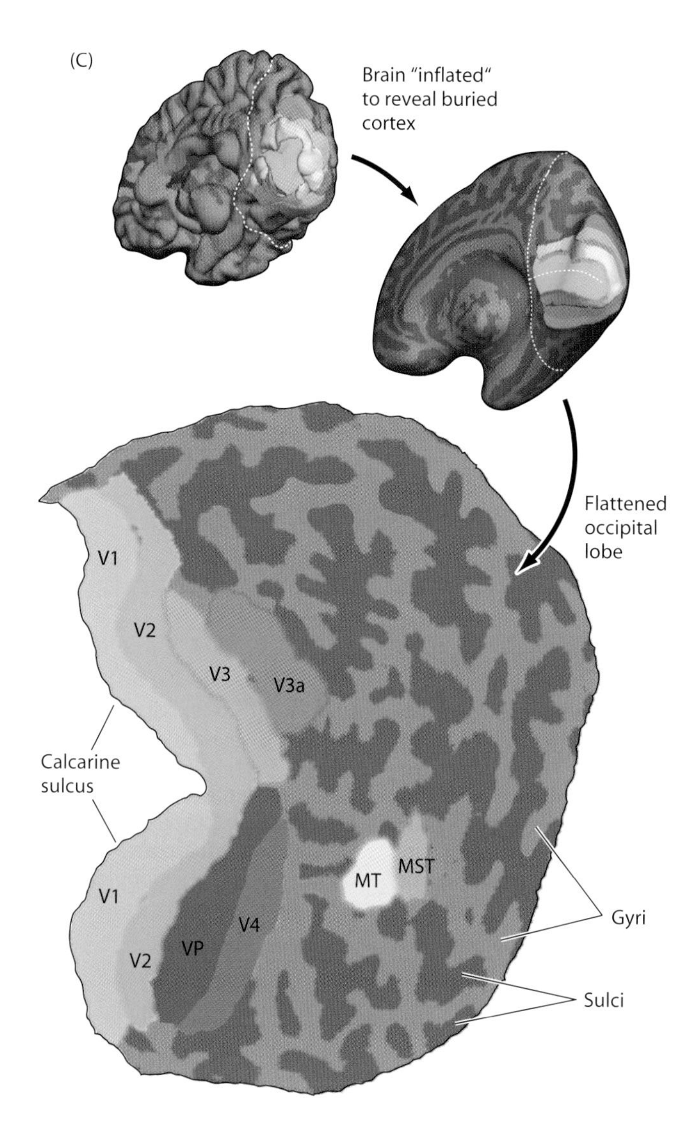

Figure A.9 Some of the major "higher order" visual areas in the human brain, determined by functional magnetic resonance imaging, a technique that allows areas activated by visual stimuli to be mapped in normal human subjects. (A, B) Lateral and medial surface views, respectively (B is comparable to the view in Figure A.8A). The primary visual cortex is indicated in green; the additional colored areas are the extrastriate areas involved in further processing visual information. (C) To better see the relation of these areas, the cortex has been computationally "inflated" to flatten its highly convoluted surface. The nomenclature is not especially important, although several of these areas are specifically referred to here and in main text. V1, V2, and V3 are the primary, secondary, and tertiary visual cortical areas (the numerical sequence should not be taken to imply sequential processing); V4 is an area important for macular vision and color processing; VP is a ventral posterior area whose function is not well understood; MST and MT (or MT+) are more anterior areas (located in the adjacent medial temporal and medial superior temporal regions, respectively) that are important in motion processing, probably among many other functions. (After Sereno et al., 1995.)

field of a visual neuron—be it in the retina, thalamus, or cortex—is defined as the region of visual space (located and measured in degrees) within which stimuli cause the cell to respond, i.e., to shift its activity from baseline levels of action potential firing to some higher (or lower) value, thus "encoding" some aspect of the information in the stimulus. This specificity of a neuron's responsiveness is generally alluded to as the "properties" of a neuron's receptive field. In the primary visual cortex, the receptive fields of cortical neurons in the region serving central vision are less than a degree of visual angle, as are the receptive fields of the corresponding retinal ganglion cells and lateral geniculate neurons (a degree of visual angle is about

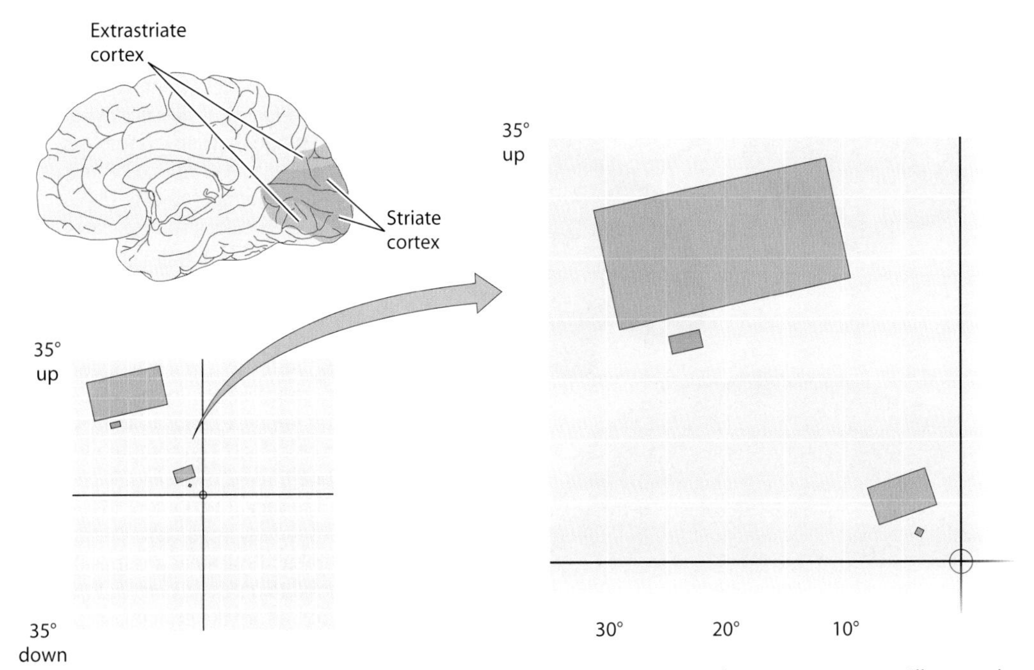

Figure A.10 Visual receptive fields in striate cortex and extrastriate cortex (illustrated here as they occur in the human brain, but based on recordings from macaque monkeys). The receptive fields in the primary visual cortex (shown in purple) are relatively small in the region that corresponds to central vision (see Figure A.8). The receptive fields of V1 neurons that serve more peripheral regions of the visual field are larger, but still relatively small. In contrast, the neuronal receptive fields mapped in extrastriate visual cortex (shown in green) are much larger; indeed, the receptive fields of some extrastriate neurons include most of the visual field (not illustrated).

half the breadth of the thumb held at arm's length). Even for cells serving peripheral vision, the receptive fields measured in the primary visual cortex are only a few degrees across. In extrastriate cortical areas, however, receptive fields often cover a substantial fraction of the entire visual field (which extends about 180° horizontally and 130° vertically). The location of retinal activity (and the corresponding topological relationships in the primary visual pathway assumed to underlie the sense of where an object is in visual field) cannot be conveyed by neurons that respond to stimuli anywhere in such a large region of space, at least not in any simple way.

A wealth of clinical and experimental evidence has shown that the extrastriate regions of the brain mediate important aspects of visual perception. Accordingly, this relative mooting of retinal topography in higher order visual cortical areas—and the diminished ability to identify the location and qualities of visual stimuli that this presumably entails—presents a problem for any rationalization of vision in terms of image reconstruction and representation. For example, processing visual qualities such as color and motion transpires in approximately defined extrastriate regions of the primate brain (including V4 for color and MT and MST for motion; see Figure A.9). One can imagine ways in which the diverse information processed in separate extrastriate areas might be "bound" back together and assigned to an appropriate location in a scene, but not easily.

Up to the level of the primary visual cortex, the organization of the visual system is patently hierarchical. At each of the stations in the primary visual pathway—the retinal ganglion cells, the dorsal lateral geniculate neurons, and the neurons at the first stages of processing in the primary visual cortex—the receptive field characteristics of the relevant neurons can be more or less understood in terms of the "lower order" cells that provide their input. Electrophysiological studies that span several decades have shown that the responses of retinal ganglion cells can be rationalized on the basis of the rods and cones that supply information to them via the bipolar and other retinal cells, that geniculate cell responses can be understood in terms of the ganglion cells that innervate them, and that the responses of at least some lower order visual cortical neurons make sense in terms of their geniculate inputs. Beyond these initial "input" levels, however, rationalizing the organization of the visual system in hierarchical terms is problematic.

Further Complications

In the primary visual cortex, stimulus qualities are organized in the plane of the cortex to form iterated columns whose overall pattern is superimposed on the topographical map described in the previous section. Some columnar maps reflect preferential responses to the orientation of contours, others to the direction of motion, and others to the eye of origin of the relevant neuronal activity. In higher order (extrastriate) visual cortex, this parallel partitioning according to visual qualities becomes even more obvious, such that entire cortical regions appear to emphasize the processing of particular stimulus qualities (e.g., color in area V4 or motion in area MT+). This parallel distribution of function is at seemingly at odds with a serial processing hierarchy.

The temporal sequence of visual processing is also problematic. "Higher" order cortical areas can become active simultaneously with, or even before, the activation of V1 (see, for example, Pascual-Leone and Walsh, 2001). To complicate matters even further, the properties of visual neurons in both thalamus and cortex are determined by descending and lateral connections to an even greater degree than by ascending inputs. Attractive though it may be to imagine that the organization of the visual system is more or less pyramidal, with perception at the apex, this framework is not warranted by the evidence.

Glossary

absorbance In the context of vision, to take up energy by chemical or physical interaction, thus reducing the intensity of light that passes through or is reflected from a medium; the reciprocal of transmittance or reflectance, and usually expressed as a logarithm.

absorption spectrum The spectral distribution of light that has passed through a medium (or been reflected from a surface) that absorbs some portion of the incident light.

accommodation The reflex response of the lens generated by convergent eye movements made to fixate on nearby objects.

achromatic Pertaining to visual stimuli (or scenes) that are perceived only in shades of gray ranging from black to white, inclusive.

action potential The electrical signal conducted along neuronal axons by which information is conveyed from one place to another in the nervous system.

acuity The ability of the visual system to accurately discriminate spatial detail, as in the standard Snellen eye chart exam; usually tested by the spatial discrimination of two points.

adaptation Sensory receptor adjustment to different levels of stimulus intensity; allows sensory systems to operate over a wide range of stimulus intensities. In vision, also used to refer to a more general adjustment to the quality of background light depth derived from the increasing fuzziness and blueness of distant objects.

aerial perspective The diminution of contrast (i.e., the increasing haziness of contour boundaries) as a function of distance from the observer; this phenomenon occurs as a result of the imperfect transmittance of the atmosphere and is a monocular cue to depth.

afferent An axon that conducts action potentials from the periphery to more central parts of the nervous system.

aftereffect The influence on visual perception that arises from looking at any sort of repetitive stimulus, such as a continuous motion stimulus (the classic example is the waterfall effect, i.e., the apparent upward movement seen after staring at falling water).

afterimage The altered image that follows the presentation and removal of a visual stimulus; e.g., the residual image observed by an individual after viewing a flash of light or after looking for many seconds at a chromatic stimulus (in which case a complementary color afterimage is seen).

albedo The maximum reflectance of light from a surface (see *highlights*).

algorithm A set of rules or procedures set down in logical notation and typically (but not necessarily) carried out by a computer.

amacrine cells One of several cell types in the vertebrate retina; amacrine means "lacking processes" (these cells are so named because of their limited number of axons and dendrites).

amblyopia Diminished visual acuity arising from a failure to establish appropriate visual cortical connections in early life as a result of visual deprivation.

analytic Relating to analysis; contrasts with *empirical* (see *analyze*).

analyze To determine the nature of something according to a set of principles, such as the features of an image.

anisotropic Characterized by a non-uniform distribution of a physical parameter such as light (see *isotropic*).

annulus A ring, typically used to describe the surround of visual neuron's receptive field.

anomalous trichromat An individual who requires long, medium, and short wavelength light to match chromatic stimuli, but who makes adjustments that deviate significantly from normal (due to anomalies in one or more of the cone opsins).

anthropoid primates Nonhuman primates that have characteristics especially similar to those of humans; generally refers to monkeys and great apes.

aperture problem The challenge of determining the speed and direction of a moving line when its ends are obscured by an opening such as a circular hole, a vertical rectangle, etc.

aperture view The presentation of visual stimulus with minimal context, usually by presenting the stimulus in a black or featureless surround (as in *colorimetry*).

apparent motion The sensation of motion elicited by presentation of a stimulus in two positions over a brief interval.

arc (visual arc) Part of the circumference of a circle; used to measure visual space in degrees (of which there are 360 in the full circumference; an arc-minute is one-sixtieth of a degree, and an arc-second is one-sixtieth of an arc minute).

area centralis Central region of the retina of most vertebrates, specialized for higher acuity vision (see *fovea*).

artificial intelligence A computational approach to mimicking brain function that generally depends on the analytical solution to problems.

artificial neural network Computer architecture for solving problems by feedback from trial and error rather than by a predetermined algorithm.

assimilation Tendency for the perception of a target to include or be influenced by the characteristics of the background.

association cortex Those regions of cerebral neocortex defined, by exclusion, as not involved in primary sensory or motor processing.

attached shadows The shadows formed when one part of an object shades another part of the same object.

attention A poorly understood phenomenon in which mental focus is preferentially accorded to a particular feature of the external or internal environment; also defined operationally as whatever occupies consciousness at any given moment.

auditory system The peripheral and central neural apparatus for hearing.

autonomic nervous system All the neural apparatus that controls visceral behavior. Includes the sympathetic, parasympathetic, and enteric systems.

awareness A synonym for the subjective, first-person aspects of consciousness.

axon The neuronal process that carries the action potential from the nerve cell body to a target.

background Referring to the part or parts of a scene that are farther away from an observer and/or less salient.

bandwidth Describing a frequency range within a spectrum.

basal ganglia A group of nuclei, lying deep in the subcortical white matter of the frontal lobes, that organize motor behavior. The caudate and putamen and the globus pallidus are major components of the basal ganglia; the subthalamic nucleus and substantia nigra are often included.

behaviorism A perspective on psychology that developed in the United States and the United Kingdom during the middle of the twentieth century. Behaviorism holds that only directly observable behavior, and not internal mental states, can be studied scientifically and that changes in behavior reflect unconscious associations of stimulus, response, and outcome, either positive or negative.

Bezold-Brücke hue shift The change in the apparent color of a stimulus as a function of luminance.

binding problem The problem presented by the uncertain way that perceptual qualities are brought together in the perception of objects. More generally, how the brain organizes related information.

binocular Referring to both eyes.

binocular disparity See *retinal disparity*.

binocular rivalry See *rivalry*.

bipolar cells A class of cells in the vertebrate retina that receive input from the photoreceptors and communicate that input directly or indirectly to retinal ganglion cells.

bistable figures Visual stimuli that elicit changes back and forth between the perception of one of two different objects (e.g., between seeing a face and a vase in a classic example).

blindsight The ability of people who are blind, usually because of damage to their cortex, to identify the properties of simple visual stimuli when forced to guess.

blobs Anatomical entities in the primary visual cortex of humans and most other subhuman primates, found in cortical layer 3; revealed by histological procedures such as cytochrome-oxidase staining. Their function is not known.

Boolean A system of symbols used to formulate logical problems (described by the nineteenth century mathematician George Boole).

bottom-up A much-abused term that loosely refers to the flow of information from sensory receptors to cerebral cortex.

brain The cerebral hemispheres and brainstem.

brain activity The electrical and metabolic changes of neurons that are associated with neural signaling and information processing.

brainstem The portion of the brain that lies between the diencephalon and the spinal cord; comprises the midbrain, pons, and medulla.

brainstem nuclei Collections of neurons in the brainstem that carry out specific functions.

brightness Technically, the apparent intensity of a source of light; more generally, a sense of the effective overall intensity of a light stimulus (see *lightness*).

brightness induction (see *simultaneous brightness contrast*).

Brodmann's Area 17 The primary visual cortex, also called the *striate cortex*.

calcarine sulcus The major sulcus on the medial aspect of the human occipital lobe; the primary visual cortex lies largely within this sulcus.

camouflage Obscuration of an object by the application of random textures or other visual noise that makes it harder to distinguish the object from the background.

candela A metric of luminance, usually expressed per unit area (the physical definition is somewhat complex, involving black body radiation at a particular temperature).

cast shadows Shadows cast on a surface by an object intervening between a light source and that surface.

cataracts Opacities in the lens that can impede normal vision.

caudal Posterior, or "tailward."

cell body The portion of a neuron that houses the nucleus.

cell The basic biological unit in plants and animals, defined by a cell membrane or wall that encloses cytoplasm and the cell nucleus.

central nervous system The brain and spinal cord of vertebrates (by analogy, the central nerve cord and ganglia of invertebrates).

central sulcus A major sulcus on the lateral aspect of the hemispheres that forms the boundary between the frontal and parietal lobes. The anterior bank of the sulcus contains the primary motor cortex; the posterior bank contains the primary sensory cortex.

cerebellum Prominent hindbrain structure concerned with motor coordination, posture, balance, and some cognitive processes. Composed of a three-layered cortex and deep nuclei; attached to the brainstem by the cerebellar peduncles.

cerebral achromatopsia Loss of color vision as a result of damage to the visual cortex.

cerebral cortex The superficial gray matter covering the cerebral hemispheres.

cerebral hemisphere One of the two halves of the forebrain.

cerebrum The largest and most rostral part of the brain in humans and other mammals, consisting of the two cerebral hemispheres.

channels Hypothetical pathways that subserve one or another aspect of a sensory modality; most often used in describing the parallel percolation of visual input qualities.

chiasm (optic chiasm) The crossing of optic nerve axons from the nasal portions of the retinas in humans and other mammals such that the temporal visual fields are represented in the contralateral cerebral hemispheres.

circuitry A general term in neurobiology referring to the connections between neurons; usually pertinent to some particular function (as in "visual circuitry").

cognition A poorly defined term referring to "higher order" mental processes.

color The subjective sensations elicited in humans (and presumably many other animals) by different spectral distributions of light.

color addition Mixing light by the superposition of different sources, typically by projection onto a surface capable of reflecting all wavelengths (e.g., a "white screen").

color blind Outmoded term for individuals who have abnormal color vision (see *color-deficient*).

color constancy The similar appearance of surfaces despite different spectral return from them; usually applied to the approximate maintenance of object appearances under different illuminants.

color contrast See *contrast*.

color-deficient Term for individuals who have abnormal color vision as a result of the absence of (or abnormalities in) one or more of the three human cone types.

color mixing (additive) Phrase used to describe the additive mixing of different spectra (see *color addition*).

color mixture (subtractive) The effect on spectral returns of mixing pigments.

color opponency Term used to refer to the experience of seeing red/green, blue/yellow, and black/white as opposites; more recently, the phrase has been extended to describe neurons whose properties suggest participation in this perceptual process by the comparison of outputs from the photoreceptors.

color shadows Demonstration of color contrast based on the remarkable effect elicited by the shadows cast on a surface illuminated by both chromatic and white light.

color space The depiction of a human color perception in diagrammatic (or mathematical) form by a space with three axes representing the perceptual attributes of hue, saturation, and brightness.

colorimetry Measurements of the human responses to uniform spectral stimuli presented in aperture view.

color-matching Referring to colorimetry experiments in which subject are asked to match color experience.

color-opponent cells Cells whose receptive fields are sensitive to opposing spectra of qualities.

competition In biology, the struggle for limited resources essential to survival or growth.

complementary colors The colors, elicited by lights, that when mixed generate a neutral sensation of some shade of gray (often applied more loosely to colors that are more or less oppositely disposed around the Newton color circle).

complex cell A cell type in the primary visual cortex of primates and carnivores whose receptive field is characterized by end-stopping and whose properties are thought to be generated by the summation of simple cell inputs.

cone opsins The three distinct photopigment proteins found in cones; the basis for color vision.

cones Photoreceptors whose organization is specialized for high visual acuity and the perception of color.

conjugate eye movements Linked movements of the two eyes in the same direction, as occurs in saccades.

connectionist theory Information process theory based on neural networks in which problems are solved by changing the "weights" of the network connections.

consciousness A contentious concept that includes the ideas of wakefulness, awareness of the world, and awareness of the self as an actor in the world.

context General term referring to the information provided by the surroundings of a "target." The division of a scene into target and surround is useful but arbitrary, since any part of a scene provides contextual information for any other part.

contour A line or edge.

contralateral On the opposite side.

contrast The physical difference between the luminance (or spectral distribution, in the case of color) of two surfaces. More specifically, the luminance difference or spectral difference between two regions of visual space (calculations of contrast vary for different models of perception).

convergence The coming together of two things, such as information from the two eyes or the two lines of sight at the point of fixation. Opposite of *divergence*.

convergence angle (vergence angle) The angle between the two lines of sight relative to parallel (i.e., to the lines of sight when looking at infinitely distant objects).

cornea Clear outer portion of the eyeball that provides the major refractive power for focusing light in human and other eyes.

Cornsweet illusion Edge effects elicited by opposing light gradients that meet along a boundary (sometimes called the Craik-O'Brien-Cornsweet illusion).

corpus callosum The large midline fiber bundle that connects the cortices of the two cerebral hemispheres.

correspondence (corresponding points) In vision, refers to the points on the two retinas that receive light rays from the same location in visual space.

correspondence problem (see also *false target problem*) The problem presented in a random dot stereogram (or any image, for that matter) of matching a dot in the retinal image in one eye to the corresponding dot in the retinal image of the other eye.

cortex The gray matter of the cerebral hemispheres and cerebellum, where most of the neurons in the brain are located.

cortical association areas See *association cortex*.

cortical columns Vertically organized groups of cortical neurons that process the same or similar information; examples are ocular dominance columns and orientation columns in the primary visual cortex.

cortical magnification The disproportionate representation of cortical space according to peripheral receptor density (such as occurs for the central representation of the fovea of the human eye).

cranial nerves The 12 pairs of nerves arising from the brainstem that carry sensory information toward (and sometimes motor information away from) the central nervous system.

critical flicker-fusion frequency The frequency at which alternating presentations of light and darkness are seen as continuous light.

critical period A restricted developmental period during which the nervous systems of humans and other mammals are particularly sensitive to the effects of experience.

crossed disparity The geometrical disparity in the projection to the two retinas of the same point(s) in visual space that lie nearer the observer than the point of fixation (or, more correctly, than the *horopter*).

cumulative probability The summed probability of some variable being less than or equal to a particular value.

cyclopean vision The normal sense, when looking at the world with both eyes, that we see it as if with a single eye.

2-deoxyglucose technique Method for marking neurons according to their level of activity by the incorporation of a radioactively tagged, non-metabolized sugar.

dark adaptation The adjustment of the sensitivity of the visual system to scotopic conditions; based on the reactivation of bleached rhodopsin.

default mode Either (1) neural processing that occurs in the absence of carrying out a specific task or (2) a pattern of brain activation that reflects passive experience.

degree Unit used to measure visual space based on the division of a circle into 360 degrees; 1 degree is approximately the width of the thumbnail held at arm's length and covers about 0.2 millimeters on the retina.

dendrite A neuronal process that receives synaptic input; usually branches near the cell body (see *axon*).

depth perception A general term used to indicate the perception of distance (either monocular or stereoscopic) from the observer.

detector In vision, a nerve cell or other device that nominally detects the presence of some particular feature of visual stimuli (e.g., luminance, orientation, etc.).

dichoptic presentation Independent presentation of a stimulus to the right and left eyes.

dichromat A color-deficient person (as well as the majority of mammals) whose color vision depends on only two cone types.

dichromatic Having only two cone types.

diencephalon Portion of the brain that lies just rostral to the brainstem; comprises the thalamus and hypothalamus.

diopter A measure of the strength of the lens defined by the reciprocal of its focal length in meters.

diplopia Seeing two images simultaneously as a result of abnormally aligned visual axes (arises from pathology or from allowing the eyes to cross in front of or behind the point of fixation).

direct light Light that reaches a surface directly from a primary source such as the sun or a source of indoor illumination (see *indirect light*).

direction The course taken by something, e.g., a point, in a frame of reference; together with speed, defines velocity.

disjunctive eye movements Movements in which the two eyes move independently, as in vergence (see *convergence*).

disparity The geometrical difference between the view from the left and right eyes in animals having frontal eyes and stereoscopic depth perception (see *retinal disparity*).

distal Farther away from a point of reference (opposite of *proximal*).

distal stimulus The light emanating from objects that reaches the eye.

divergent fusion Fusing the right and left eye images when the two lines of sight intersect beyond the object of interest (typically a pair of stereo images).

dorsal Toward the back. Opposite of *ventral*.

dorsal lateral geniculate nucleus The thalamic nucleus that relays information from the retina to the cerebral cortex; usually referred to as the *lateral geniculate nucleus* or just the geniculate.

dorsal stream A partially segregated visual processing network passing from primary visual cortex through extrastriate areas of the parietal cortex and thought to be concerned primarily with spatial aspects of visual processing. Sometimes referred as the "direct" visual pathway.

double opponent neurons Neurons whose receptive field centers are activated by one spectral distribution and inhibited by its opposite, and conversely (i.e., the surround is also inhibited by the spectrum that activates the center, and the center is inhibited by the opposite spectrum). Such cells are deemed important for seeing color contrast.

drift The slow, ongoing eye movements that occur even when the direction of gaze is fixed.

eccentricity A measure of the distance away from the center; in vision, typically refers to the distance in degrees away from the line of sight.

edge effects A class of perceptual phenomena in which the qualities at an edge affect the perception of the qualities (e.g., brightness of color) of the adjoining territory (or territories) (see *Cornsweet illusion*).

efferent An axon that conducts information away from the central nervous system.

efferent neuron An axon that conducts information away from the central nervous system.

electromagnetic radiation The full spectrum of radiation in the universe, of which light comprises only a tiny portion.

electrophysiology Study of the nervous system by means of electrical recording.

empirical Derived on the basis of past experience, effectively by trial and error (contrasts with *analytical*).

epiphenomenon An effect taken to be an incidental consequence of some more basic property or principle.

equiluminant (*isoluminant*) Having the same luminance.

Euclidean space The three-dimensional space of ordinary geometry.

excitatory neuron A neuron whose activity depolarizes (excites) the target cells it contacts.

excitatory response (excitation) A cellular response involving an increased rate of action potential generation per unit time.

executive control The cognitive function associated with altering thought and behavior in a goal-directed, context-dependent, and flexible manner.

extrastriate Referring to the regions of visual cortex that lie outside the primary (or striate) visual cortex.

extrastriate visual areas Include areas such as V4, MT, and MST, which are taken to be particularly pertinent to the processing of one or another categories of visual information (e.g., color in V4, motion in MT and MST).

false target problem The problem arising from the fact that the binocular disparity generated by any point in visual space can have been generated by a pair of points at a different distance from the observer (see *correspondence problem*).

feature detection The idea that vision entails the detection and subsequent processing of features in the retinal image.

filling-in The perceptual attribution of a property or properties to a region of visual space when the information from that space is either absent or physically different from what is actually seen.

fixation Looking steadily a particular point in visual space; normally the point of fixation is where the lines of sight from the left and right eyes intersect.

forebrain The anterior portion of the brain that includes the cerebral hemispheres (the telencephalon and diencephalon).

Fourier analysis Mathematical procedure for representing any function as the sum of a set of sinusoids.

Fourier's theorem The idea, established by Jean-Batiste Fourier in the late eighteenth century, that any periodic function can be decomposed into a series of sine or cosine waves that are harmonically related.

fovea Area of the human retina specialized for high acuity; contains a high density of cones and few rods. Most mammals do not have a well defined fovea, although many have an area of central vision (called the *area centralis*) in which acuity is higher than in more eccentric retinal regions.

foveola The capillary- and rod-free zone in the center of the fovea, at which acuity is highest.

free fusion Technique of bringing together two separate images by having the lines of sight from the two eyes intersect in front of or behind the plane of the stimulus.

frequency How often something occurs in a unit of time and/or space.

frequency distribution Histogram or other graphical representation showing the relative frequency of occurrence of an event.

frontal eye fields A region of the prefrontal cortex in human and nonhuman primates, often associated with Brodmann's area 8a, that plays a key role in voluntary visual orienting movements.

frontal lobe One of the four lobes of the brain; includes all the cortex that lies anterior to the central sulcus and superior to the lateral fissure.

fronto-parallel plane A plane orthogonal to the line of sight.

functional imaging Technique of noninvasive brain imaging that depends on the metabolic activity of the relevant brain tissue to reveal the location of a neural function; usually comprises *positron emission topography* and *functional magnetic resonance imaging*.

functional magnetic resonance imaging (fMRI) A technique that uses standard MRI scanners to measure changes in local blood deoxygenation (an index of neuronal metabolism) that can be used to create spatial maps of brain function.

ganglion cell An output neuron from the vertebrate retina whose axons form the optic nerve.

ganzfeld A completely uniform and featureless visual field (from the German word for "whole field").

gene A hereditary unit located on a chromosome and defined by encoding the information for a particular protein; the genetic information is carried by linear sequences of nucleotides in DNA that code for corresponding sequences of amino acids.

genetic algorithms A computer-based scheme for simulating evolution using artificial neural networks.

genome The complete set of an animal's genes.

genotype The genetic makeup of an individual.

geometrical illusions Discrepancies between a visual stimulus and the resulting percept based on geometrical measurements (i.e., measurements of length, angle, etc.).

Gestalt laws Generalizations about the rules of perception as applied to stimulus categories; not widely accepted today, but of historical interest.

gestalt psychology A school of psychology founded by Max Wertheimer in the early twentieth century in which the overall qualities of a scene are taken to be determinants of its perception; "Gestalt" in German means "an integrated perceptual whole."

gradient Monotonic variation of the amount of some variable (e.g., light).

gray matter General term describing regions of the central nervous system rich in neuronal cell bodies; includes the cerebral and cerebellar cortices, the nuclei of the brain, and the central portion of the spinal cord.

growth factors Molecules that promote the survival and growth of cells (nerve cells in the present context).

hardware The machinery of a computer.

Hering illusion A classical geometrical effect in which parallel lines placed on a background of radiating lines look bowed.

hertz (Hz) Measure of frequency (1 Hz = 1 cycle/second).

heuristic A rule of procedure derived from past experience that can be used to solve a problem when an algorithm for getting the answer is not known; in vision, such rules are often taken to be the determinants of perception.

hierarchy A system of higher and lower ranks; in vision, the idea that neurons in stages of the visual system determine the properties of higher order neurons.

higher-order Descriptive term for processes and/or areas taken to be further removed from the input stages of a system; in neuroscience, this phrase is sometimes used as a synonym for "cognitive."

higher-order neurons Neurons that are relatively remote from peripheral sensory receptors or motor effectors.

highlights Photometric maxima that arise due to the interaction of illuminants and reflective surfaces.

hippocampus A specialized cortical structure located in the medial portion of the temporal lobe; in humans, it is concerned with short-term declarative memory, among many other functions.

historical Determined by events rather than logical or other a priori factors.

history What happened in the past.

homunculus Greek for "little man," referring to the shape of the primary sensory or motor cortical map; also used to refer (often derisively) to the dualist notion of an "I" that stands above ordinary neural processing.

horizontal cells A class of retinal neurons that mediate lateral interactions between photoreceptor terminals and the dendrites of bipolar cells.

horopter The plane (actually a manifold) in visual space defined by the fused appearance of objects seen by the two eyes at any given viewing distance.

hue That aspect of color sensation (brightness and saturation being the others) that pertains specifically to judgments about a color being different from, or similar to, shades of red, green, blue, or yellow. Usually diagrammed as a series of circles that define this aspect of perceptual color space.

hypercomplex cells (end-stopped cells) Cells in the primary visual cortex whose receptive field properties are sensitive to the length of the stimulus; originally thought to be determined by convergent innervation from complex cells.

hypothalamus A collection of small but critical nuclei in the diencephalon that lies just inferior to the thalamus; governs reproductive, homeostatic, and circadian functions.

illuminant A source of illumination.

illumination The light that falls on a scene or surface.

illusions Discrepancies between the physically measured properties of a stimulus and what is actually perceived. Use of the term is problematic since all visual percepts entail such discrepancies.

image In vision, the projection of a scene on the retina; also used to refer to the "representation" of a scene in perception.

image formation The result of focusing the light rays diverging from a collection of adjacent points on object surfaces onto another surface (e.g., a screen or the retina) to form a corresponding set of points that represents the three-dimensional sources on a two-dimensional plane.

image processing Improving (or otherwise changing) images by application of one or more algorithms.

indirect light Light that reaches a surface only after reflection from another object or surface.

indirect pathway (see *ventral pathway*)

information The systematic arrangement of a parameter such that an observer (or a receiver) can , in principle, extract a signal from the background noise.

information theory Theory of communication channel efficiency first elaborated by Claude Shannon in the late 1940s.

inhibition Decrease in neuronal signaling.

inhibitory neuron A neuron whose activity decreases the likelihood that the target cells it contacts will in turn be activated.

inhibitory response (inhibition) A cellular response involving a decrease in the rate of action potential generation per unit time.

innervate Establish synaptic contact with another neuron or target cell.

innervation Referring to all the synaptic contacts made with a target.

input The information supplied to a neural processing system.

integration In neuroscience, the summation of excitatory and inhibitory synaptic conductance changes by postsynaptic cells.

intercellular communication Signaling interactions between cells; if the cells are neurons, it typically occurs by means of synaptic transmission.

interneuron Literally, a neuron in the pathway between primary sensory and primary effecter neurons; more generally, a neuron that branches locally to innervate other neurons.

interocular transfer The effect in one eye when stimulus is presented to the other eye.

inverse optics problem The impossibility of knowing the world directly by virtue of light patterns projected onto the retina.

ipsilateral On the same side.

iris Circular, pigmented membrane behind the cornea; perforated by the pupil.

irradiance Radiometric measure of light striking a surface.

isoluminant Having the same luminance (synonymous with *equiluminance*).

isotropic A physical property that is evenly distributed in space.

lambertian Describing a surface that reflects light uniformly in all directions; named after the eighteenth century German mathematician Johann Lambert.

lamina (pl., laminae) One of the cell layers that characterize the neocortex, hippocampus, cerebellar cortex, spinal cord, or retina.

laminated Layered.

lateral geniculate nucleus See *dorsal lateral geniculate nucleus*.

lateral inhibition Inhibitory effects extending laterally in the plane of the tissue, e.g., the retina or the visual cortex; widely assumed to play a major role in perceptual phenomenology.

learning The acquisition of novel information and behavior through experience.

lens The transparent and more or less spherical part of the eye whose thickening or flattening under neural control allows the light rays emitting from objects at different distances to be focused on the retina. More generally, any object that refracts light.

light The wavelength range in the electromagnetic spectrum that elicits visual sensations in humans (i.e., photons having a wavelength of about 400–700 nm).

light adaptation See *adaptation*.

lightness The apparent reflectance of a surface (or transmittance of a medium), experienced as relative achromatic values ranging from white through grays to black (see *brightness*).

lightness constancy The similar appearance of a surface despite differences in spectral return, typically as a function of illumination (see *color constancy*).

line A set of points connected by a common property (e.g., straightness); an extension in length without thickness.

line of sight An imaginary line from the center of the fovea through the center of the point of fixation.

linear perspective Geometrical changes arising from the projection of three-dimensional objects onto a two-dimensional surface.

lobes The four major divisions of the cerebral cortex (frontal, parietal, occipital, and temporal).

local circuit neuron A neuron whose local connections contribute to processing circuitry.

logic A set of formal principles, the use of which allows valid inferences to be made.

logical Operating according to the rules of formal logic.

lowlights Photometric minima that occur as the result of the physical interaction of direct and indirect light with reflective surfaces.

luminance The physical (photometric) intensity of light returned to the eye (or some other detector) adjusted for the sensitivity of the average human observer.

luminance gradients Gradients of light intensity.

luminosity function Curve representing the differential sensitivity of the average human observer to light of different wavelengths.

Mach bands Perceptual bands of brightness maxima and minima that occur at the onset and offset of luminance gradients, described by the physicist Ernst Mach in 1865.

machine Any human-made device or, more broadly, any apparatus that accomplishes a purpose by the operation of a series of causally connected parts.

macula (macula lutea) Central region of the retina that includes the fovea (the term derives from the yellowish appearance of this region in ophthalmoscopic examination).

magnification factor See *cortical magnification*.

magnocellular Primary visual processing channel comprising relatively large neurons; specialized for detecting motion and stimulus change.

magnocellular system Component of the primary visual pathway specialized, among other functions, for the perception of motion; so named because of the relatively large cells involved.

mammal An animal the embryos of which develop in a uterus, and the young of which suckle at birth (technically, a member of the class Mammalia).

map A systematic arrangement of information in space. In neurobiology, the ordered projection of axons from one region of the nervous system to another, by which the organization of a relatively peripheral part of the body (e.g., the retina) is reflected in the organization of the nervous system (e.g., the primary visual cortex).

mapping The corresponding arrangement of the peripheral and central components of a sensory or motor system, or between two other sets of information.

matte Describing surfaces that appear dull because they reflect relatively little light and do so more or less isotropically (see *lambertian*).

M-cells Retinal ganglion cells initiating information flow in the magnocellular system.

medial Located nearer to the midsagittal plane of an animal (the opposite of lateral).

medium In the context of vision, a substance (e.g., the atmosphere or a filter) between the observer and the object or objects in a scene.

mesopic Light levels at which both the rod and cone systems are active.

metamers Light spectra that have different wavelength distributions but nonetheless elicit the same color percepts.

Michelson contrast See *Rayleigh contrast*.

microelectrode A recording device (typically made of wire or a glass tube pulled to a point and filled with an electrolyte) used to monitor electrical potentials from individual or small groups of nerve cells.

microspectrophotometry The detection of spectral qualities, typically of retinal photoreceptors based on a microscopic approach to spectrometry.

midbrain The most rostral portion of the brainstem; identified by the superior and inferior colliculi on its dorsal surface and the cerebral peduncles on its ventral aspect.

mind The state of a person's thoughts and feelings, which thus includes the full spectrum of cognitive functions at any point in time.

modality A category of function; e.g., vision, hearing, and touch are different sensory modalities.

module A general term used to refer to a cortical unit (e.g., ocular dominance columns, orientation columns, or *blobs*).

Mondrian A collage of papers or other surfaces used to explore lightness or color perception; named after the early twentieth century abstract artist Piet Mondrian, who produced many works of this sort (current usage has anglicized his name).

monochromatic light Light comprising (nominally) a single wavelength; in practice, often a narrow band of wavelengths generated by an interference filter.

monochromats Color-deficient individuals who have only one or no cone opsins and therefore have no color vision.

monocular Pertaining to one eye.

monocular cues Information about depth present in the view of a single eye.

motion The changing position of an object; defined by speed and direction in a frame of reference.

motion parallax The different degree of movement of near and far objects as a function of moving the head or body while observing a scene.

motor Pertaining to movement.

motor cortex The region of the cerebral cortex in humans and other mammals, lying anterior to the central sulcus, concerned with motor behavior in humans; includes the primary motor cortex in the precentral gyrus and associated cortical areas in the frontal lobe.

motor neuron A nerve cell that innervates skeletal muscle.

motor system A broad term used to describe all the central and peripheral structures that support motor behavior.

MST Extrastriate cortical region in primates specialized for motion processing.

MT Related extrastriate cortical region in primates specialized for motion processing. In humans, MST and MT are often referred to as MT+ because the distinction between them is not well defined.

Mueller-Lyer illusion A geometrical effect in which the length of a line terminated by arrowheads appears shorter than the same line terminated by arrow tails; first described by the nineteenth century German philosopher and sociologist F. D. Mueller-Lyer.

myotatic reflex (stretch reflex) A spinal reflex comprising the motor response to afferent sensory information arising from muscle sensors (called muscle spindles).

nasal (nasal division) Referring to the half of the visual field, seen by each eye, that lies in the direction of the nose (can also refer to the half of the retina nearer the nose).

near point The nearest distance at which the lens can no longer accommodate sufficiently to bring objects into focus.

near reflex Set of responses induced by changing binocular fixation from a further to a closer target; comprises convergence, accommodation, and pupillary constriction.

Necker cube A depicted wire frame cube that can be seen in two different configurations; first described by the nineteenth century Swiss naturalist L. A. Necker.

neocortex The six-layered cortex that covers the bulk of the cerebral hemispheres.

nerve A collection of peripheral axons that are bundled together and travel a common route.

neural circuit A collection of interconnected neurons dedicated to a neural processing goal.

neural network Typically refers to an artificial network of interconnected nodes whose connections change in strength as a result of experience.

neural plasticity The ability of the nervous system to change as a function of experience; typically applied to the changes in the fully formed adult nervous system.

neural processing A general term used to describe operations carried out by neural circuitry.

neural system A collection of peripheral and central neural circuits dedicated to a particular function (e.g., the visual system, the auditory system, etc.).

neuron Cell specialized for the conduction and transmission of electrical signals in the nervous system.

neuronal receptive field The area of the sensory periphery (e.g., visual field, skin, etc.) that elicits activity in a sensory neuron.

neuroscience Study of the structure and function of the nervous system.

neurotransmitter A chemical agent released at synapses that affects the signalling activity of the postsynaptic target cells.

neurotransmitter receptor A molecule, embedded in the membrane of a postsynaptic cell, that binds a neurotransmitter.

nodal point The point in the eye where light rays entering the pupil cross en route to the retina; located approximately in the center of the lens.

noise A signal that does not carry information.

nucleus (pl., nuclei) Collection of nerve cells in the brain that are anatomically discrete and that typically serve a particular function.

objects The physical entities that give rise to visual stimuli by reflecting illumination (or by emitting light if, as more rarely happens, they are themselves generators of light).

occipital cortex Part of the brain, nearest the back of the head, mainly containing visual processing areas.

occipital lobe The posterior of the four lobes of the human cerebral hemisphere; primarily devoted to vision.

occlusion Blockage of the background of a visual scene by an object in the foreground.

ocular dominance columns The segregated termination patterns of thalamic inputs representing the two eyes in primary visual cortex of some species of primates and carnivores.

ontogeny The developmental history of an individual animal; also used as a synonym for development.

opponency See *color opponency*.

opponent cell See *color opponency*.

opponent colors Colors that appear as perceptual opposites, sited across from each other on the Newton color circle (e.g., red/green or blue/yellow).

opsins Proteins in photoreceptors that absorb light (in humans, rhodopsin and the three specialized cone opsins).

optical axis See *line of sight*.

optic chiasm See *chiasm*.

optic disk The region of the retina where the axons of retinal ganglion cells exit to form the optic nerve.

optic nerve The nerve (cranial nerve II) containing the axons of retinal ganglion cells; extends from the eye to the optic chiasm.

optic radiation Portion of the internal capsule that contains the axons of lateral geniculate neurons that carry visual information to the striate cortex.

optic tectum The first central station in the visual pathway of many vertebrates (homologous to the superior colliculus in mammals).

optic tract The axons of retinal ganglion cells after they have passed through the region of the optic chiasm en route to the lateral geniculate nucleus of the thalamus.

organelle A subcellular component visible by means of a light or electron microscope (e.g., nucleus, ribosome, endoplasmic reticulum, etc.).

orientation The arrangement of an object in the three dimensions of Euclidean space.

orientation selectivity Describing neurons that respond selectively to edges presented over a relatively narrow range of stimulus orientations.

orientation tuning curve The curve derived when a neuron's receptive field is tested with stimuli at the different orientations to demonstrate the range of responses.

orthogonal Making a right angle with another line or surface; also called "normal."

Panum's fusional area The distance in front of and beyond the horopter within which disparate views of the two eyes are nonetheless seen as fused and single; named after the mid-nineteenth century German psychologist P. L. Panum.

parallel processing Processing visual or other information simultaneously by different components or pathways in a sensory (or other) system.

parietal lobe The lobe of the human brain that lies between the frontal lobe anteriorly and the occipital lobe posteriorly.

parvocellular Referring to relatively small neurons in the primary visual system specialized for processing color and form.

parvocellular system Referring to the component of the primary visual pathway in primates specialized for the detection of detail and color.

penumbra The fuzzy borders that occur at the edges of shadows cast by sources that are extended in space (e.g., the sun).

perception Typcially considered the subjective awareness (usually taken to be conscious awareness) of any aspect of the external or internal environment.

peripheral nervous system All the nerves and neurons that lie outside the brain and spinal cord.

perspective The representation of three-dimensional objects on a two-dimensional surface by projection through a lens. As a result, the same objects subtend different angles on the retinal surface as a function of distance.

phase Temporal description of a repeating wave.

phasic Transient firing of action potentials in response to a prolonged stimulus; the opposite of *tonic*.

phenomenology Description of the behavior of something (e.g., the phenomenology of visual perceptions).

phenotype The visible (or otherwise discernible) characteristics of an animal that arise during development.

photometry The measurement of light.

photopic Referring to normal levels of light, in which the predominant information is provided by cones (see *scotopic*).

photopic system The components of the visual system activated at relatively high light levels.

photopic vision Vision at relatively high light levels.

photoreceptors Cells in the retina specialized to absorb photons and thus to generate neural signals in response to light stimuli.

phylogeny The evolutionary history of a species or other taxonomic category.

phylum A major division of the plant or animal kingdom that includes classes having a common ancestry.

physiological blind spot The locus in visual space of the optic nerve head on the retina, which has no overlying photoreceptors and thus provides no visual information.

pigment epithelium Pigmented layer of cells underlying the retina; important in the normal turnover of photoreceptor pigments.

pigments Substances that both absorb and reflect light, such that they are perceived as colored.

pixel (picture element) Member of the array of discrete elements that comprises a two-dimensional digital image.

plasticity See *neural plasticity*.

Poggendorff illusion A geometrical effect that entails seeing an angled, collinear line that is occluded as being non-collinear; first described by the mid-nineteenth century German psychologist J. C. Poggendorff.

point The geometrical concept of a dimensionless location in space.

point of gaze The point in space on which the eyes are fixed.

polymodal Pertaining to more than one sensory modality.

Ponzo illusion A geometrical effect that entails seeing two identical horizontal lines as being unequal in length when they are placed between two converging lines; first described by the early twentieth century Italian psychologist Mario Ponzo.

positron emission topography (PET) A technique for measuring brain function based on the local concentration of an injected radioisotope that binds to some brain metabolite (e.g., oxygen or glucose).

postcentral gyrus The gyrus that lies just posterior to the central sulcus; contains the primary somatic sensory cortex.

posterior Toward the back; sometimes used as a synonym for *caudal*.

power The amount of work performed per unit time.

power law A relationship described by an exponential function.

prägnanz A word used by gestalt psychologists to indicate the simplest completion of a stimulus (a partly occluded letter for example) according to expectation; the meaning of the German word is "terseness" or "precision."

prefrontal cortex Cortical regions in the frontal lobe that are anterior to the primary motor and premotor cortices; thought to be involved in planning complex cognitive behaviors and in the expression of personality and appropriate social behavior.

primal sketch The initial generation of the internal representation postulated by David Marr and others; suggested to organize feature detection and information.

primary colors The four qualities or categories of human color vision (red, green, blue, and yellow) that are defined by a unique color sensation. In art and elsewhere, the term may have other definitions; computer images, for instance, are based on three colors (red, green, and blue).

primary motor cortex A major source of descending projections to motor neurons in the spinal cord and cranial nerve nuclei; located in the precentral gyrus (Brodmann's area 4) and essential for the voluntary control of movement.

primary sensory cortex Any one of several cortical areas in direct receipt of the thalamic input for a particular sensory modality.

primary visual cortex See *striate cortex*.

primary visual pathway (retino-geniculocortical pathway) Pathway from the retina via the lateral geniculate nucleus of the thalamus to the primary visual cortex; carries the information that allows conscious visual perception.

primate Order of mammals that includes lemurs, tarsiers, marmosets, monkeys, apes, and humans (technically, a member of this order).

priming A phenomenon in which the effect of an initial exposure is expressed by improved performance at a later time.

probability The likelihood of an event, usually expressed as a value from 0 (will never occur) to 1 (will always occur).

probability distribution Probability of a variable having a particular value, expressed as a function of all the possible values of that variable.

processing A general term that refers to the neural activity associated with carrying out some function.

proximal Closer to a point of reference (the opposite of *distal*).

proximal stimulus The light that actually interacts with photoreceptors (and therefore corresponds to a pattern of retinal activity).

psychology The study of mental processes in humans and other animals.

psychophysics The study of mental processes by quantitative methods, typically involving reports by human subjects of the sensations elicited by carefully measured stimuli.

pupil The perforation in the iris that controls the amount of light reaching the retina.

pupillary light reflex Decrease in the diameter of the pupil that follows stimulation of the retina by light.

Purkinje shift Change in sensitivity as a function of overall light level as the visual system shifts from primarily rod function to cone function.

radiance The electromagnetic energy emitted by something.

random dot stereograms Method of binocular stimulation using patterns of random dots to camouflage a hidden object.

range The distance of a point in space from an observer or a measuring device.

range image A digital image that includes information about the range of every constituent pixel.

rate code The proposition that neural information is encoded by the firing rate of neuronal action potentials.

ray Conventional term used to indicate the passage of photons from source to a target or detector in a straight line.

Rayleigh scattering The scattering of sunlight by molecules and very small particles in the earth's atmosphere, the amount of scatter being inversely proportional to the fourth power of the light's wavelength (meaning that short-wavelength light is scattered more than long-wavelength light; thus, we see the sky as blue because the atmosphere is effectively a source of isotropically scattered light in which short wavelengths predominate).

real world Phrase used to convey the idea that there is an external world that determines what we see, even though that world is unknowable by means of light stimuli.

receptive field Region of the retinal surface and the specific characteristics of the stimuli that cause a sensory nerve cell (or axon) to respond by increasing or decreasing its baseline activity.

receptor Nerve cells specialized to transduce physical energy into neural signals.

receptor cells The cells in a sensory system that transduce environmental energy into neural signals (e.g., photoreceptors in the retina, hair cells in the inner ear, etc.).

reflectance The percentage of incident light reflected from a surface (see *reflectance efficiency function*).

reflectance efficiency function Function describing the degree to which a surface reflects different wavelengths of light; more precisely, the ratio of the light flux reflected from the surface to the flux reflected from a perfectly reflecting diffuser under the same conditions of measurement.

reflection The return of light hitting a surface as a result of its failure to be absorbed or transmitted.

reflex A stereotyped response elicited by a defined stimulus. Usually taken to be restricted to "involuntary" actions (an assumption that is not necessarily appropriate).

reflex arc Referring to the circuitry that connects a sensory input to a motor output.

refraction The altered direction and speed of light as a result of passing from one medium to another (e.g., from air to the substance of the cornea).

representation In vision, the idea that the visual system reconstructs the retinal image or its real-world source, either literally or figuratively.

resolution The ability to distinguish two points in space (see *acuity*).

retina Laminated neural component of the eye that contains the photoreceptors (rods and cones) and the initial processing circuitry for vision.

retinal disparity The geometrical difference between the same points in the images projected on the two retinas, measured in degrees with respect to the fovea.

retinal ganglion cells The output neurons of the retina, whose axons form the optic nerve.

retinal image The image focused on the retina by the optical properties of the eye.

retinex theory Edwin Land's algorithm for explaining color contrast and constancy.

retinotectal system The pathway between retinal ganglion cells and the optic tectum in vertebrates such as frogs and fish.

retinotopic map A map in which neighbor relationships at the level of the retina are maintained at higher stations in the visual system.

retinotopy The maintenance of the neighbor relationships at progressively higher stations in the visual system.

rhodopsin The photopigment found in vertebrate rods.

rivalry (see *binocular rivalry*) The unstable visual experience that occurs when the right and left eye are presented with incompatible or conflicting images.

rods System of photoreceptors specialized to operate at low light levels.

rostral Anterior, or "headward." Opposite of *caudal*.

rotation A physical movement defined by an angular change in the position of a point or points in a frame of reference (see *translation*).

saccades Ballistic, conjugate eye movements that change the point of binocular foveal fixation; normally occur at a rate of approximately 3 to 4 per second.

saturation The aspect of color sensation pertaining to the subjective sense of the perceptual distance of the color from neutrality (thus, an unsaturated color is one that approaches a neutral gray).

scale An ordering of quantities according to their magnitudes.

scaling Psychophysical technique for measuring the magnitude of a sensation.

scatter Dispersion of light that degrades an image.

scene The arrangement of objects and their illumination with respect to the observer; gives rise to the visual stimulus synonymous with "source")

sclera The external connective tissue coating of the eyeball.

scotoma A defect in the visual field as a result of injury or disease to some component of the primary visual pathway.

scotopic Referring to vision in dim light, where only the rods are operative.

scotopic system Components of the visual system operating at relatively low levels of light (see *photopic system*).

sensation The subjective experience elicited by energy in one form or another impinging on an organism's sensory receptors (a word that has little or no meaning when differentiated from *perception*).

sensitivity The degree of ability to respond to the energy in a sensory stimulus.

sensory Pertaining to sensation.

sensory coding Referring to the manner in which information is generated in sensory systems.

sensory neuron Any neuron involved in sensory processing.

sensory stimuli Any pattern of energy impinging on a sensory receptor sheet such as the retina, skin, or basilar membrane.

sensory system Term used to describe all the components of the central and peripheral nervous system concerned with sensation (or a modality such as vision).

sensory transduction Process by which energy in the environment is converted into electrical signals by sensory receptors.

shadows Regions of diminished light that occur when objects are interposed between light sources such as the sun and a surface potentially in receipt of that light.

simple cell A cell in visual cortex that receives direct information from the lateral geniculate nucleus; center surround receptive fields are organized as if constructed from the characteristics of geniculate neurons.

simultaneous brightness contrast The ability of contextual information to alter the perception of luminance (i.e., the lightness or brightness) of a visual target.

simultaneous color contrast The ability of contextual information to alter the perception of the spectral distribution ("color") of a visual target.

simultaneous lightness contrast See *simultaneous brightness contrast*.

single-unit recording Method of studying the activity of single neurons using an electrode and amplifiers.

sinusoid Pattern defined by a sine (or cosine) function.

size constancy An abused phrase indicating the sense that familiar objects extend to maintain their perceived size despite being seen at different distances from an observer.

size–distance relationship The diminished size of a retinal projection as a function of distance from the observer.

skylight The isotropic light reflected from the atmosphere that reaches Earth's surface; it appears bluish because more short- than long-wavelength light is permitted through by the filtering properties of the atmosphere.

smooth pursuit Slow tracking movements of the eyes designed to keep a moving stimulus aligned with the fovea.

software The programs that run computers.

spatial frequency The spatial interval over which a pattern repeats, usually measured in cycles per degree (or cycles per millimeter). More specifically, the number of cycles of luminance variation by some measure in a given direction over 1° of visual angle.

species A taxonomic category subordinate to genus; members of a species are defined by extensive similarities and the ability to interbreed.

spectral differences Differences in the distribution of spectral power in a visual stimulus that give rise to perceptions of color.

spectral sensitivity The sensitivity of a photoreceptor (or other detecting device) to light at different wavelengths.

spectrograph The depiction of the frequency of a signal plotted over time (density of the graph signifies power).

spectrophotometer A device for measuring the spectral power distribution in light (i.e., electromagnetic radiation with wavelength between about 400 and 700 nanometers (the actual range varies as a function of the specific purpose for which the machine is used).

spectrum (pl. spectra) The power distribution of a given light at different wavelengths; more generally, the distribution of a continuous variable.

specular Referring to surfaces that reflect light preferentially at the angle of incidence (e.g., a mirror).

speed The rate of change that, together with direction, defines the velocity of a moving object.

stereogram A pair of pictures of the same scene taken from slightly different angles that, when fused, create a sensation of stereoscopic depth.

Stereopsis The special sensation of depth that results from fusion of the two eyes' views of relatively nearby objects.

stereoscope A viewing devise that mimics with two-dimensional pictures or photographs the normally different views of the right and left eyes.

Stiles-Crawford effect Observation that photons entering the center of the pupil are more effective for vision than those entering at the edge, particularly in photopic vision.

striate cortex The primary visual cortex in the occipital lobe in humans and other primates (also called *Brodmann's area 17* or *V1*). So named because the prominence of layer IV in myelin-stained sections gives this region a striped appearance.

sulcus (pl. sulci) Valleys that arise from the enfolding of the cerebral hemisphere between gyral ridges.

superior colliculus Laminated structure that forms part of the roof of the midbrain; important in orienting movements of the head and eyes.

surface Any physical interface separable from the medium in which it resides.

symbolic A formal system of notation whereby a series of steps or relationships is spelled out.

synapse Specialized apposition between a neuron and a target cell; transmits information by release and reception of a chemical transmitter agent.

synaptic potentials Membrane potentials generated by the action of chemical transmitter agents.

tectum A general term referring to the dorsal region of the brainstem (from the Latin word for "roof").

temporal coding Computational principle in which information is encoded by the precise timing of neuronal activity.

temporal lobe The hemispheric lobe that lies inferior to the lateral fissure.

texture An interactive quality of surfaces that can be indicated by a variety of characteristics and can provide information about the orientation or distance of objects in space.

thalamus A collection of nuclei that forms the major component of the diencephalon. Although its functions are many, a primary role of the thalamus is to relay sensory information from the periphery to the cerebral cortex.

threshold The lowest energy level of a stimulus that causes a perceptual response; also, the level of membrane potential at which an action potential is generated.

T-illusion A geometrical effect in which a vertically oriented line looks longer than a horizontally oriented line of the same length.

tonic Sustained activity in response to an ongoing stimulus; the opposite of *phasic*.

top-down A term that refers to the effects of what are taken to be "*higher order*" mental processes (usually considered a consequence of experience) on primary sensory or other "*bottom-up*" information (usually considered independent from experience).

top-down processing The idea that cognitive influences arising from "*higher order*" cortical regions influence "lower order" cortical or sub-cortical processing.

transduction The cellular and molecular process by which energy is converted into neural signals.

translation A physical movement in which a point or points move without rotation at the same velocity in a frame of reference.

translucent An object or medium that allows some, but not all, of the light incident upon it to pass through.

transmittance The amount of light that reaches the eye or some other detector from a surface, compared with the amount that is initially reflected from it (or emanated by it), and expressed as a percentage. More precisely, the ratio of transmitted flux to incident flux under specified conditions.

transparent As object or medium that allows all light incident upon it to pass through.

trichromacy theory The theory that human color vision generally is explained on the basis of three different cone types.

trichromatic Referring to the three different cone types in the human retina that absorb long, medium, and short wavelengths of light, respectively.

tuning curve Result of an electrophysiological test in which the receptive field properties of neurons are gauged; the maximum sensitivity (or responsiveness) is defined by the peak of the tuning curve.

uncrossed disparity Geometrical disparity arising from objects that are more distant from the observer than the *horopter*.

unique hues One of the four particular hues around the Newton color circle that seen as having no admixture of another hue (i.e., unique red, green, blue, and yellow).

unit recording See *single unit recording*.

univariance Principle that individual receptors cannot signal the wavelength of the light they have absorbed.

V1 The primary visual cortex.

V2 The secondary visual cortex.

V4 Area of extrastriate cortex important in color vision.

variable A measurement that can in principle assume any value within some appropriate range.

velocity Vector defined by the speed and direction of a moving object.

ventral Referring to the belly; the opposite of *dorsal*.

ventral stream (direct pathway) The stream of visual information directed toward the temporal lobe that is especially pertinent to object recognition.

ventricles The spaces in the vertebrate brain that represent the lumen of the embryonic neural tube.

vergence movements Disjunctive movements of the eyes (convergent or divergent) that normally align the fovea of each eye with targets located at different distances from the observer.

vertebrate An animal with a backbone (technically, a member of the subphylum Vertebrata).

vertical disparity Geometrical disparity arising from right and left eye images displaced along a vertical axis of visual space.

vestibular system The sensory system dedicated to generating information about the position of the head in space and the acceleration of the body.

vestibulo-ocular reflex Involuntary movement of the eyes in response to displacement of the head; allows retinal images to remain stable while the head is moved.

Vieth-Mueller circle A circle containing the point of fixation and the nodal point of both eyes. Contains all the points in the plane of the circle that project to corresponding points on the two retinas, thus defining this part of the geometrical horopter.

vision The process by which the eye and brain translate information conveyed by light into perceptions appropriate for visually guided behavior.

visual acuity See *acuity*.

visual angle The angle between two lines that extend from the observer's eye to different points on an object in space.

visual association cortices The neocortex of the occipital lobe and the adjacent regions of the parietal and temporal lobes devoted to 'higher order' visual processing.

visual field The area of visual space normally seen by one or both eyes (referred to, respectively, as the monocular and binocular fields).

visual perception The manifestation in consciousness of the empirical significance of visual stimuli (and not, therefore, a necessary accompaniment of vision, since vision often occurs without any particular awareness of what is being seen).

visual percepts Mental constructs that represent the empirical significance of light stimuli, which thus allow the observer to think about visual experience (or scenes).

visual pigment A pigment (in humans, rhodopsin or one of the three cone opsins) that absorbs light and initiates the process of vision.

visual processing Transformations (known or unknown) carried out on information in the retinal stimulus.

visual qualities The mental descriptors of visual percepts (e.g., brightness, color, depth, form, and motion).

visually guided responses The observer's actions undertaken to deal with the object or objects in the scene giving rise to the visual stimuli.

wavelength The interval between two wave crests or troughs in any periodic function; for light, the standard way of measuring the energy of different photons (measuring the frequency of photon vibration is another way).

Weber's law Principle that the just noticeable difference in a stimulus increment is a constant fraction of the stimulus; named after the 19th C. German physiologist and anatomist Ernst Weber.

white light Broadband light that is perceived as lacking color (i.e., as neutral).

white matter A general term that refers to large axon tracts in the brain and spinal cord; the phrase derives from the fact that axonal tracts have a whitish cast when viewed in the freshly cut material.

White's illusion A classic brightness illusion that has posed a major problem in explaining visual processing of luminance.

Young-Helmholtz theory Synonym for *trichromacy theory*.

References

Abney WD (1910). On the changes in hue of spectrum colours by dilution with white light. *Proc R Soc (Lond)* 83: 120–127.

Adelson EH (1993). Perceptual organization and the judgment of brightness. *Science* 262: 2042–2044.

Adelson EH (2000). Light perception and lightness illusions. In: *The new cognitive neurosciences,* (2nd edition) (Gazzaniga MS, ed.), pp. 339–351. Cambridge, MA: MIT Press.

Adelson EH, Bergen JK (1985). Spatiotemporal energy models for the perception of motion. *J Opt Soc Am* 2: 284–299.

Adelson EH, Movshon JA. (1982) Phenomenal coherence of moving visual patterns. *Nature* 300: 523–525.

Alais D, O'Shea RP, Mesana-Alais C, Wilson IG (2000). On binocular alternation. *Perception* 29: 1437–1445.

Allman JM (1999). *Evolving brains.* New York: WH Freeman.

Anderson BL (1998). Stereoscopic occlusion and the aperture problem for motion: A new solution. *Vision Res* 39: 1273–1284.

Anderson JA, Rosenfeld E (2000). *Talking nets: An oral history of neural networks.* Cambridge, MA: MIT Press.

Andrews DP (1967). Perception of contour orientation in the central fovea. Part I: Short lines. *Vision Research* 7: 975–997.

Andrews TJ, Halpern SD, Purves D (1997). Correlated size variations in human visual cortex, lateral geniculate nucleus and optic tract. *J Neurosci* 17: 2859–2868.

Andrews TJ, Purves D (1997). Similarities in normal and binocularly rivalrous viewing. *Proc Natl Acad Sci USA* 94: 9905–9908.

Andrews TJ, White LE, Binder D, Purves D (1996). Temporal events in cyclopean vision. *Proc Natl Acad Sci USA* 93: 3689–3692.

Appel K, Haken W (1976). Proof of 4-color theorem. *Discretionary Mathematics* 16: 179–180.

Arend LE, Goldstein R (1987). Lightness models, gradient illusions, and curl. *Percep Psychophys* 54: 446–456.

Aschenbrenner CM (1954). Problems in getting information into and out of air photographs. *Photogrammetric Engineering* 20: 398–401.

Asher H (1953). Suppression theory of binocular vision. *Brit J Ophthal* 37: 37–49.

Atick JJ, Redlich AN (1992). What does the retina know about natural scenes? *Neural Computation* 4: 196–210.

Barlow HB (1953). Summation and inhibition in the frog's retina. *J Physiol (Lond)* 119: 69–88.

Barlow HB (1961). Possible principles underlying the transformation of sensory messages. In: *Sensory communication* (Rosenblith WA, ed.) pp. 217–236. Cambridge, MA: MIT Press.

Barlow HB (1995). The neuron doctrine in perception. In: *Cognitive neurosciences* (Gazzaniga MS, ed.), pp. 415–436. Cambridge, MA: MIT Press.

Barlow HB, Blakemore C, Pettigrew JD (1967). The neural mechanism of binocular depth discrimination. *J Physiol (Lond)* 193: 327–342.

Barlow HB, Blakemore C, Pettigrew JD (1967). The neural mechanism of binocular depth perception. *J Physl (Lond)* 193: 327–342.

Barlow HB, Pettigrew JD (1971). Lack of specificity of neurones in the visual cortex of young kittens. *J Physiol (Lond)* 218: 98–100.

Bartels A, Zeki S (2000). The architecture of the colour center in the human visual brain: new results and a review. *European J Neurosci* 12: 172–190.

Basole A, White LE, Fitzpatrick D (2003). Mapping multiple features in the population response of visual cortex. *Nature* 423: 986–990.

Bayes, T (1763). An essay toward solving a problem in the doctrine of chances. *Philosophical Transactions of the Royal Society* 53: 370–418.

Bedford RE, Wyszecki G (1958). Wavelength discrimination for point sources. *J Opt Soc Am A* 48: 129–135.

Bell AJ, Sejnowski TJ (1997). The "independent components" of natural scenes are edge filters. *Vision Research* 37: 3327–3338.

Benary W (1924). Beobachtungen zu einem Experiment uber Helligkeitskontrast. *Psychol Forsch* 5: 131–142.

Bennett BM, Hoffman DD, Nicola JE, Prakash, C (1989). Structure from two orthographic views of rigid motion. *J Opt Soc Am A* 6: 1052–1069.

Berkeley G (1709/1975). *Philosophical works including works on vision.* (Ayers MR, ed.) London: Everyman/ J.M. Dent.

Berry MJ, Brivanlou IH, Jordan TA, Meister M (1999). Anticipation of moving stimuli by the retina. *Nature* 398: 334–338.

Biehler W (1896). *Beitrage zur Lehre vom Augenmass fur Winkel.* Dissert. Freiburg.

Billock VA, Tsou BH (2010). Pushing the perceptual envelope: inducing specific hallucinations and forbidden colors. *Scientific American* (in press).

Billock VA, Gleason GA, Tsou BH (2001). Perception of forbidden colors in retinally stabilized equiluminant imges: An indication of softwired cortical color opponency? *J Opt Soc Am A* 18 (10): 2398–2403.

Blake R (2000). A primer on binocular rivalry, including current controversies. *Brain and Mind* 2: 5–38.

Blakemore C, Carpenter RHS, Georgeson MA (1970). Lateral inhibition between orientation detectors in the human visual system. *Nature* 228: 37–39.

Blakemore C, Cooper GF (1970). Development of the brain depends on the visual environment. *Nature* 228: 477–478.

Blakemore C, Tobin EA (1972). Lateral inhibition between orientation detectors in the cat's visual cortex. *Exp Brain Res* 15: 439–440.

Blakeslee B, McCourt ME (1997). Similar mechanisms underlie simultaneous brightness contrast and grating induction. *Vision Res* 37: 2849–2869.

Blakeslee B, McCourt ME (1999). A multiscale spatial filtering account of the White effect, simultaneous brightness contrast and grating induction. *Vision Res* 39: 4361–4377.

Blakeslee B, McCourt ME (2001). Multiscale spatial filtering account of the Wertheimer-Benary effect and the corrugated Mondrian. *Vision Res* 41: 2487–2502.

Blasdel GG, Salama G (1986). Voltage sensitive dyes reveal a modular organization in monkey striate cortex. *Nature* 321: 579–585.

Bolles RC (1969). The role of eye movements in the Müller-Lyer illusion. *Percept Psychophys* 6: 175–176.

Bonhoeffer T, Grinvald A (1993). The layout of iso-orientation domains in area 18 of the cat visual cortex: Optical imaging reveals a pinwheel-like organization. *J Neurosc* 13: 4157–4180.

Bonhoeffer T, Grinvald A (1996). Optical imaging based on intrinsic signals: The methodology. In: *Brain mapping: The methods* (Toga A, ed.), pp. 55–97. New York: Academic Press.

Boots B, Nundy S, Purves D (2007). Evolution of visually-guided behavior in artificial agents. *Network: Computation in Neural Systems* 18 (1): 11–34.

Bouma H, Andriessen JJ (1968). Perceived orientation of isolated line segments. *Vision Res* 8: 493–507.

Bouma H, Andriessen JJ (1970). Induced changes in the perceived orientation of line segment. *Vision Res* 10: 333–349.

Bowmaker JK (1984). Microspectrophotometry of vertebrate photoreceptors. *Vision Res* 24: 1641–1650.

Bowmaker JK (1998). The evolution of color vision in vertebrates. *Eye* 12: 541–547.

Boynton RM, Gordon J (1965). Bezold–Brücke hue shift measured by color-naming technique. *J Opt Soc Am A* 55(1): 78–86.

Braddick, O (1974). A short-range process in apparent motion. *Vision Res* 14: 519–528.

Brainard DH, Freeman WT (1997). Bayesian color constancy. *J Opt Soc Am* 7: 1393–1411.

Brainard DH, Wandell BA (1986). Analysis of the retinex theory of color vision. *J Opt Soc Am A* 3:1651–1661.

Brainard DH, Freeman WT (1997). Bayesian color constancy. *J Opt Soc Am A* 7: 1393–1411.

Breese BB (1899). On inhibition. *Psychological Monographs* 3: 1–65.

Brenner E, Smeets JBJ (2000). Motion extrapolation is not responsible for the flash-lag effect. *Vision Res* 40: 1645–1648.

Bross M, Blair R, Longtin P (1978). Assimilation theory, attentive fields, and the Mueller-Lyer illusion. *Perception* 7: 297–304.

Brown PK, Wald G (1964). Visual pigments in single rods and cones of the human retina. *Science* 144: 44–51.

Brown RO, MacLeod DIA (1997). Color appearance depends on the variance of surround colors. *Curr Biol* 7: 844–849.

Bruce CJ, Desimone R, Gross CG (1981). Visual properties of neurons in a polysensory area in superior temporal sulcus of the macaque. *J Neurophys* 46: 369–384.

Brünswik E (1956/1997). *Perception and the psychological design of representative experiments,* (2nd edition). Berkeley: University of California Press.

Brünswik E (1952/1955). The conceptual framework of psychology. In: *International Encyclopedia of Unified Science* (Neurath O, Carnap R, Morris C, eds.) pp. 665–760. Chicago: University of Chicago Press.

Buchsbaum G, Gottschalk A (1983). Trichromacy, opponent colour coding and optimum colour information transmission in the retina. *Phil T Roy B* 220: 89–113.

Buck S (1997). Influence of rod signals on hue perception: evidence from successive scotopic contrast. *Vision Res* 37: 1295–1301.

Burns BD, Pritchard R (1971). Geometrical illusions and the response of neurons in the cat's visual cortex to angle patterns. *J Physiol (Lond)* 213: 599–616.

Burr D, Ross J (1986). Visual processing of motion. *Trends Neurosci* 9: 304–307.

Burton GJ, Moorhead IR (1987). Color and spatial structure in natural scenes. *Applied Opt* 26: 157–170.

Campbell FW, Robson JG (1968). Application of Fourier analysis to the visibility of gratings. *J Physiol* 197: 551–566.

Carpenter RHS, Blakemore C (1973). Interactions between orientations in human vision. *Exp Brain Res* 18: 287–303.

Carr HA (1935). *An introduction to space perception*. New York, NY: Longmans, Green.

Castet E, Charton V, Dufour A (1999). The extrinsic/intrinsic classification of two-dimensional motion signals with barber-pole stimuli. *Vis Res* 39: 915–932.

Cayley A (1878). On the colouring of maps. *Proc Lond Math Soc* 9: 148.

Changizi MA, Widders D (2002). Latency correction explains the classical geometrical illusions. *Perception* 31: 1241–1262.

Chevreul ME (1839). De la loi du contraste simultané des couleurs et de l'assortiment des objets colorés. - translated into English by Charles Martel as The principles of harmony and contrast of colours (1854) (1st French edition 1838; Martel C, transl.) London: Longmans.

Chiao CC, Cronin TW, Osorio D (2000). Color signals in natural scenes: characteristics of reflectance spectra and effects of natural illuminants. *J Opt Soc Am A Opt Image Sci Vis* 17: 218–224.

Chichilnisky EJ, Wandell BA (1995). Photoreceptor sensitivity changes explain color appearance shifts induced by large uniform backgrounds in dichoptic matching. *Vision Res* 35: 239–254.

Chubb C, Sperling G, Solomon JA (1989). Texture interactions determine perceived contrast. *Proc Natl Acad Sci USA* 86: 9631–9635.

Clarke R and Lotto RB (2009). Visual processing of the bee innately encodes higher-order image statistics when the information is consistent with natural ecology. *Vision Res* 49: 1455–1464.

Cole BL (1993). Does defective colour vision really matter? In: *Colour vision deficiencies XI* (Drum B, ed.), pp. 67–86. Amsterdam: Kluwer Academic Publishers.

Conway BR, Livingstone MS (2006). Spatial and Temporal Properties of Cone Signals in Alert Macaque Primary Visual Cortex (VI). *Journal of Neuroscience* 26(42): 10826–10846.

Coppola DM, Purves H, McCoy A, Purves D (1998). The distribution of oriented visual contours. *Proc Natl Acad Sci USA* 95: 4002–4006.

Coppola DM, White LE, Fitzpatrick D, Purves D (1998). Unequal representation of cardinal and oblique contours in ferret visual cortex. *Proc Natl Acad Sci USA* 95: 2621–2623.

Coren S (1971). A size–contrast illusion without physical size differences. *American J Psychology* 84: 565–566.

Coren S, Enns JT (1993). Size contrast as a function of conceptual similarity between test and inducers. *Percept Psychophys* 54: 579–588.

Coren S, Porac C, Ward LM (1999). *Sensation and perception*, 5th edition. New York: Harcourt Brace.

Cormack EO, Cormack RH (1974). Stimulus configuration and line orientation in the horizontal–vertical illusion. *Percept Psychophys* 16: 208–212.

Corney D, Haynes J, Rees G, Lotto, RB (2009). The brightness of colour. *PLoS ONE* 4 (3): e5091. doi: 10.1371/journal.pone.0005091.

Corney D, Lotto RB (2007). What are lightness illusions and why do we see them? *PLoS Comput Biol* 3 (9): e180. doi: 10.1371/journal.pcbi.0030180.

Cornsweet TN (1970). *Visual perception*. New York: Academic Press.

Cornsweet TN (1985). A simple retinal mechanism that has complex and profound effects on perception. *Am J Opt Psycho Optics* 62: 427–438.

Craik KJW (1948/1966). *The nature of psychology: A selection of papers, essays and other writings* (Sherwood SL, ed.). Cambridge, England: Cambridge University Press.

Crane HD, Piantanida TP (1983). On seeing reddish-green and yellowish-blue. *Science* 221: 1078–1079.

Craven BJ (1993). Orientation dependence of human line-length judgments matches statistical structure in real-world scenes. *Proceedings of Royal Society of London Series B, Biological Sciences* 253: 101–106.

Creutzfeldt B, Lange-Malecki B, Dreyer E (1990). Chromatic induction and brightness contrast: a relativistic color model. *J Opt Soc A* 7: 1644–1653.

Creutzfeldt B, Crook J, Kastner S, Li C-Y, Pei X (1991a). The neurophysiological correlates of colour and brightness in lateral geniculate neurons. *Exp Brain Res* 87: 3–21.

Creutzfeldt B, Kastner S, Pei X, Valberg A (1991b). The neurophysiological correlates of colour and brightness contrast in lateral geniculate neurons. *Exp Brain R* 87: 22–45.

Cumming BG, Parker AJ (2000). Local disparity not perceived depth is signaled by binocular neurons in cortical area V1 of the macaque. *J Neurosc* 20: 4758–4767.

D'Zmura M (1998). Color contrast gain control in color vision. In: *Color vision* (Backhausm WGK, Kliegl R, Werner JS, eds.). pp. 251–266. Berlin: Walter de Gruyter and Co.

D'Zmura M, Lennie P (1986). Mechanisms of color constancy. *J Opt Soc Am A* 3: 1662–1672.

D'Zmura M, Lennie P, Tiana C (1997). Color search and visual field segregation. *Percept Psychophys* 59: 381–388.

Das A, Gilbert CD (1999). Topography of contextual modulations mediated by short-range interactions in primary visual cortex. *Nature* 399: 655–661.

Daw NW (1968). Colour-coded ganglion cells in the goldfish retina: extension of their receptive fields by means of new stimuli. *J Physiol (Lond)* 197: 567–592.

Daw NW (1984). The psychology and physiology of color vision. *Trends Cogn Sci* 9: 330–335.

Daw NW (1995). *Visual development*. New York: Plenum Press.

Day RH (1962). Inappropriate constancy explanation of spatial distortions. *Nature* 207: 891–893.

De Valois R, Albrecht DG, Thorell L (1978). Cortical cells: bar detectors or spatial filters? In: *Frontiers of visual science* (Cool S, Smith E, eds.). New York: Raven Press.

De Valois RL, De Valois KK (1988). *Spatial vision*. New York: Oxford University Press.

De Valois RL, Albrecht DG, Thorell LG (1982a). Spatial frequency selectivity of cells in macaque visual cortex. *Vision Res* 22 (5): 545–559.

De Valois RL, Yund EW , Hepler, N (1982b). The orientation and direction selectivity of cells in macaque visual cortex. *Vision Res* 22 (5): 531–544.

DeAngelis GC, Cumming BG, Newsome WT (1998). Cortical area MT and the perception of stereoscopic depth. *Nature* 394: 667–680.

DeAngelis GC, Freeman RD, Ohzawa I (1994). Length and width tuning of neurons in the cat's primary visual cortex. *J Neurophys* 71: 347–374.

DeAngelis GC, Ohzawa I, Freeman RD (1993a). Spatiotemporal organization of simple-cell receptive fields in the cat's striate cortex. I. General characteristics and postnatal development. *J Neurophys* 69: 1091–1117.

DeAngelis GC, Ohzawa I, Freeman RD (1993b). Spatiotemporal organization of simple-cell receptive fields in the cat's striate cortex. II. Linearity of temporal and spatial summation. *J Neurophys* 69: 1118–1135.

Delboeuf JLR (1892). Sur une nouvelle illusion d'optique. *Bulletin de l'Academie royale de Belgique* 24: 545–558.

Descartes R (1637/1965). *Discourse on method, optics, geometry, and meteorology* (Olscamp P, transl.), Optics, pp. 65–162. Indianapolis: Bobbs-Merrill.

Desimone R, Gross CG (1979). Visual areas in the temporal cortex of the macaque. *Brain Res* 178: 363–380.

Desimone R, Albright TD, Gross CG, Bruce CJ (1984). Stimulus-selective properties of inferior temporal neurons in the macaque. *Neuroscience* 4: 2051–2062.

Dewar RE (1967). The effect of angle between the oblique lines on the decrement of the Müller-Lyer illusion with extended practice. *Percept Psychophys* 2: 426–428.

Diaz-Caneja E (1928). Sur l'alternace binoculaire. Annales d'Oculistique pp. 721–731. Alais, D., O'Shea, R. P., Mesana-Alais, C., & Wilson, I. G. (2000). *Translation of Diaz-Caneja* (1928) Dunedin, New Zealand: Department of Psychology, University of Otago. http://psy.otago.ac.nz/r_oshea/br_DJtrans.html.

Dixon MW, Proffitt DR (2002). Overestimation of heights in virtual reality is influenced more by perceived distal size than by the 2-D versus 3-D dimensionality of the display. *Perception* 31: 103–112.

Dobbins AC, Jeo RM, Fiser J, Allman JM (1998). Distance modulation of neural activity in the visual cortex. *Science* 281: 552–555.

Dong DW, Atick JJ (1995). Statistics of natural time-varying images. *Network: Computation in Neural Systems* 6: 345–358.

Douglas RJ, Martin KAC (2007). Mapping the matrix: the ways of neocortex. *Neuron* 56: 226–238.

Dragoi V, Sur M (2000). Dynamic properties of recurrent inhibition in primary visual cortex: Contrast and orientation dependence of contextual effects. *J Neurophys* 83: 1019–1030.

Eagleman, DM (2001). Visual Illusions and Neurobiology. *Nat Rev Neurosci* 2: 920–926.

Eagleman DM, Sejnowski TJ (2000a). Motion integration and postdiction in visual awareness. *Science* 287: 2036–2038.

Eagleman DM, Sejnowski TJ (2000b). The position of moving objects. Response. *Science* 289: 1107a.

Eagleman DM, Sejnowski TJ (2002). Untangling spatial from temporal illusions. *Trends Neurosci* 25: 293.

Earlebacher A, Sekuler R (1969). Explanation of the Müller-Lyer illusion: Confusion theory examined. *J Exp Psychol* 80: 462–467.

Edelman S (1999). *Representation and recognition in vision*. Cambridge, MA: MIT Press.

Engel SA, Furmanski CS (2001). Selective adaptation to color contrast in human primary visual cortex. *J Neurosc* 21: 3949–3954.

Epstein W, Landauer AA (1969). Size and distance judgments under reduced conditions of viewing. *Percept Psychophys* 6: 269–272.

Erkelens CJ, van Ee R (2002). Multi-coloured stereograms unveil two binocular colour mechanisms in human vision. *Vision Res* 42: 1103–1112.

Evans RM (1948). *An introduction to color*. New York: John Wiley.

Evans RM, Swenholt BK (1968). Chromatic strength of colors. II. *J Opt Soc Am* 58: 580–584.

Fechner GT (1860/1966). *Elements der psychophysik*. Leipzig: Brietkopf und Hartel. (Vol. 1 translated as *Elements of psychophysics* by Adler HE). New York: Holt, Rinehart & Winston.

Feldman, J (2001). Bayesian contour integration. *Percept Psychophys* 63 (7): 1171–1182.

Fennema CI, Thompson WB (1979). Velocity determination in scenes containing several moving objects. *Comput Graph Image Process* 9: 301–315.

Field DJ (1994). What is the goal of sensory coding? *Neural Computation* 6: 559–601.

Fisher GH (1969). An experimental study of angular subtension. *Q J Exp Psychol* 21: 356–366.

Fisher GH (1970). An experimental and theoretical appraisal of the perspective and size-constancy theories of illusions. *Q J Exp Psychol* 22: 631–652.

Fitzpatrick D (2000). Seeing beyond the receptive field in primary visual cortex. *Curr Opin Neurobiol* 10: 438–443.

Foley JM (1985). Binocular distance perception: egocentric distance tasks. *J Exp Psychol: Human Perception & Performance* 11: 133–149.

Freeman WT (1994). The generic viewpoint assumption in a framework for visual perception. *Nature* 368: 542–545.

Ganz L (1966). Mechanism of the figural after-effects. *Psychol Rev* 73: 128–150.

Geisler WS, Kersten D (2002). Illusions, perception and Bayes. *Nat Rev Neurosci* 5 (6): 508–510.

Geisler WS, Perry JS, Super BJ, Gallogly DP (2001). Edge co-occurrence in natural images predicts contour grouping performance. *Vision Res* 41: 711–724.

Ghazanfar AA, Nicolelis MAL (1999). Spatiotemporal properties of layer V neurons of the rat primary somatosensory cortex. *Cereb Cortex* 4: 348–361.

Gibson JJ (1950a). *The perception of the visual world.* Boston, MA: Houghton Mifflin.

Gibson JJ (1950b). The perception of visual surfaces. *Am J Psychol* 63: 367–384.

Gibson JJ (1966). *The senses considered as perceptual systems.* Boston, MA: Houghton Mifflin.

Gibson JJ (1979/1986). *The ecological approach to visual perception.* Hillsdale, New Jersey: Lawrence Erlbaum.

Gilbert CD, Wiesel TN (1990). The influence of contextual stimuli on the orientation selectivity of cells in primary visual cortex of the cat. *Vision Res* 30: 1689–1701.

Gilchrist AL (1977). Perceived lightness depends on perceived spatial arrangement. *Science* 195: 185–187.

Gilchrist AL (1980). When does perceived lightness depend on perceived spatial arrangement? *Percept Psychophys* 28: 521–538.

Gilchrist AL (1994). *Lightness, brightness and transparency.* Hillsdale, NJ: Lawrence Erlbaum.

Gilchrist, AL, Kossifydis, C, Bonato, F, Agostini, T, Cataliotti, J, Li, X, Spehar, B, Annan, V, and Economou, E (1999). An anchoring theory of lightness percepion. *Psychol Rev* 106: 795–834.

Gilchrist AL (2006). *Seeing in black and white.* Oxford: Oxford University Press.

Gillam B, Blackburn S, Nakayama K (1999). Stereopsis based on monocular gaps: metrical encoding of depth and slant without matching contours. *Vision Res* 39: 493–502.

Gillam B (1971). A depth processing theory of the Poggendorff illusion. *Percep Psychophys.* 10(4A): 211–216.

Gillam B (1995). The perception of spatial layout from static optical information. In: *Perception of space and motion* (Epstein W, Rogers SJ, eds), pp. 23–67. San Diego: Academic Press.

Gillam B (1998). Illusions at century's end. In: *Perception and cognition at century's end*, 2d edition (Hochberg J, ed.), pp. 98–136. New York: Academic Press.

Girgus JS, Coren S (1975). Depth cues and constancy scaling on the horizontal–vertical illusion: the bisection error. *Can J Psychol* 29: 59–65.

Girgus JS, Coren S, Agdern M (1972). The interrelationship between the Ebbinghaus and Delboeuf illusions. *J Exp Psychol* 95: 453–455.

Goethe JW (1840/1967). *Goethe's theory of colours* (Frank Cass & Co., Ltd., Eastlake CL, transl.). London: John Murray, Albemarle Street.

Gogel WC (1965). Equidistance tendency and its consequences. *Psychological Bulletin* 64: 153–163.

Gogel WC, Tietz JD (1979). A comparison of oculomotor and motion parallax cues of egocentric distance. *Vision Res* 19: 1161–1170.

Goldstein EB (2000). *Sensation and perception*, 5th edition. Pacific Grove, CA: Brooks/Cole.

Gottschalk A, Buchsbaum G (1983). Information theoretic aspects of color signal processing in the visual system. *IEEE Transactions on Systems, Man, and Cybernetics. Special Issue on Sensory Communication*, SMC-13/57, pp. 864–873.

Green RT, Stacey BG (1966). Misapplication of the misapplied constancy hypothesis. *Life Sciences* 5: 1871–1880.

Greene E (1994). Collinearity judgment as a function of induction angle. *Percept Motor Skill* 78: 655–674.

Gregory RL (1963). Distortion of visual space as inappropriate constancy scaling. *Nature* 199: 678–780.

Gregory RL (1968). Perceptual illusions and brain models. *Proceedings of Royal Society of London Series B, Biological Sciences* 171: 279–296.

Gregory RL (1974). *Concepts and mechanisms of perception.* London: Duckworth.

Gregory RL (1998). *Eye and brain: The psychology of seeing*, 5th edition. Oxford: Oxford University Press.

Gregory RL, Harris J, Heard P (1995). *The artful eye.* New York: Oxford University Press.

Griggs R (1974). Constancy scaling theory and the Mueller-Lyer illusion: More disconfirming evidence. *Bulletin of the Psychonomic Society* 4: 168–170.

Grimson WEL (1981). *From images to surfaces: a computational study of the human early vision system.* Cambridge, MA: MIT Press.

Grossberg S (1987). Cortical dynamics of three-dimensional form, color, and brightness perception: I. Monocular theory. *Percept Psychophys* 41: 87–116.

Hammond P (1978). Directional tuning of complex cells in area 17 of the feline visual cortex. *J Phys (Lond)* 285: 479–491.

Hartline HK (1940). The receptive fields of optic nerve fibers. *Am J Physiol* 130: 690–699.

Hartline HK (1969). Visual receptors and retinal interaction. *Science* 164: 270–278.

Hartline HK, Graham CH (1932). Nerve impulses from single receptors in the eye. *J Cell Comp Physiol* 1: 277–295.

Helmholtz HLF von (1866/1924–1925). *Helmholtz's treatise on physiological optics*, (3rd German edition, Vols. I–III, 1909; Southall JPC, transl. Rochester, NY: The Optical Society of America.

Helson H (1938). Fundamental principles in color vision. I. The principle governing changes in hue, saturation and lightness of non-selective samples in chromatic illumination. *J Exp Psychol* 23: 439–471.

Helson H (1964). *Adaptation-level theory*. New York: NY: Harper & Row.

Hendley C, Hecht S (1949). The colors of natural objects and terrains, and their relation to visual color deficiency. *J Opt Soc A* 39: 870–873.

Hering E (1861). *Beiträge zur Physiologie*. Leipzig: Engelman.

Hering E (1868/1964). Outlines of a theory of the light sense. In: *Grundzüge der Lehre vom Lichtsinn* (Hurvich LM, Jameson D, transl.). Cambridge, MA: Harvard University Press.

Hershenson M (1999). *Visual space perception.* Cambridge, MA: MIT Press.

Heyer D, Mausfeld R (eds.) (2002). *Perception and the physical world: psychological and philosophical issues in perception*. New York: John Wiley and Sons.

Heywood S, Chessell K (1977). Expanding angles? Systematic distortions of space involving angular figures. *Perception* 6: 571–582.

Hildreth E (1984). *The measurement of visual motion.* Cambridge, MA: MIT Press.

Hodgkinson IJ, O'Shea RP (1994). Constraints imposed by Mach bands on shape from shading. *Computer Graphics* 18: 531–536.

Hoffman CS (1962). Comparison of monocular and binocular color matching. *J Opt Soc A* 52: 75–80.

Hooke R (1665). *Micrographia: or, some physiological descriptions of minute bodies made by magnifying glasses [microform]: with observations and inquires thereupon*. Pp.188–187. Edinburgh: WF Clay.

Horn BKP, Schunck BG (1981). Determining optical flow. *Artif Intell* 17: 185–203.

Horton JC, Adams DL (2005).The cortical column: a structure without a function. *Phil Trans R Soc B* 360: 837–862.

Hotopf WH, Robertson SH (1975). The regression to right angles tendency, lateral inhibition, and the transversals in the Zollner and Poggendorff illusions. *Percept Psychophys* 18: 453–459.

Howard IP (1982). Human visual orientation. New York: John Wiley.

Howard IP, Rogers BJ (1995). *Binocular vision and stereopsis,* Oxford Psychology Series No. 29. New York: Clarendon Press.

Howard RB, Wagner M, Mills RC (1973). The superiority of the pair-comparisons method for scaling visual illusions. *Percept Psychophys* 13: 507–512.

Howe CQ, Purves D (2002). Range image statistics can explain the anomalous perception of length. *Proc Natl Acad Sci USA* 99: 13184–13188.

Howe CQ, Purves D (2004). Size contrast and assimilation explained by the statistics of natural scene geometry. *J Cognitive Neurosci* 16: 90–102.

Howe CQ, Purves D (2005a). *Perceiving geometry: geometrical illusions explained by natural scene statistics.* New York: Springer Press.

Howe CQ, Purves D (2005b). Natural scene geometry predicts the perception of angles and line orientation. *Proc Natl Acad Sci USA* 102: 1228–1233.

Howe CQ, Purves D (2005c). The Müller-Lyer illusion explained by the statistics of image-source relationships. *Proc Natl Acad Sci USA* 102: 1234–1239.

Howe CQ, Yang Z, Purves D (2005d). The Poggendorff illusion explained by natural scene geometry. *Proc Natl Acad Sci USA* 102: 7707–7712.

Howe CQ, Lotto RB, Purves D (2006). Comparison of Bayesian and empirical ranking approaches to visual perception. *J Theoretical Biology* 241: 866–875.

Hoyer PO, Hyvärinen A (2000). Independent component analysis applied to feature extraction from colour and stereo images. *Network: Computation in Neural Systems* 11: 191–210.

Hubel DH (1982). Exploration of the primary visual cortex, 1955–1978. *Nature* 299: 515–524.

Hubel DH (1988). *Eye, brain and vision*. New York: WH Freeman.

Hubel DH, Wiesel TN (1959). Receptive fields of single neurons in the cat's striate cortex. *J Physiol* 148: 574–591.

Hubel DH, Weisel TN (1962). Receptive fields, binocular interaction and functional architecture in the cat's visual cortex. *J Physiol (Lond)* 160: 106–154.

Hubel DH, Wiesel TN (1963). Receptive fields of cells in striate cortex of very young, visually inexperienced kittens. *J Neurophys* 26: 994–1002.

Hubel DH, Wiesel TN (1968). Receptive fields and functional architecture of monkey striate cortex. *J Physiol* 195: 215–243.

Hubel DH, Wiesel TN (1974). Sequence regularity and geometry of orientation columns in the monkey striate cortex. *J Comp Neurol* 158: 267–294.

Hubel DH, Wiesel TN (1977). Ferrier lecture: functional architecture of macaque monkey visual cortex. *Phi T Roy B* 198: 1–59.

Hubel DH, Wiesel TN, LeVay S (1977). Plasticity of ocular dominance columns in the monkey striate cortex. *Phi T Roy B* 278: 377–409.

Hubel DH, Wiesel TN (2005). *Brain and visual perception*. New York: Oxford.

Hunt RWG (1952). Light and dark adaptation and the perception of color. *J Opt Soc Am* 42: 190–199.

Hurlbert A (1996). Colour vision: putting it in context. *Curr Biol* 6: 1381–1284.

Hurvich L (1981). *Color vision*. Sunderland, MA: Sinauer Associates.

Hurvich, L M, Jameson, D. (1955). Brightness, saturation, and hue in normal and dichromatic vision. *J Opt Soc Am* 45: 602–616.

Ichikawan M, Egusa H (1993). How is depth perception affected by long term wearing of left–right spectacles? *Perception* 22: 971–984.

Ikeda H, Obonai T (1955). Figural after-effect, retroactive effect and simultaneous illusion. *Japanese J Psychol* 26: 235–246.

Ikeda M, Nakashima Y (1981). Wavelength difference for binocular color fusion. *Vision Res* 20: 693–697.

Ikeda M, Yaguchi H, Sagawa K (1982). Brightness efficiency functions for 2deg and 10deg fields. *J Opt Soc A* 72: 1660–1665.

Indow T (1991). A critical review of Luneburg's model with regard to global structure of visual space. *Psychol Rev* 98: 430–453.

Jacobs GH (1993). The distribution and nature of colour vision among the mammals. *Biol Rev* 68: 413–471.

Jacobs GH, Neitz M, Deegan JF, Neitz J (1996). Trichromatic colour vision in New World monkeys. *Nature* 382: 156–158.

Jaeger T, Grasso K (1993). Contour lightness and separation effects in the Ebbinghaus illusion. *Percept Motor Skill* 76: 255–258.

Jaeger T, Lorden R (1980). Delboeuf illusions: contour or size detector interactions? *Percept Motor Skill* 50: 376–378.

Jaeger T, Pollack RH (1977). Effect of contrast level and temporal order on the Ebbinghaus circles illusion. *Percept Psychophys* 21: 83–87.

Jameson D, Hurvich L (1989). Essay concerning color constancy. *Ann Rev Psychol* 40: 1–22.

Jordan K, Uhlarik J (1986). Length contrast in the Mueller-Lyer figure: functional equivalence of temporal and spatial separation. *Percept Psychophys* 39: 267–274.

Judd DB (1933). The 1931 ICI standard observer and coordinate system for colorimetry. *J Opt Soc Am* 23: 359–374.

Judd DB (1940). Hue saturation and lightness of surface colors with chromatic illumination. *J Opt Soc Am* 30: 2–32.

Judd DB (1960). Appraisal of Land's work on two primary color projections. *J Opt Soc Am* 50: 254–268.

Julesz B (1959). Binocular depth perception of computer-generated patterns. *Bell System Technical Journal* 39: 1125–1162.

Julesz B (1971). *Foundations of cyclopean perception*. Chicago: Chicago University Press.

Julesz B (1995). *Dialogues on perception*. Cambridge, MA: MIT Press.

Kaiser PK, Boynton RM (1996). Human Color Vision. *Opt Soc Am*. Washington D.C.

Kapadia MK et al. (1995). Improvement in visual sensitivity by changes in local context: parallel studies in human observers and in V1 of alert monkeys. *Neuron* 15: 843–856.

Katz D (1911/1935). *The world of colour*. London: Paul, Trench and Trubner.

Kay LM, Laurent G (1999). Odor- and context-dependent modulation of mitral cell activity in behaving rats. *Nat Neurosc* 11: 1003–1009.

Kersten D (2000). High-level vision as statistical inference. In: *The New Cognitive Neurosciences* (Gazzaniga MS, ed.), pp. 353–363. Cambridge, MA: MIT Press.

Kersten D, Yuille A (2003). Bayesian models of object perception. *Current Opinion in Neurobiology* 13: 150–158.

Khurana B, Nijhawan R (1995). Extrapolation or attention shift? Reply. *Nature* 378: 555–556.

Khurana B, Watanabe K, Nijhawan R (2000). The role of attention in motion extrapolation: are moving objects "corrected" or flashed objects attentionally delayed? *Perception* 29: 675–692.

Knierim JJ, Van Essen DC (1992a). Neuronal responses to static texture patterns in area V1 of the alert macaque monkey. *J Neurophys* 67: 961–980.

Knierim JJ, Van Essen DC (1992b). Visual cortex: cartography, connectivity, and concurrent processing. *Current Opinion in Neurobiology* 2: 150–155.

Knill DC (1998). Surface orientation from texture: ideal observers, generic observers and the information content of texture cues. *Vision Res* 38: 1655–1682.

Knill DC (1998). Discrimination of planar surface slant from texture: human and ideal observers compared. *Vision Res* 38: 1683–1711.

Knill DC et al. (1996). Introduction: a Bayesian formulation of visual perception. In *Perception as Bayesian inference* (Knill DC, and Richards W., eds.), pp. 1–21. Cambridge, England: Cambridge University Press.

Knill DC, Kersten D (1991). Apparent surface curvature affects lightness perception. *Nature* 351: 228–230.

Knill DC, Pouget A (2004). The Bayesian brain: The role of uncertainty in neural coding and computation. *Trends Neurosci* 27: 712–719.

Knill DC, Richards W (eds.) (1996). *Perception as Bayesian inference*. Cambridge, England: Cambridge University Press.

Koffka K (1935). *Principals of gestalt psychology*. New York: Harcourt, Brace.

Kohler W (1947). *Gestalt psychology: An introduction to new concepts in modern psychology*. New York: Liveright.

Komatsu H (1998). Mechanisms of central color vision. *Curr Opin Neurol* 8: 503–508.

Kraft JM, Brainard DH (1999). Mechanisms of color constancy under nearly natural viewing. *Proc Natl Acad Sci USA* 96: 307–312.

Kruskal JB (1964). Nonmetric multidimensional scaling: A numerical method. *Psychometrika* 29: 1–27; 115–129.

Kruskal JB, Wish M (1977). *Multidimensional scaling*. Beverly Hills, CA: Sage Publications.

Kuehni RG, Schwarz (2008). *Color ordered: a survey of color order systems from antiquity to the present*. Oxford: Oxford University Press.

Kuennapas TM (1957). The vertical–horizontal illusion and the visual field. *J Exp Psychol* 53: 405–407.

Kuffler SW (1953). Discharge patterns and functional organization of mammalian retina. *J Neurophys* 16: 37–68.

Kuffler SW (1973). The single-cell approach in the visual system and the study of receptive fields. *Invest Ophthal* 12: 794–813.

Laming DRJ (1997). *The measurement of sensation*. Oxford: Oxford Univerity Press.

Lamme VAF (1995). The neurophysiology of figure-ground segregation in primary visual cortex. *J Neurosc* 15: 1605–1615.

Land EH (1959a). Color vision and the natural image, Part I. *Proc Natl Acad Sci USA* 45: 116–129.

Land EH (1959b). Color vision and the natural image. Part II. *Proc Natl Acad Sci USA* 45: 636–644.

Land EH (1986). Recent advances in retinex theory. *Vision Res* 26: 7–21.

Land EH, McCann JJ (1971). Lightness and retinex theory. *J Opt Soc Am* 61: 1–11.

Leibowitz H, Toffey S (1966). The effect of rotation and tilt on the magnitude of the Poggendorff illusion. *Vision Res* 6: 101–103.

Lennie P (1998). Single units and visual cortical organization. *Perception* 27: 889–935.

Lennie P, Fairchild MD (1994). Ganglion cell pathways for rod vision. *Vision Res* 34: 477–482.

Levine MW (2000). *Levine & Shefner's fundamentals of sensation and perception*. 3rd edition. Oxford: Oxford University Press.

Lewen G, Bialek W, Steveninck RR (2001). Neural coding of natural motion stimuli. *Network* 12: 317–329.

Lewis EO (1909). Confluxion and contrast effects in the Mueller-Lyer illusion. *British J Psychol* 3: 21–41.

Li CY, Li W (1994). Extensive integration field beyond the classical receptive field of cat's striate cortical neurons: classification and tuning properties. *Vision Res* 34: 2337–2355.

Limb JO, Murphy JA (1975). Estimating the velocity of moving images in television signals. *Comput Graph Image Process* 4: 311–327.

Livingstone MS, Hubel DH (1987). Psychophysical evidence for separate channels for the perception of form, color, movement, and depth. *J Neurosci* 7: 3416–3468.

Livingstone MS, Hubel DH (1988). Segregation of form, color, movement, and depth: Anatomy, physiology, and perception. *Science* 240: 740–749.

Logothetis NK, Leopold DA, Sheinberg DL (1996). What is rivaling during binocular rivalry? *Nature* 380: 621–624.

Long F, Purves D (2003). Natural scene statistics as the universal basis for color context effects. *Proc Natl Acad Sci, USA* 100: 15190–15193.

Long F, Yang Z, Purves D (2006). Spectral statistics in natural scenes predict hue, saturation, and brightness. *Proc Natl Acad Sci, USA* 103: 6013–6018.

Loomis JM, Da Silva JA, Philbeck JW, Fukusima SS (1996). Visual perception of location and distance. *Current Directions in Psychological Sci, USA* 5: 72–77.

Lotto RB, Chittka L (2005). Seeing the light: Illumination as a contextual cue to color choice behavior in bumblebees. *Proc Natl Acad Sci, USA* 102: 3852–3856.

Lotto RB, Purves D (1999). The effects of color on brightness. *Nat Neurosc* 2: 1010–1014.

Lotto RB, Purves D (2000). An empirical explanation of color contrast. *Proc Natl Acad Sci USA* 97: 12834–12839.

Lotto RB, Purves D (2001). An empirical explanation of the Chubb illusion. *J Cogn Neur* 13: 1–9.

Lotto RB, Purves D (2002). A rationale for the structure of color space. *Trends Neurosci* 2: 84–88.

Lotto RB, Williams SM, Purves D (1999a). Mach bands as empirically derived associations. *Proc Natl Acad Sci USA* 96: 5245–5250.

Lotto RB, Williams SM, Purves D (1999b). An empirical basis for Mach bands. *Proc Natl Acad Sci USA* 96: 5239–5244.

Lu ZL, Sperling G (1995). The functional architecture of human visual motion perception. *Vision Res* 35: 2697–2722.

Luckiesh M (1922). *Visual illusions. Their causes, characteristics and applications*. New York: D. Van Nostrand Company.

Luneburg RK (1947). *Mathematical analysis of binocular vision*. Princeton, NJ: Princeton University Press.

Lynch DK, Livingston WC (1995). *Color and light in nature*. Cambridge, England: Cambridge University Press.

MacEvoy SP, Paradiso MA (2001). Lightness constancy in primary visual cortex. *Proc Nat Acad Sci USA* 98: 8827–8831.

Mach E (1886/1959). The analysis of sensations and the relation of the physical to the psychical (1st German edition, Williams CM, transl.). New York: Dover.

Mach E (1865). Über die Wirkung der raümlichen Verthelung des Lichtreizes auf die Netzhaut. I. Sitzungsberichte der mathematisch-naturwissenschaftlichen Classe der kaiserlichen. *Akademie der Wissenschaften* 52: 303–322. (For an authoritative translation of this and other papers on vision by Mach see Ratliff, 1965.)

Mach E (1914). *The analysis of sensations*. Chicago: Open Court Publishing Company.

MacKay DM (1958). Perceptual stability of a stroboscopically lit visual field containing self-luminous objects. *Nature* 181: 507–508.

Maclean IE, Stacey BG (1971). Judgment of angle size: An experimental appraisal. *Percept Psychophys* 9: 499–504.

Maffei L, Fiorentini A (1976). The unresponsive regions of visual cortical receptive fields. *Vision Res* 16: 1131–1139.

Maloney LT (1986). Evaluation of linear models of surface spectral reflectance with small numbers of parameters. *J Opt Soc Am* 3: 1673–1683.

Maloney LT (2002). Statistical decision theory and biological vision. In: *Perception and the physical world: psychological and philosophical issues in perception* (Heyer D and Mausfeld R., eds.), pp. 145–189. New York: John Wiley and Sons.

Maloney LT, Schirillo JA (2002). Color constancy, lightness constancy, and the articulation hypothesis. *Perception* 31: 135–139.

Mamassian P, et al. (2002). Bayesian modeling of visual perception. In: *Probabilistic models of the brain: perception and neural function* (Rao RPN. et al., eds.), pp. 13–36. Cambridge, MA: MIT Press.

Marks WB, Dobelle WH, MacNichol EF (1964). Visual pigments of single primate cones. *Science* 143: 1181–1183.

Marr D (1982). *Vision: A computational investigation into human representation and processing of visual information*. San Francisco: W.H. Freeman.

Marr D, Poggio T (1976). Cooperative computation of stereo disparity. *Science* 194: 283–287.

Marr D, Poggio T (1979). A computational theory of human stereo vision. *Phi T Roy B* 204: 283–287.

Marr D, Ullman S (1981). Directional selectivity and its use in early visual processing. *Proc Roy Soc London Ser B*, 211: 151–180.

Massaro DW, Anderson NH (1971). Judgmental model of the Ebbinghaus illusion. *J Exp Psychol* 89: 147–151.

Maxwell JC (1855). Experiments on colour, as perceived by the eye, with remarks on colour-blindness. *Trans R Soc Edinburgh* 21: 275–298.

Maxwell JC (1861). On the theory of compound colors, and the relations of the colours of the spectrum. *Phil T Roy B* 150: 57–84.

Mayhew JEW, Frisby JP (1981). Psychophysical and computational studies towards a theory of human stereopsis. *Art Intell* 17: 349–385.

MacEvoy SP, Paradiso MA (2001). Lightness constancy in primary visual cortex. *Proc Nat Acad Sci, USA* 98: 8827–8831.

McCulloch WS, Pitts W (1943). A logical calculus of the ideas immanent in nervous activity. *Bulletin of Mathematical Biophysics* 5: 115–133.

McManus IC (1978). The horizontal-vertical illusion and the square. *British J Psychol* 69: 369–370.

Metelli F (1970). An algebraic development of the theory of perceptual transparency. *Ergonomics* 13: 59–66.

Metelli F (1974). Achromatic color conditions in the perception of transparency. In: *Perception: essays in honor of James J. Gibson* (MacLeod RB, Pick HL, eds.), pp. 95–116. New York: Cornell University Press.

Metelli F, da Pos O, Cavedon A (1985). Balanced and unbalanced, complete and partial transparency. *Percept Psychophys* 38: 354–366.

Miller SM (2001). Binocular rivalry and the cerebral hemispheres. *Brain and Mind* 2: 119–149.

Miller SM, Liu GB, Ngo TT, Hooper G, Riek S, Carson RG, Pettigrew JD (2000). Interhemispheric switching mediates perceptual rivalry. *Curr Bio* 10: 383–392.

Minnaert MGJ (1937/1992). *Light and color in the outdoors*. New York: Springer.

Mollon JD (1991). The uses and evolutionary origins of primate colour vision. Visual Dysfunction (Cronly-Dillon J, Gregory RL, eds.), Vol. 2, pp. 306–319. London: Macmillan.

Mollon JD (1995). See colour. In: *Colour: art & science*. (Lamb T, Bourriau J, eds.), pp. 127–150. Cambridge, England: Cambridge University Press.

Mollon JD, Regan BC (1999). The spectral distribution of primate cones and of the macular pigment: matched to properties of the world? *J Opt Technol* 66: 847–851.

Morgan MJ, Adam A, Mollon JD (1992). Dichromats break colour-camouflage of textural boundaries. *Proc Roy Soc B* 248: 291–295.

Morrison JD, Whiteside TC (1984). Binocular cues in the perception of distance of a point source of light. *Perception* 13: 555–566.

Morrison LC (1977). Inappropriate constancy scaling as a factor in the Muller-Lyer illusion. *British J Psychol* 68: 23–27.

Movshon JA (1975). The velocity tuning of single units in cat striate cortex. *J Physiol* 249: 445–468.

Movshon JA, Adelson EH, Gizzi MS, Newsome WT (1986). The analysis of moving visual patterns. In: *Pattern recognition mechanisms* (Chagas C, Gattass R, Gross C, eds.), pp. 148–163. New York: Springer-Verlag.

Movshon JA, Thompson ID, Tolhurst DJ (1978). Receptive field organization of complex cells in the cat's striate cortex. *J Physiol* 283: 79–99.

Müller RU, Kubie JL (1987). The effects of change in environment on the spatial firing of hippocampal complex-spike cells. *J Neurosc* 7: 1951–1968.

Müller-Lyer FC (1889). Optische Urteilstäuschungen. *Archiv für Anatomie und Physiologie Supplement-Band*: 263–270.

Murakami I (2001a). A flash-lag effect in random motion. *Vision Res* 41: 3101–3119.

Murakami I (2001b). The flash-lag effect as a spatiotemporal correlation structure. *Journal of Vision* 1: 126–136.

Nakagawa D (1958). Mueller-Lyer illusion and retinal induction. *Psychologia* 1: 167–174.

Nakayama K, Silverman GH (1988). The aperture problem: II Spatial integration of velocity information along contours. *Vision Res* 28: 747–753.

Nakayama K (1998). Vision Fin de Siecle: A reductionistic explanation of perception for the 21st century? In: *Perception and cognition at century's end: history, philosophy theory* (Hochberg J, ed.). pp. 307–331. San Diego: Academic Press.

Nakayama K, Silverman GH (1988). The aperture problem. I: Perception of nonrigidity and motion direction in translating sinusoidal lines. *Vision Res*. 28, 739–746.

Necker LA (1832). Observations on some phaenomena seen in Switzerland; and an optical phaenomenon, which occurs on viewing on a crystal or geometrical solid. *Phil Mag A* 1: 329–337.

Nelson JI, Frost BJ (1978). Orientation-selective inhibition from beyond the classic visual receptive field. *Brain Res* 139: 359–365.

Neumeyer C (1991). Evolution of colour vision. In: *Uses and evolutionary origins of primate color vision; evolution of the eye and visual system*, from the series *Vision and visual dysfunction*, Vol. 2, pp. 284–305, (Cronly-Dillon JR, Gregory RL, eds.). Boca Raton, FL: CRC Press.

Newell A, Shaw JC, Simon HA (1958). Chess-playing programs and the problem of complexity. *IBM J Res Dev* 2: 320.

Newton I (1704/1952). *Opticks: or, a treatise of the reflexions, refractions, inflexions and colours of light* (based on the 4th London edition). New York: Dover.

Nichols MJ, Newsome WT (2002). Middle temporal visual area microstimulation influences veridical judgments of motion direction. *J Neurosci* 22 (21): 9530–9540.

Nicolelis MAL, Ghazanfar AA, Fagin B, Votaw S, Oliveira LMO (1997). Reconstructing the engram: Simultaneous, multiple site, many single neuron recordings. *Neuron* 18: 529–537.

Nijhawan R (1994). Motion extrapolation in catching. *Nature* 370: 256–257.

Nijhawan R (1997). Visual decomposition of colour through motion extrapolation. *Nature* 386: 66–69.

Nijhawan R (2007). Visual prediction: Psychophysics and neurophysiology of compensation for time delays. *Behavioral and Brain Sciences* 31: 179–198.

Nilsson D-E, Land MF (2001). *Animal eyes*. Oxford: Oxford University Press.

Noe, A (2009). *Out of our heads*. New York: Hill and Wang.

Nundy S, Lotto B, Coppola D, Shimpi A, Purves D (2000). Why are angles misperceived? *Proc Natl Acad Sci USA* 97: 5592–5597.

Nundy S, Purves D (2002). A probabilistic explanation of brightness scaling. *Proc Natl Acad Sci* 99 (22): 14482–14487.

Obermeyer K, Blasdel G (1993). Geometry of orientation and ocular dominance columns in monkey striate cortex. *J Neurosci* 13: 4114–4129.

Obonai T (1954). Induction effects in estimates of extent. *J Exp Psychol* 47: 57–60.

O'Brien V (1958). Contour perception, illusion and reality. *J Opt Soc Am* 48: 112–119.

O'Brien V (1959). Contrast by contour-enhancement. *Am J Psychol* 72: 299–300.

Ogle KN (1964). *Researches in binocular vision*. New York: Hafner.

Oja E (1982). A simplified neuron model as a principal component analyzer. *J Math Biol* 15: 267.

O'Keefe J (1979). A review of the hippocampal place cells. *Prog Neurobio* 13: 419–439.

O'Keefe J, Dostrovsky T (1971). The hippocampus as a spatial map. Preliminary evidence from unit activity in the freely moving rat. *Brain Res* 34: 171–175.

Olshausen BA, Field DJ (1996). Emergence of simple-cell receptive field properties by learning a sparse code for natural images. *Nature* 381: 607–609.

Olshausen BA, Field DJ (1997). Sparse coding with an overcomplete basis set: a strategy employed by V1? *Vision Res* 37: 3311–3325.

Olzak LA, Laurinen PI (1999). Multiple gain control processes in contrast-contrast phenomena. *Vision Res* 39: 3983–3987.

Ooi TL, Wu B, He ZJ (2001). Distance determined by the angular declination below the horizon. *Nature* 414: 197–200.

Owens DA, Leibowitz HW (1976). Oculomotor adjustments in darkness and the specific distance tendency. *Percept Psychophys* 20: 2–9.

Oyama T (1960). Japanese studies on the so-called geometrical-optical illusions. *Psychologia* 3: 7–20.

Panum PL (1858). *Physiolgische untersuchungen über das sehen mit zwei augen*. Kiel: Schwers.

Paradiso MA, Carney T (1988). Orientation discrimination as a function of stimulus eccentricity and size: nasal/temporal retinal asymmetry. *Vision Res* 28: 867–874.

Paradiso MA, Hahn S (1996). Filling-in percepts produced by luminance modulation. *Vision Research* 36: 2657–2663.

Parkhurst DJ, Niebur E (2003). Scene content selected by active vision. *Spatial Vision* 16: 125–154.

Pascual-Leone A, Walsh V (2001). Fast backprojections from the motion to the primary visual area necessary for visual awareness. *Science* 292: 510–512.

Pavlov IP (1927/1960). Conditioned reflexes: an investigation of the physiological activity of the cerebral cortex (Anrep GV, transl. and ed.). New York: Dover.

Pearce D, Matin L (1969). Variation of the magnitude of the horizontal–vertical illusion with retinal eccentricity. *Percept Psychophys* 6: 241–243.

Pettigrew JD, Miller SM (1998). A "sticky" interhemispheric switch in bipolar disorder? *Phi T Roy B* 265: 2141–2148.

Philbeck JW, Loomis JM (1997). Comparison of two indicators of perceived egocentric distance under full-cue and reduced-cue conditions. *J Exp Psychol: Human Perception & Performance* 23: 72–85.

Poggio GF (1995). Mechanisms of stereopsis in monkey visual cortex. *Cereb Cort* 3: 193–204.

Poggio GF, Gonzalez F, Krause F (1988). Stereoscopic mechanisms in monkey visual cortex: binocular correlation and disparity selectivity. *J Neurosci* 8: 4531–4550.

Poggio GF, Poggio T (1984). The analysis of stereopsis. *Ann Rev Neurosci* 7: 379–412.

Poggio T, Torre V, Koch C (1985). Computational vision and regularization theory. *Nature* 317: 314–319.

Pokorny J, Smith V (2004). Chromatic discrimination. In: *The visual neurosciences* (Chalupa, L M, Werner JS, eds.), pp. 908–923. Cambridge, MA: MIT Press.

Polat U, Mizobe K, Pettet MW, Kasamatsu T, Norcia AM (1998). Collinear stimuli regulate visual responses depending on cell's contrast threshold. *Nature* 391: 580–584.

Pollock WT, Chapanis A (1952). The apparent length of a line as a function of its inclination. *Q J Exp Psy* 4: 170–178.

Polonsky A, Blake R, Braun J, Heeger DJ (2000). Neuronal activity in human primary visual cortex correlates with perception during binocular rivalry. *Nat Neurosc* 3: 1153–1159.

Polyak S (1957). *The vertebrate visual system*. Chicago: University of Chicago Press.

Ponzo M (1928). Urteilstäuschungen über Mengen. *Archiv für die gesamte Psychologie* 65.

Pouget A, Dayan P, Zemel R (2000). Information processing with population codes. *Nature Rev Neurosci* 1: 125–132.

Prazdny K (1985). Detection of binocular disparities. *Biol Cybern* 52: 93–99.

Predebon J (1983). Illusion of extent in simple angular figures. *Perception* 12: 571–580.

Pressey AW (1967). A theory of the Mueller-Lyer illusion. *Percept Motor Skill* 25: 569–572.

Prinz AA, Bucher D, Marder E (2004). Similar network activity from disparate circuit parameters. *Nat Neurosci* 7: 1345–52.

Prinzmetal W, Gettleman L (1993). Vertical–horizontal illusion: one eye is better than two. *Percept Psychophys* 53: 81–88.

Purdy DM (1931). Spectral hue as a function of intensity. *Am J Psych* 43: 541–559.

Purkinje JE, Kruta V (1823/1969). *Physiologist: A short account of his contributions to the progress of physiology, with a bibliography of his work* (Kruta V, transl.), pp. 75–131. Prague: Academia, Publishing House of the Czechoslovak Academy of Sciences.

Purushothaman G, Patel SS, Bedell HE, Ogmen H (1998). Moving ahead through differential visual latency. *Nature* 396: 424.

Purves D (1994). *Neural activity and the growth of the brain*. Cambridge, England: Cambridge University Press.

Purves D, Lichtman JW (1985). *Principles of neural development*. Sunderland, MA: Sinauer Associates.

Purves D, Lotto RB (2003). *Why we see what we do: an empirical theory of vision*. Sunderland, MA: Sinauer Associates.

Purves D, Lotto RB, Polger T (2000). Color vision and the four-color-map problem. *J Cogn Neurosci* 12: 233–237.

Purves D, Lotto RB, Williams SM, Nundy S, Yang Z (2001). Why we see things the way we do: evidence for a wholly empirical strategy of vision. *Phi T Roy B* 356: 285–297.

Purves D, Riddle D, LaMantia A (1992). Iterated patterns of brain circuitry (or how the cortex gets its spots). *Trends Neurosci* 15: 362–368.

Purves D, Shimpi A, Lotto RB (1999). An empirical explanation of the Cornsweet effect. *J Neurosci* 19: 8542–8551.

Purves D et al. (2008a). *Neuroscience*, 4th edition. Sunderland, MA: Sinauer Associates.

Purves D et al. (2008b). *Principles of Cognitive Neuroscience*. Sunderland, MA: Sinauer Associates.

Quiroga RQ, Reddy L, Kreiman GG, Koch C, Fried I (2005). Invariant visual representation by single neurons in the human brain. *Nature* 435: 1102–1107.

Rao RPN, Olshausen BA, Lewicki MS (eds.) (2002). *Probabilistic models of the brain: Perception and neural function*. Cambridge, MA: MIT Press.

Ratliff F (1965). *Mach bands: quantitative studies on neural networks in the retina*. San Francisco, CA: Holden-Day (contains translations of Mach's original papers also).

Read JC, Cumming BG (2007). Sensors for impossible stimuli may solve the stereo correspondence problem. *Nat Neurosci* 10: 1322–1328.

Recanzone GH, Wurtz RH, Schwarz U (1997). Responses of MT and MST neurons to one and two moving objects in the receptive field. *J Neurophys* 78: 2904–2915.

Reinagel P (2001). How do neurons respond to the real world? *Curr Opin Neurol* 11: 437–442.

Restle F, Merryman CT (1968). An adaptation-level theory account of a relative-size illusion. *Psychonomic Science* 12: 229–230.

Roberson AR (1970). A new determination of lines of constant hue. *AIC Proceedings Color* 69: 395–402.

Robinson JO (1998). *The psychology of visual illusions*. New York: Dover (corrected republication of the 1972 edition published by Hutchinson and Co. in England).

Rock I (1995). *Perception*. New York, NY: Scientific American Library.

Rodieck RW (1998). *First steps in seeing*. Sunderland, MA: Sinauer Associates.

Rossi AF, Paradiso MA (1999). Neural correlates of brightness in the responses of neuorons in the retina, LGN, and primary visual cortex. *J Neurosc* 19: 6145–6156.

Rossi AF, Rittenhouse CD, Paradiso MA (1996). The representation of brightness in primary visual cortex. *Science* 273: 1104–1107.

Ruderman DL, Bialek W (1994). Statistics of natural images: scaling in the woods. *Physical Review Letters* 73: 814–817.

Ruderman DL, Cronin TW, Chiao CC (1998). Statistics of cone responses to natural images: implications for visual coding. *J Opt Soc Am A* 15: 2036–2045.

Rust NC, Mante V, Simoncelli EP, Movshon JA (2006). How MT cells analyze the motion of visual patterns. *Nat Rev Neurosci* 9: 1421–1431.

Sagawa K (1982). Dichoptic color fusion studied with wavelength discrimination. *Vision Res* 2: 945–952.

Schein SJ, Desimone R (1990). Spectral properties of V4 neurons in the macaque. *J Neurosc* 10: 3369–3389.

Schiffman HR, Thompson JG (1975). The role of figure orientation and apparent depth in the perception of the horizontal–vertical illusion. *Perception* 4: 79–83.

Schlessinger E, Bentley PJ, Lotto RB (2005). Evolving visually guided agents in an ambiguous virtual world. *GECCO* 2005: 115–120.

Schnapf JL, Kraft TW, Baylor DA (1987). Spectral sensitivity of human cone photoreceptors. *Nature* 325: 439–441.

Schwartz O, Hsu A, Dayan P (2007). Space and time in visual context. *Nat Rev Neurosci* 8 (7): 522–535.

Schwartz O, Sejnowski TJ, Dayan P (2009). Perceptual organization in the tilt illusion. *J Vision* 9: 1–20.

Seckel A (2000). *The art of optical illusions*. London: Carlton Books Ltd.

Seckel A (2002). *More optical illusions*. London: Carlton Books Ltd.

Sedgwick HA (1986). Space perception. In: *Handbook of Perception and Human Performance: Vol. 1. Sensory processes and perception* (Boff KR, Kaufman L, Thomas JP, eds), pp. 21.21–21.57. New York: Wiley.

Sereno MI, Dale AM, Reppas JB, Kwong KK, Belliveau JW, Bradley TJ, Rosen BR, Tootell RBH (1995). Borders of multiple visual areas in humans revealed by functional magnetic resonance imaging. *Science* 268: 889–893.

Shannon CE (1950). Programming a computer for playing chess. *Phil Mag* 41: 256.

Shapley R, Enroth-Cugell C (1984). Visual adaptation and retinal gain control. *Prog Retinal Res* 3: 263–343.

Shepard RN (1962). The analysis of proximities: Multidimensional scaling with an unknown distance function. *Psychometrika* 27: 125–140 (part I); 219–246 (part II).

Shepard RN (1992). *L'oeil qui pense. visions, illusions, perceptions*. Paris: Le Seuil, coll. Points Sciences, S140.

Shepard RN, Carroll JD (1966). Parametric representation of non-linear data structures. In: *International symposium on multivariate analysis* (Krishnaiah PR, ed.). New York: Academic Press.

Shepard RN, Cooper LA (1992). Representation of colors in the blind, color-blind and normally sighted. *Psychol Sci* 3 (2): 97–103.

Sherrington CS (1904). On binocular flicker and the correlation of activity of corresponding retinal points. *Brit J Psychol* 1: 26–60.

Sherrington CS (1906a). On reciprocal action in the retina as studied by means of some rotating discs. In: *Reciprocal action in the retina* (reprinted from *J Physiol (Lond)*, February 1897), pp. 33–54.

Cambridge, England: Cambridge University Press.

Sherrington CS (1906b/1947). *The integrative action of the nervous system*, 2nd edition. New Haven, CT: Yale University Press.

Shimojo S, Nakajima Y (1981). Adaptation to the reversal of binocular depth cues: effects of wearing left–right reversing spectacles on stereoscopic depth perception. *Perception* 10: 391–402.

Shimojo S, Nakayama K (1990a). Real world occlusion constraints and binocular rivalry. *Vision Res* 30: 69–80.

Shimojo S, Nakayama K (1990b). Amodal representation of occluded surfaces: role of invisible stimuli in apparent motion correspondence. *Perception* 19: 285–299.

Shimojo S, Silverman GH, Nakayama K (1989). Occlusion and the solution to the aperture problem for motion. *Vision Res* 29: 619–626.

Shipley T (1971). The first random-dot texture stereogram. *Vision Res* 11: 1491–1492.

Shipley WC, Mann BM, Penfield MJ (1949). The apparent length of tilted lines. *J Exp Psychol* 39: 548–551.

Signac P (1921/1992). *D'Eug ne Delacroix au néo-impressionnisme*. (Silverman W, Ratliff F, transl and eds.), pp. 193–285. New York: The Rockefeller UP.

Sillito AM, Grieve KL, Jones HE, Cudeiro J, David J (1995). Visual cortical mechanisms detecting focal orientation discontinuities. *Nature* 378: 492–496.

Simoncelli EP, Olshausen BA (2001). Natural image statistics and neural representation. *Annu Rev Neurosci* 24: 1193–1216.

Simoncelli EP, Schwartz O (1999). Modeling surround suppression in V1 neurons with a statistically-derived normalization model. In: *Advances in neural information processing systems #11*. (Kearns MS, Solla SA, Cohn DA, eds.), pp. 153–159. Cambridge, MA: MIT Press,

Sleight RB, Austin TR (1952). The horizontal–vertical illusion in plane geometric figures. *J Psychol* 33: 279–287.

Solomon SG, Lennie P (2007). The machinery of colour vision. *Nature Rev Neurosci* 8: 276–286.

Sterling P (1997). The retina. In: *Synaptic organization of the brain* (Shepherd GM, ed.), pp. 205–253. Oxford: Oxford University Press.

Stevens SS (1966). Concerning the measurement of brightness. *J Opt Soc Am* 56: 1135–1136.

Stevens SS (1975). *Psychophysics*. New York: John Wiley.

Stiles WS, Crawford BF (1933). The luminous efficiency of rays entering the eye pupil at different points. *Phi T Roy B* 112: 428–450.

Stiles WS (1972). The line element in colour theory: A historical review. In: *Color metrics* (Vos, JJ, Fridle, LFC., Walraven, PL, eds.), pp. 1–25. Soesterberg: AIC/Holland.

Stocker AA, Simoncelli EP (2006). Noise characteristics and prior expectations in human visual speed perception. *Nat Neurosci* 9: 578–585.

Stockman A, Plummer DJ, Montag ED (2005). Spectrally-opponent inputs to the human luminance pathway: slow +M and -L cone inputs revealed by intense long-wavelength adaptation. *J Physiol* 566: 61–76.

Stryker MP, Sherk H (1975). Modification of cortical orientation selectivity in the cat by restricted visual experiences: a re-examination. *Science* 190: 904–906.

Stumpf P (1911). Über die Abhangigkeit der visuellen Bewegungsrichtung und negativen Nachbildes von den Reizvorgangen auf der Netzhaut. *Zeitschrift fur Psychologie* 59: 321–330. A gem from the past: Pleikart Stumpf's 1911 anticipation of the aperture problem, Reichardt detectors, and perceived motion loss at equiluminance. (Todorovic D, transl.) *Perception* (1996) 25: 1235–1242.

Sugrue LP, Corrado GS, WT Newsome (2005). Choosing the greater of two goods: neural currencies for valuation and decision making. *Nature Rev Neurosci* 6: 363–375.

Sumner P, Mollon JD (2000). Catarrphine photopigments are optimised for detecting targets against a foliage background. *J Exp Biol* 203: 1963–1986.

Sung K, Wojtach WT, Purves D (2009). An empirical explanation of aperture effects. *Proc Natl Acad* Sci 106: 298–303.

Svaetichin G, MacNichol EF (1958). Retinal mechanisms for chromatic and achromatic vision. *Ann NY Acad Sci* 74: 385–404.

Switkes E, Mayer MJ, Sloan JA (1978). Spatial frequency analysis of the visual environment: anisotropy and the carpentered environment hypothesis. *Vision Res* 18: 1393–1399.

Todd JT, Oomes AH, Koenderink JJ, Kappers AM (2001). On the affine structure of perceptual space. *Psychological Sci* 12: 191–196.

Todorovic D (1997). Lightness and junctions. *Perception* 26: 379–394.

Tong F, Engel SA (2001). Interocular rivalry revealed in the human cortical blindspot representation. *Nature* 411: 195–199.

Tootell RBH, Silverman MS, Switkes E, De Valois RL (1982). Deoxyglucose analysis of retinoptopic organization in primate striate cortex. *Science* 218: 902–904.

Tootell RBH, Switkes E, Silverman MS, Hamilton SL (1988). Functional anatomy of macaque striate cortex, II. Retinotopic organization. *J Neurosc* 8: 1531.

Tootell RBH, Silverman MS, Hamilton SL, De Valois RL, Switkes E (1988). Functional anatomy of macaque striate cortex. III. Color. *J Neurosci* 8: 1569–1593.

Torgerson WS (1952). Multidimensional scaling: I. Theory and method. *Psychometrika* 17: 401–419.

Turiel A, Parga N, Ruderman DL, Cronin TW (2001). Multiscaling and information content of natural color images. *Physical Review E* 62: 1138–1148.

Turing AM (1936). On computable numbers, with an application to the Entscheidungs problem. *Proc London Mathematical Soc* 42: 230–265.

Turing AM (1937). On computable numbers, with an application to the Entscheidungs problem: a correction. *Proc London Mathematical Soc* 43: 544–6.

Turing A (1950). Computing machinery and intelligence. *Mind* 59: 433–460.

Turner RS (1994). *In the eye's mind: vision and the Helmholtz-Hering controversy*. Princeton: Princeton University Press.

Tynan PD, Sekuler R (1982). Motion processing in peripheral vision: Reaction time and perceived velocity. *Vision Res* 22: 61–68.

Ullman S (1979). *The interpretation of visual motion*. Cambridge, MA: MIT Press.

Ullman S (1987). Computational vision: A brief introduction. In: *Frontiers of visual science: proceedings of the 1985 symposium, Committee on Vision, Commission on Behavioral and Social Sciences and Education, National Research Council*, pp. 63–79. Washington, DC: National Academy Press.

Ullman S (1996). *High-level vision*. Cambridge, MA: MIT Press.

Ullman S, Sali E (2000). *Biologically motivated computer vision*. (Lee SW, Blülthoff HH, Poggio T, eds.), pp. 73–87. Berlin: Springer-Verlag.

Van Essen DC, Gallant JL (1994). Neural mechanisms of form and motion processing in the primate visual system. *Neuron* 13: 1–10.

van Hateren JH, van der Schaaf A (1998). Independent component filters of natural images compared with simple cells in primary visual cortex. *Proc R Soc London Ser B, Biological Sciences* 265: 359–366.

van Santen JP, Sperling G (1984). Temporal covariance model of human motion perception. *J Opt Soc Am A* 1: 451–473.

Verhoeff FH (1935). A new theory of binocular vision. *Arch Ophth* 13: 151–175.

Vinje WE, Gallant JL (2000). Sparse coding and decorrelation in primary visual cortex during natural vision. *Science* 287: 1273–1276.

von Collani G (1985). The horizontal–vertical illusion in photographs of concrete scenes with and without depth information. *Percept Motor Skill* 61: 523–531.

von Kries J (1905). Die Gesichtsempfindugen. In: *Handbuch der Physiologie des Menschen* (Nagel W, ed.) pp. 109–282. Braunschweig: Viewveg.

von Neumann J, Morgenstern O (1947). *Theory of games and economic behavior*. Princeton: Princeton University Press.

Vos JJ (1978). Colorimetric and photometric properties of a 2 fundamental observer. *Col Res App* 3: 125–128.

Wachtler T, Lee TW, Sejnowski TJ (2001). Chromatic structure of natural scenes. *J Opt Soc Am A Opt Image Sci Vis* 18: 65–77.

Wade NJ (1987). On the late invention of the stereoscope. *Perception* 16: 785–818.

Wagner M (1985). The metric of visual space. *Percept Psychophys* 38: 483–495.

Wald G, Wooten BR (1973). Color-vision in mechanisms in the peripheral retinas of normal and dichromatic observers. *J Gen Physiol* 61: 125–145.

Wallach H (1935/1996). Über visuell wahrgenommene bewegungsrichtung. *Psychologische Forscheung* 20: 325–380. On the visually perceived direction of motion by Hans Wallach: 60 years later (Wuerger S, Shapley R, Rubin N, transl.). *Perception* 25: 1317–1367.

Wallach, H (1939). On the constancy of speed. *Psychol Rev* 46: 541–552.

Wallach H, O'Leary A (1982). Slope of regard as a distance cue. *Percept Psychophys* 31: 145–148.

Wallis JD, Anderson KC, Miller EK (2001). Single neurons in prefrontal cortex encode abstract rules. *Nature* 411: 953–956.

Walls GL (1948). The vertebrate eye and its adaptive radiation. *Optic J and Rev Optometry* 85: 33–43.

Walls GL (1956). The G. Palmer story. *J Hist Med* 11: 66–96.

Walsh V (1995). Adapting to change. *Curr Biol* 5: 703–705.

Wandell BA (1995). *Foundations of vision*. Sunderland MA: Sinauer Associates.

Watanabe M, Rodieck RW (1989). Parasol and midget ganglion cells of the primate retina. *J Comp Neurol* 289: 434–454.

Watson JB (1913). Psychology as a behaviorist views it. *Psychol Rev* 20: 158–177.

Weber EH (1834/1996). *E.H. Weber on the tactile senses*, 2nd edition (Ross HE, Murray DJ, eds.). Hove: Erlbaum (UK) Taylor & Francis.

Webster MA, Mollon JD (1991). Changes in colour appearance following post-receptoral adaptation. *Nature* 349: 235–238.

Webster MA, Mollon JD (1995). Colour constancy influenced by contrast adaptation. *Nature* 373: 694–698.

Weintraub DJ, Krantz DH (1971). The Poggendorff illusion: amputations, rotations, and other perturbations. *Percept Psychophys* 10: 257–264.

Weiss Y, Adelson E (1998). Slow and smooth: a Bayesian theory for the combination of local motion signals in human vision. *Technical Report*

1624, AI Laboratory, Massachusetts Institute of Technology, Cambridge, MA.

Weiss Y, Simoncelli EP, Adelson EH (2002). Motion illusions as optimal percepts. *Nat Rev Neurosci* 5: 598–604.

Weliky M, Bosking WH, Fitzpatrick D (1996). A systematic map of direction preference in primary visual cortex. *Nature* 279: 725–728.

Wenderoth P, Parkinson A, White D (1979). A comparison of visual tilt illusions measured by the technique of vertical setting, parallel matching and dot alignment. *Perception* 9: 47–57.

Werner JS (1998). Color vision in art and science. In: *Color vision: perspectives from different disciplines* (Backhaus WGK, Kliegel R, Werner JS, eds.) pp. 1–39. Berlin: Walter de Gruyter.

Wertheimer M (1912/1950). Laws of organization in perceptual forms. In: *A sourcebook of gestalt psychology* (Ellis WD, transl. and ed.), pp. 71–88. New York: Humanities Press.

Westfall RS (1976). *Never at rest: A biography of Isaac Newton*. Cambridge, England: Cambridge University Press.

Wheatstone C (1838). Contributions to the physiology of vision. I. On some remarkable and hitherto unobserved phenomena of binocular vision. *Phi T Roy B* 128: 371–394.

White IM, Wise SP (1999). Rule-dependent neuronal activity in the prefrontal cortex. *Exp Brain Res* 126: 315–335.

White M (1979). A new effect of pattern on perceived lightness. *Perception* 8: 413–416.

White LE, Basole A, Fitzpatrick D (2001a). Effect of speed on responses of V1 neurons to motion of contour and terminator cues. *Soc Neurosci* Abstract 27, Program 164.4.

White LE, Bosking WH, Fitzpatrick D (2001b). Consistent mapping of orientation preference across irregular functional domains in ferret visual cortex. *Vis Neurosci* 18: 65–76.

White LE, Coppola DM, Fitzpatrick D (2001). The contribution of sensory experience to the maturation of orientation selectivity in ferret visual cortex. *Nature* 411: 1049–1052.

Whitney D, Murakami I (1998). Latency difference, not spatial extrapolation. *Nat Rev Neurosci* 1: 656–657.

Whitney D, Cavanagh P, Murakami I. (2000b). Temporal facilitation for moving stimuli is independent of changes in direction. *Vision Res* 40: 3829–3839.

Whitney D, Murakami I, Cavanagh P (2000a). Illusory spatial offset of a flash relative to a moving stimulus is caused by differential latencies for moving and flashed stimuli. *Vision Res* 40: 137–149.

Whittle P (1992). Brightness, discriminability and the crispening effect". *Vision Res* 32: 1493–1507.

Wiesel TN (1982). Postnatal development of the visual cortex and the influence of environment. *Nature* 299: 583–591.

Wiesel TN, Hubel DH (1966). Spatial and chromatic interactions in the lateral geniculate body of the rhesus monkey. *J Neurophys* 29: 1115–1156.

Williams SM, McCoy AN, Purves D (1998a). The influence of depicted illumination on perceived brightness. *Proc Natl Acad Sci USA* 95: 13296–13300.

Williams SM, McCoy AN, Purves D (1998b). An empirical explanation of brightness. *Proc Natl Acad Sci USA* 95: 13301–13306.

Wojtach WT, Sung K, Truong S, Purves D (2009). An empirical explanation of the flash-lag effect. *Proc Natl Acad Sci, USA* 105: 16338–16343.

Wojtach WT, Sung K, Purves D (2009). An empirical explanation of the speed–distance effect. *PLoS ONE* 4 (8): e6771. doi: 10.1371/journal.pone.0006771.

Worgotter F, Eysel UT (2000). Context, state and the receptive fields of striatal cortex cells. *Trends Neurosci* 23: 497–503.

Wuerger S, Shapley R, Rubin N (1996). On the visually perceived direction of motion" by Hans Wallach: 60 years later. *Perception* 25, 1317–1367 (translated 1996; original 1935).

Wundt W (1862). *Beiträge zur Theorie der Sinneswahrnehmung*. Leipzig und Heidelberg: C. F. Winter'sche Verlagshandlung.

Wundt W (1862/1961). Contributions to the theory of sensory perception. (English translation is taken from Wundt W [1961]). In: *Classics in psychology* (Shipley T, ed.), pp. 51–78. New York: Philosophical Library.

Wundt W (1902). *Beitrage zue Theorie der Sinnewahrnelmung*. Leipzig: Wilhelm Engelmann. Trans Judd CH in *Outlines of Psychology*, reprinted in 1999. Bristol UK:Thoemmes.

Wyszecki G (1986). Color appearance. In: *Handbook of perception and human performance*, Chapter 9 (Boff KR, Kaufman L, Thomas JP, eds.). New York: John Wiley.

Wyszecki G, Stiles WS (1982). *Color science: concepts and methods, quantitative data and formulae* (Wyszecki G, Stiles WS, eds.). New York: John Wiley.

Yang T, Shadlen MN (2007). Probabilistic reasoning by neurons. *Nature* 447: 1075–1082.

Yang Z, Purves D (2003). A statistical explanation of visual space. *Nat Rev Neurosci* 6: 632–640.

Yang Z, Shimpi A, Purves D (2001). A wholly empirical explanation of perceived motion. *Proc Natl Acad Sci USA* 9: 5252–5257.

Yang Z, Shimpi A, Purves D (2002). Perception of objects that are translating and rotating. *Perception* 31 (8): 925–942.

Yang Z, Purves D (2004). The statistical structure of natural light patterns determines perceived light intensity. *Proc Natl Acad Sci* 101: 8745–8750.

Yarbus AL (1959). *Eye movements and vision*. (Haigh B, transl; Riggs LA, ed.)., New York: Plenum Press.

Young T (1801/1923). Zwei Abhandlungen von Thomas Young (Rohr M, ed.). *Zeitschrift fur opthalmologische Optik* 11: 102–155.

Young T (1802). An account of some cases of the production of colours, not hitherto described. *Phi T Roy B* 92: 387–397.

Yuille A, Ullman S (1990). Rigidity and smoothness of motion: Justifying the smoothness assumption in motion measurement. Image understanding. *Adv Computational Vision* 3: 163–184.

Yuille AL, Grzywacz NM (1998). A theoretical framework for visual motion. In *High-level motion processing: computational, neurobiological, and psychophysical perspectives* (Watanabe, T., ed.), pp. 187–211. Cambridge, MA: MIT Press.

Zeki SM (1983a). Colour coding in the cerebral cortex: The reaction of cells in monkey visual cortex to wavelengths and colours. *Neurosci* 9: 741–765.

Zeki SM (1983b). Colour coding in the cerebral cortex: The responses of wavelength-selective and colour-coded cells in monkey visual cortex to changes in wavelength composition. *Neurosci* 9: 767–781.

Zigler E (1960). Size estimates of circles as a function of size of adjacent circles. *Percept Motor Skill* 11: 47–53.

Zipser K, Lemme VAF, Schiller PH (1996). Contextual modulation in primary visual cortex. *J Neurosci* 16: 7376–7389.

Index

Page numbers sets in *italics* indicate there is information in a figure.